In this important first book in the series Cambridge Studies in Probability, Induction, and Decision Theory, Ellery Eells explores and refines current philosophical conceptions of probabilistic causality. In a probabilistic theory of causation, causes increase the probability of their effects rather than necessitate their effects in the ways traditional deterministic theories have specified. Philosophical interest in this subject arises from attempts to understand population sciences as well as indeterminism in physics.

Taking into account issues involving spurious correlation, probabilistic causal interaction, disjunctive causal factors, and temporal ideas, Professor Eells advances the analysis of what it is for one factor to be a positive causal factor for another. A salient feature of the book is a new theory of token-level probabilistic causation in which the evolution of the probability of a later event from around the time of an earlier event is central.

This will be a book of crucial significance to philosophers of science and metaphysicians; it will also prove stimulating to many economists, psychologists, and physicists.

Probabilistic causality

Cambridge Studies in Probability, Induction, and Decision Theory

General editor: Brian Skyrms

Advisory editors: Ernest W. Adams, Ken Binmore, Persi Diaconis, William L. Harper, John Harsanyi, Richard C. Jeffrey, Wolfgang Spohr, Patrick Suppes, Amos Tversky, Sandy Zabell

This new series is intended to be the forum for the most innovating and challenging work in the theory of rational decision. It focuses on contemporary developments at the interface between philosophy, psychology, economics, and statistics. The series addresses foundational theoretical issues, often quite technical ones, and therefore assumes a distinctly philosophical character.

Probabilistic causality

Ellery Eells

University of Wisconsin–Madison

CAMBRIDGE
UNIVERSITY PRESS

CAMBRIDGE UNIVERSITY PRESS
Cambridge, New York, Melbourne, Madrid, Cape Town, Singapore, São Paulo

Cambridge University Press
The Edinburgh Building, Cambridge CB2 8RU, UK

Published in the United States of America by Cambridge University Press, New York

www.cambridge.org
Information on this title: www.cambridge.org/9780521392440

First published 1991
Reprinted 1996
This digitally printed version 2008

A catalogue record for this publication is available from the British Library

ISBN 978-0-521-39244-0 hardback
ISBN 978-0-521-06132-2 paperback

To Joanne, Justin, Erika

Contents

Preface

My interest in probabilistic causality arose naturally from my earlier interest, as a graduate student and after, in the area of the philosophical foundations of decision theory. I was especially interested in the decision-theoretical puzzle known as Newcomb's paradox and the idea of causal decision theory (Eells 1982). Causal decision theory was designed to accommodate the fact that the *evidential,* or "average" *probabilistic,* significance of one factor or event for another need not coincide with the *causal* significance of the first factor or event for the other – a fact vividly illustrated by the Newcomb problem. Causal decision theory involves ideas and techniques quite similar to the ideas and techniques involved in untangling and understanding the relations between probabilistic and causal significance in the theory of probabilistic causality.

Most of the recent philosophical literature in this area has seemed to concentrate on what I call here type-level probabilistic causation, though some authors have either noted or developed theories of what I call here token-level probabilistic causation. The first five chapters of this book are about type-level probabilistic causation. The last, very long, chapter is on token-level probabilistic causation. It is probably Chapter 6, which gives a new theory of token-level probabilistic causation, that contains the most novel proposals of this book. In this book, I distinguish two conceptions of probability change: the *conditional probability-comparison conception* (which is the usual idea) and an *"actual probability-trajectory"* conception (on which, roughly, the profile of the actual evolution, in time, of the probability of the relevant later token event, is what is crucial). Chapters 1–5 refine the first idea (in some usual and some not so usual ways) into a probability-increase theory of type-level probabilistic causation. Chapter 6 describes the second idea and refines it into a probability-increase theory of token-level probabilistic causation.

In learning about probabilistic causality, I have benefited especially from the writings of the philosophers I. J. Good, Patrick Suppes, Nancy Cartwright, Wesley Salmon, and Brian Skyrms – as indicated by my frequent reference to, and discussion of, their works. As to more recent help on my thinking about probabilistic causality, I thank most of all Elliott Sober – for helping in the first place to make vivid to me the difference between the type and token levels of probabilistic causation, for his (at least initial) skepticism about the possibility of a theory of token-level probabilistic causation, for his encouragement, and for his ongoing comments and criticisms in the course of this project. Next, I must thank Malcolm Forster and Dan Hausman, for detailed and comprehensive comments on earlier drafts of this book; these comments were influential in many parts of the book. Malcolm Forster attended a seminar in which I presented some of the main ideas in this book. And Dan Hausman provided generous comments on all the chapters of a draft of this book.

Probably the most significant specific criticisms and comments that shaped the theory of Chapter 6 came from Bill Tolson and Malcolm Forster, in discussions following presentations at Northern Illinois University and at the University of Wisconsin–Madison, respectively. Their criticisms (counterexamples, in fact, to earlier versions of my theory of token-level probabilistic causation) inspired important, though natural, qualifications on the basic probability-increase idea for token-level probabilistic causation. Their important contributions to the theory are acknowledged in footnotes in Chapter 6.

I have also benefited a great deal from instructive comments and criticisms from students in seminars on probability and causation and on probability and explanation that I offered in the spring semester of 1988 and the fall semester of 1989 at the University of Wisconsin–Madison. In this connection, I am especially indebted to Marty Barrett, Greg Mougin, Aladdin Yaqub, and Tony Peressini.

There are others who have helped me in this project – in discussion, correspondence, and other ways. I thank them as well, and I am embarrassed if their help is not acknowledged in the text or notes. Of course, the people who have helped me in this project are not unanimous about the ideas advanced in this book. I would say these people are mixed in this connection, on different mixtures of the ideas advanced here.

For financial support, I thank the Research Committee of the Graduate School of the University of Wisconsin–Madison (for various grants, including summer support and a generous Romnes Faculty Fellowship), the John Simon Guggenheim Foundation, and the National Science Foundation. This material is based on work supported by the National Science Foundation under grant number SES-8605440. Any opinions, findings, and conclusions or recommendations expressed in this material are those of the author and do not necessarily reflect the views of the National Science Foundation.

Finally, I should note that parts of the material in the chapters that follow have been adapted from previously published articles. (See Bibliography for full references to the articles mentioned below.) Part of Section 1.2 comes from my "Objective Probability Theory Theory" (1983, *Synthese*). Part of Section 2.3 comes from my "Probabilistic Causality: Reply to John Dupré" (1987a, *Philosophy of Science*). Chapter 3 leans heavily on my "Probabilistic Causal Interaction" (1986, *Philosophy of Science*) and my "Probabilistic Causal Interaction and Disjunctive Causal Factors" (1988a, in *Probability and Causality,* Reidel). Section 4.3 draws from Eells and Sober, "Probabilistic Causality and the Question of Transitivity" (1983, *Philoso-*

phy of Science). Part of Section 5.3 leans on Eells and Sober, "Old Problems for a New Theory: Mayo on Giere's Theory of Causation" (1987, *Philosophical Studies*). Finally, an early form of the theory of Chapter 6 was given in my "Probabilistic Causal Levels" (1988b, in *Causation, Chance, and Credence,* Kluwer). In some cases, especially in the last, some of the ideas advanced here are somewhat different from, or advance, or replace, some of the ideas in the articles just cited.

Ellery Eells

Introduction

In the past 30 years or so, philosophers have become increasingly interested in developing and understanding probabilistic conceptions of causality – conceptions of causality according to which causes need not necessitate their effects, but only, to put it very roughly, raise the probabilities of their effects. This philosophical project is of interest not only because the problem of the nature of causation is itself so central in philosophy, and not only because of the nature of causation as well as physical indeterminism in current scientific theory. The theory of probabilistic causation also has applications in other philosophical problems, such as the nature of scientific explanation and the nature of probabilistic laws in a variety of sciences, as well as the character of rational decision. And the theory has applications in these areas whether or not determinism is assumed. In this book, however, very little is said about such applications. I focus on the theory of probabilistic causation itself.

In philosophy, the development of the probabilistic view of causality owes much to the work of I. J. Good (1961–2), Patrick Suppes (1970), Wesley Salmon (1971, 1978, 1984), and Nancy Cartwright (1979) (as well as others). In this book, I articulate and defend a conception of probabilistic causation that owes much to, but differs in important details from, the work of these and other authors. I also examine and appraise several alternatives to the ideas advanced here.

It is only putting it very roughly that, according to probabilistic conceptions of causality, *causes raise the probabilities of their effects*. As I see it, there are two main reasons why this

1

can only be a very rough characterization of the theory of probabilistic causation. The first has to do with a possibility known as Simpson's paradox (discussed in detail in Chapter 2). Whether or not there is a correlation between two factors, and whether a correlation is positive or negative, can be different in an "overall" population from the way things are in subpopulations, and things can vary in this respect from subpopulation to subpopulation. For example, an overall positive correlation can be reversed or disappear in all subpopulations, when the overall population is carved up in certain ways.

A frequently discussed example of this is sometimes called the Berkeley discrimination case (see Chapter 2). Suppose there is an overall positive correlation between getting admitted to the graduate school and being male: A higher percentage of male applicants are admitted than female applicants. This suggests discrimination against women, that being a man (or checking "male" on the application form) is a positive causal factor for getting admitted. However, admissions decisions are made at the departmental level. And if it is found, on a department-by-department examination of the data, that, within each department, there is no correlation between being male and getting admitted (the same percentage of male applicants is admitted as female applicants), then the graduate school may rightly be exonerated of the charge of discrimination. The overall positive correlation is consistent with statistical independence (or even negative correlation) in the departmental subpopulations. It might be that, for whatever reason, women tend to apply to departments that are harder to get into, while men tend to apply to departments that are easier to get into. In this example, it would seem that it is within the departmental subpopulations (the subpopulations of individuals applying to particular departments), and not in the overall population (of all applicants), that probability comparisons coincide with the causal facts.

2

In another frequently discussed example (also of a kind treated in Chapter 2), smoking is a positive causal factor for heart disease. This causal fact gives a "positive component" to the overall correlation between smoking and heart disease. However, suppose also that, in this hypothetical example, it is mainly the exercisers who tend to be smokers (say this is because there is a genetic common cause of smoking and exercising, or that exercising somehow causes smoking). This gives a "negative component" to the overall correlation between smoking and heart disease, because exercising prevents heart disease. As to the relative contributions of these two components of the overall correlation between smoking and heart disease, two of the possibilities are these: The negative component may dominate the positive component (resulting in an *overall negative* correlation), or the positive and negative components may exactly balance (resulting in *no overall* correlation). That is, overall, there may be a negative correlation, or no correlation at all, between smoking and heart disease, while at the same time, both among exercisers and among nonexercisers, there is a positive correlation between smoking and heart disease. Again, it is within the appropriate subpopulations that correlations coincide with causation.

Thus, the *first* main kind of qualification that has to be made, on the basic probability-increase idea for probabilistic causality, involves describing appropriate "causal background contexts." These are the subpopulations within which the direction of probability change should coincide with kind of causal significance (positive, negative, neutral, for example). In the Berkeley discrimination case, the appropriate causal background contexts would seem to be the different pools of applicants to the different departments. And in the smoking–heart disease–exercising example, the appropriate causal background contexts are the exercisers and the nonexercisers. In the chapters that follow, many examples like these, illustrating various reasons for "holding fixed" various kinds of factors in causal background contexts, will be analyzed.

The *second* main reason why the basic probability-increase idea for probabilistic causality is only very rough has to do with how we should understand the idea of probability change in the first place. In the examples above, probability change was understood as a *comparison of conditional probabilities:* the probability of getting admitted to the graduate school given that an applicant is male, compared to the probability of getting admitted given that an applicant is female; and the probability of heart disease given that one is a smoker, compared to the probability of heart disease given that one is not a smoker. Within the appropriate subpopulations, it is this kind of comparison that is supposed to reveal the causal relations. However, there another way in which to understand probability change: We can look at *how the probability of the exemplification of a factor in a particular individual changes over time.*

To vaguely suggest the difference between the two probability-change ideas, consider the following kind of scenario (a concrete example of this kind is described later in this Introduction and analyzed in detail in Chapter 6). Suppose that the exemplification of a causal factor C usually (say 90 percent of the time in which it is exemplified) confers a very low probability on the exemplification, by the same individual, of a factor E. On the other occasions in which factor C is exemplified (10 percent of the time that it is exemplified), it gives a very high probability to E. Suppose also that, all things considered (including perhaps a low probability of C), E is intermediate or somewhat high in probability. In this example, whether a particular exemplification of C will be positive or negative for E is a purely chancy matter. However, the probability of E conditional on C is lower than the probability of E conditional on not-C. On the other hand, in 10 percent of the exemplifications of C, the probability of E is increased to a very high value across the time of the exemplification of C. In such cases, we may want to say that the particular exemplification of C is causally positive for the later exemplification of E, despite the (negative) conditional probability comparisons.

I will argue that the *conditional probability comparison* understanding of probability change is appropriate for understanding what I call "type-level," or "property-level," or "population-level," probabilistic causal relations, while the *temporal-evolution, or "probability-trajectory,"* idea of probability change is appropriate for understanding probabilistic causal relations on what I call the "token-level." Conditional probability comparisons are for the analysis of type-level probabilistic causal relations, and not for the analysis of token-level probabilistic causal relations; and probability trajectories are for the analysis of token-level probabilistic causal relations, and not for the analysis of type-level probabilistic causal relations. This kind of sensitivity to the difference between the type and token levels of probabilistic causality, in terms of the different senses of probability change described above, is the second of the two main qualifications of the very rough probability increase idea mentioned at the outset.

For both levels of probabilistic causality, we must be careful to identify the appropriate causal background contexts. The first qualification applies to both levels identified in the second qualification. Both of these two kinds of qualifications on the basic probability-increase idea, as well as my own versions or formulations of these qualifications, will be motivated by many examples in the chapters that follow, some formal, some intuitive, and most of them hypothetical.

Between the two main kinds of qualifications of the basic probability-increase idea for probabilistic causation, it is the distinction between the type and token levels that I would most like to emphasize here. To understand probabilistic causality, and to assess theories of probabilistic causation correctly, requires understanding two main issues. There is the problem of the nature of probabilistic *type* causation (a relation

5

between event types, or factors, or properties), and there is the problem of the nature of probabilistic *token* causation (a relation between particular, actually occurring, token events). Philosophers have for the most part been concerned with the problem of type-level probabilistic causation.[1] Accordingly, I have two main aims in this book. In the first five chapters, I explain, extensively refine, and defend one common current way of understanding type-level probabilistic causation. In Chapter 6, I advance a theory of probabilistic token causation. In the final sections of Chapters 5 and 6, I examine alternative proposals. As will become clear, it is crucial to the assessment of a probabilistic theory of causation that the distinction between the two levels of causation be kept clearly in mind and that it be kept clearly in mind which of the two problems the theory is (or should be) intended to address.

In the rest of this Introduction, I will outline just a rough understanding of each of these two levels of probabilistic causation, paying attention mainly to ways in which the two levels differ. The problems of type and token probabilistic causation turn out to be quite distinct. Causal claims made on one of the two levels of causation turn out to be quite independent of claims made on the other. And the two levels of probabilistic causation require quite different kinds of theories.

I argue that (1) very little (if anything) about what happens on the token level can be inferred from type-level probabilistic causal claims, and that (2) very little (if anything) about type-level probabilistic causal relations can be inferred from token-level probabilistic causal claims. Of course, (1) and (2) are not independent. Roughly, if the probabilistic causal facts on one level can be anything given any way the probabilistic causal facts are on the other level, then the same goes the other way around. In this Introduction, I motivate (1) and (2) with the use of just a few examples and a little intuitive clarification of the two kinds of

[1]A notable exception is I. J. Good (1961–2), who both draws the distinction and offers a theory for each of the two ways of understanding probabilistic causality.

causal claims. I simplify things a little here by considering only type- and token-level claims of *positive causation,* as opposed to prevention or causal neutrality, for example. The discussion here is mainly for the purpose of distinguishing the two problems mentioned above and for conveying a rough understanding of the two levels of causation. We must know at least roughly what we have in mind before trying to develop a theory of it. Later in the book, before we deal with token-level probabilistic causation in detail, claims (1) and (2), as well as some of the examples used in this Introduction, will have to be considered again, in a little more detail.

Turning first to (1), let me note first something that is perhaps obvious about how token causation should be understood. In order for one token event to be a cause (also called a token cause) of a second, the two events must *both actually occur,* each at a definite time and place.[2] In addition, of course, the relevant causal relation must obtain between the two events.

It will be easy to see, first, that one type may be a probabilistic cause of a second type, while not every token of the first type is a cause, probabilistic or deterministic, of any token of the second. But also, on the way I shall understand statements of positive type level probabilistic causation, they do not even imply that *any* token of the cause type *ever* causes a token of the effect type, nor do they imply that the cause type is *ever exemplified,* nor do they even imply that if the cause type were *universally* exemplified then the effect type would have to be exemplified *at least once.* Eventually, it will emerge that type-level probabilistic causal claims are consistent with the token-level facts being even more "unexpected" than the possibilities just described. Type-level causal claims have little to say about what happens causally at the token level. Let us

---

[2]This may be helpful: In a theory of *possible* token events, in order for a possible token *c* to be a cause of a possible token *e* in a world *w*, *c* and *e* must each occur in world *w*.

turn to some concrete examples illustrating these possibilities, and then to the question of how type-level probabilistic causal claims should be understood.

The surgeon general says that smoking is a positive causal factor for lung cancer. This, of course, is a type-level causal claim, about the *properties* of being a smoker and of developing lung cancer. And it is consistent with various pertinent possibilities regarding *token events,* and the token causal relations between them. The claim is consistent, for example, with Harry's being a heavy smoker and never contracting lung cancer. It is also consistent with Larry's being a heavy smoker and a victim of lung cancer, where Larry's lung cancer was not caused by his smoking, but rather (for example) by overexposure to asbestos fibers at work. Also, as I argue below, the surgeon general's type-level causal claim, properly understood, is consistent with the possibility that nobody is a smoker, with the possibility that no smokers (even if there are many smokers) develop lung cancer, and even with the possibility that if everyone were to smoke, still nobody would contract lung cancer.

The surgeon general's claim implies only some kind of a probabilistic connection between smoking and lung cancer; nobody believes that smoking (even heavy smoking) *necessitates* lung cancer. Obviously, there would be no necessitation if determinism were false and we were indeterministic systems. And, of course, the hypothesis of physical indeterminism makes the surgeon general's type-level causal claim consistent with all the possibilities mentioned above for what happens (or might happen) in singular cases. But even without the assumption of physical indeterminism, the surgeon general's claim is still consistent with the ones about Harry and Larry, and with the other possibilities mentioned above. Even if Harry and Larry and the rest of us were all deterministic systems, the possibilities mentioned above would still be consistent with the type-level claim that smoking is a positive causal factor for lung cancer. Let us now see why.

8

As J. L. Mackie (1974) has emphasized, singular (or token) causal claims (either asserting or denying the presence of a causal connection) should be understood as depending for their truth on an implicitly assumed physical context, which he (following John Anderson 1938) calls a "causal field."[3] A causal field is, roughly, a constellation of factors, other than the token cause in question, that are causally relevant to the effect in question and that are actually present in the singular situation in question. An example commonly used to illustrate the idea is this: The claim, "It was the striking of the match that caused the explosion," is true only given the background assumption of the physical context that included the presence of gasoline vapor and oxygen. The point is that singular causal claims are usually elliptical, in that the causal event mentioned is only one part of the full constellation of conditions that is claimed to bring about the effect, where the rest of the full constellation is implicitly assumed to be present, in the causal field.

Likewise, the *denial* of a token causal connection generally depends for its truth on the assumption of an operative causal field within which the alleged noncause is a noncause, for it is often true that in other causal fields, the alleged noncause *would* be a cause. Thus, assuming determinism, the operative causal field in Harry's situation evidently included factors (possibly involving exercise, diet, genetic makeup, and environmental factors, for example) that counteracted his smoking's impact on his prospects for lung cancer. And in the case of Larry, not only is this true, but also the causal field included the presence of the factor of overexposure to asbestos, which itself caused his lung cancer, and, in combination with the other factors in the causal field, rendered smoking causally inert in his contracting lung cancer.

[3]The idea of physical contexts plays a crucial role also in the theory of type-level, probabilistic causal connection, which does not assume, but is compatible with, determinism. This idea is also central to the explication of probabilistic token causation, in Chapter 6.

How are the other possibilities described above consistent with the surgeon general's claim? First, as mentioned above, I interpret the type-level claim as asserting the existence of a certain relation between the properties of being a smoker and of contracting lung cancer, a relation that holds, of course, because of the nature of human physiology. As I interpret this claim, the relation between the two properties (and human physiology) can hold regardless of whether either of the two properties ever happen to be exemplified. In part, this interpretation is stipulative, but I think it has strong intuitive support, and, as I argue below, it has an advantage over an alternative kind of interpretation that may initially seem quite plausible.

As to the consistency of the surgeon general's claim with the possibility that nobody is a smoker (or, for that matter, ever contracts lung cancer), I merely cite an entirely parallel case that, I think, should be quite intuitively compelling. Drinking a quart of plutonium is a positive causal factor for a certain kind of death. This is true even though (or even if) nobody will ever drink so much plutonium or ever die in quite that way. If we insisted that the relevant types have to be exemplified in order for a type-level causal claim to be true, then we would have to deny this clear type-level causal truth, about the relation between the relevant properties and human physiology. The type-level causal claim is true in virtue of a *potential* physiological connection between exemplifications of the two properties. Of course, in the smoking example, it may be that the surgeon general would never have made his claim, or had any evidence for it, if nobody smoked or nobody ever contracted lung cancer; but, of course, questions of relevance and justification are quite different from the questions of the truth and the meaning of the claim.

What about the possibility that there are many smokers, but nobody contracts lung cancer? Consistent with determinism, the surgeon general's claim can be true in this case if nobody's causal field were quite right for the exemplification of the potential physiological connection between smoking and lung cancer. For example, it could be that all the smokers are like Harry, in the example described above, and it is only for the nonsmokers that smoking could have had a chance of making a difference. All this is consistent with determinism, and with the claimed relation between the two relevant properties that holds in virtue of the nature of human physiology but is never exemplified.

Consider now the possibility that even if everyone were to smoke, still nobody (or at least no smokers) would contract lung cancer. Consistent with human physiology being just as it actually is (so that the surgeon general's claim is still true), is the possibility that everybody's causal field happens (improbably enough) to be such that, if they were to become smokers, they would, just before the time lung disease had a chance to develop, die, from some other cause that, given the causal field, is deterministically token causally related to smoking. Perhaps some individuals are careless with matches and would perish in a house fire if they became smokers, while others are easily distracted when driving and would suffer a fatal automobile crash after lighting up while driving if they became smokers, and so on. For each individual, the operative causal field, together with being a smoker, leads deterministically to such a fate just before lung disease has a chance to develop. Again, this kind of distribution of causal fields to members of the human population is consistent with human physiology being just as it actually is, namely, such that there exists the kind of relation between the properties of smoking and of contracting lung cancer in virtue of which it is true that smoking is a positive causal factor for lung cancer in humans.

The following objection to my use of this example naturally arises: It is part of the example that everybody's causal field is such that, were anyone to smoke, he or she would fail to contract lung cancer; so, given the actual causal fields individu-

11

als in fact have in this example, it seems that we should not say that smoking is a positive causal factor for lung cancer after all. On the one hand, I agree that the objection contains a valid point, if the point is formulated just right. But on the other hand, it is clear that if we concluded simply that smoking is not a positive causal factor for lung cancer in the human population of the example, then we would allow an important part of the causal truth – truth in virtue of human physiology – to escape what can be characterized in terms of type-level causal claims about the population. It is worth explaining here roughly how this kind of dilemma will be resolved.

Chapter 1 shows that type-level probabilistic causal claims must be made relative to given, token, populations. And in Section 1.2, I sketch an interpretation of probability that makes use of the idea of hypothetical populations: hypothetical "versions" of an actual population. Probability is characterized in terms of (to put it somewhat vaguely here) limiting frequencies in such hypothetical populations. Of course, the relevant hypothetical populations must be relevantly related to (they must be "versions of") the actual population: they must be, roughly speaking, "of the same kind" as the actual population. But any given token population is a token of many types, so that *which* hypothetical populations are relevant will depend on which *type* the given token population is considered to be a token of. For the surgeon general's claim, the relevant token population is the human population. This token population is of the type (among others) *human*. And in the example just described, this population is also of kind Q, where the description of Q is a detailed physical description of the causal fields of the individuals in the population (in virtue of which the individuals do not develop lung cancer, and would not develop lung cancer even if they were smokers, and so on, as in the example). The relevant token population is a token of each of these two types.[4]

[4]In Chapter 1, I argue that there are some kinds (of population) such that it is never appropriate to assess probabilistic or causal relations, among the factors in question

If we consider the actual population as a token of the first type (simply as the human population), then the corresponding hypothetical populations should not be homogeneous with respect to the factors (Q) of being physically disposed not to develop lung cancer and being physically determined to die prior to the development of lung disease if one were to smoke. The possession of these properties is not part of what it is to be human, and it was just an accident that these factors were universally exemplified in the actual population of the example. Also, it is clear that the surgeon general's claim, in the example, should be understood as relative to the population considered as an instance of the kind *human population*, and not as relative to the population considered as an instance of the other kind, Q.

Suppose probability is defined in terms of frequencies in such hypothetical populations, so that probabilities will be determined according to how factors like those mentioned above are distributed in such hypothetical populations (see Section 1.2). Then we can expect that, if the causal claim in question is true, then the probability of lung cancer will be higher given smoking than given not smoking – even if nobody actually develops lung cancer, and even if it is true of all actual individuals that even if they were to smoke they would still not get lung cancer.

Setting aside for the moment several fine points and qualifications that will have to be addressed in the coming chapters, we might expect that in an appropriate hypothetical population of the simply human type, every individual is either such that smoking increases (or would increase) the probability of lung cancer, or such that smoking leaves unchanged (or would leave unchanged) the probability of lung cancer,

in a given problem, relative to populations considered to be of those kinds. For example, it is inappropriate to assess such relations relative to a population considered to be of a kind in which the frequencies of the factors in question are such and such. Neither of the two kinds under discussion here are inappropriate in this way, however.

where there are individuals of each of these two kinds.[5] We have an increase in the case of individuals who, unlike all actual individuals in the example, lack the physical disposition to die before any onset of lung disease if they were to smoke; and we may have no probability change for those who, like the actual individuals, have that disposition. It is (very roughly) these probabilistic conditions that make the surgeon general's claim true, on the interpretation of type-level probabilistic causal claims that I shall elaborate in the chapters that follow.[6] And note that this (rough) interpretation of the surgeon general's type-level causal claim is compatible with determinism and with all the token-level possibilities for Harry, Larry, and the rest of us, given above.

Thus, to summarize my response to the natural objection: (1) It is true that in the population of humans considered to be of the kind Q, smoking is neutral for lung cancer; but (2) this is consistent with its being true that in the population of humans considered to be simply of the kind *human*, smoking is a positive (or a Pareto-positive, or a "mixed") causal factor for lung cancer; and (3) the intent of the population-level causal claim would seem to be to refer to the actual population as of the kind simply *human*. Part of the rationale for points (1) and (2) can be expressed like this: In the actual human population considered to be of the kind simply *human*, the actual frequency of the factors involved in the description of Q is 1, but the probability is not 1.

There is another possible interpretation of type-level probabilistic causal claims, quite different from the one sketched

[5]In this case, we have an example of "Pareto-positive" type-level causation, discussed in Section 2.3 below.

[6]I argue in Section 2.3 that *Pareto*-positive causal factors (see note 5) are usually (when there is both positive and neutral probabilistic significance) best thought of not as *positive* causal factors at all, but rather as having a *mixed* causal significance for their effect factors. Another possibility for this example worth noting is that, for some actual or hypothetical individuals, smoking may *prevent* lung cancer, by causing death before lung disease has a chance to develop, where, were it not for this early death, lung cancer might have developed. In this case, the causal significance of smoking for lung cancer is even more mixed.

14

above, that I should mention and distinguish from the kind of interpretation I am endorsing here. Assuming Mackie's idea of a causal field, a rough but perhaps natural way interpreting the surgeon general's type-level causal claim is this: Relative to the causal fields of *some* (or perhaps *many* or *most*) people, smoking is (or would be?) a (deterministic?) *token* cause of lung cancer. This, of course, suggests the possibility of analyzing type-level probabilistic causal claims in terms of token-level causal claims (or perhaps in terms of token-level causal claims plus counterfactuals). This alternative interpretation explicates type-level probabilistic causal claims as generalizations over instances of token causation. I have three brief points to make about this kind of suggestion.

First, it is consistent with a type-level probabilistic causal claim, as I understand such claims, that the cause and effect factors involved never in fact happen to be exemplified. Thus, as I understand type-level probabilistic causal claims, they are not generalizations over instances of token causation

Second, in motivating the kind of interpretation of type-level probabilistic causal claims that I am endorsing, I have described examples, possible situations, in which the surgeon general's type-level claim is intuitively true, yet in which there are no cases in which a token of the cause type ever causes, or even would cause, a token of the effect type. In these cases, there are no instances of the relevant kind of token causation to generalize over.

Finally, there is a problem for the suggestion that type-level causal claims be understood as generalizations over instances of token causation: What are the formal properties of the generalizations? In order for a type C to be a type-level cause of a type E, must a token of E be "token caused" in *all* cases in which a token of type C occurs – or only in most such cases, or only many, or only some? Also, could C be a positive causal factor for E if instances of C sometimes cause the *absence* of (that is, sometimes *prevent*) E? On the one hand, until all this is made precise, we have only a very minimal

15

theory; and on the other hand, it seems that any way of making it precise makes the theory either too weak, too strong, or arbitrary.[7]

Given the way I shall be interpreting type-level probabilistic claims, it is clear that the problem of the nature of type-level causation is quite distinct from the problem of the nature of token-level causation, both in the case of physical determinism and in the case of physical indeterminism. Because of the variety of possibilities for what can be true at the token-level consistent with the truth of a type-level probabilistic causal claim, it seems hopeless to try to characterize type-level probabilistic causation in terms of token-level causal relations. Thus, when the theory of type-level probabilistic causation is laid out in detail in the first five chapters of this book, no appeal will be made to causal relations at the token level.[8]

Turning to the theory of token causation, the idea naturally arises of trying to make use of the theory of type-level probabilistic causation in characterizing probabilistic causation at the token level. However, the explication of probabilistic causation at the two levels requires quite different kinds of theories, involving quite different concepts, though there are interesting analogies between the two theories (see Chapter 6). Given the conceptual independence of token-level causal facts from type-level causal facts, it should not be surprising that what is true at the type level is equally conceptually independent of what is true at the token level, and that token-level causation cannot be straightforwardly understand in terms of type-level causal relations.

For one thing, any token event is a token of many types. So there is first of all the issue of whether there are certain "relevant" or "appropriate" types for token causes and effects, if the

[7]This kind of difficulty, for a related kind of proposal, is discussed in more detail in Section 2.3.

[8]In Section 4.1, I criticize a recent argument to the effect that token causal facts *must* be appealed to in the explication of type-level causation.

16

idea of one token's causing another is to be explicated in terms of causal relations among "corresponding" types. (This issue arises anyway if probability is to play a role in the explication.) Also, the basic idea of probabilistic type causation is, subject to a number of important qualifications, that the cause type raises the probability of the effect type. Of course, it is very plausible that in some cases of token causation, a "relevant" type of the token cause raises the probability of a "relevant" type of the token effect. But, as detailed in Chapter 6, it is also possible for one token event to be a token cause of a second token event even though a relevant or appropriate type of which the first is a token *lowers the probability* of a relevant or appropriate type of which the second is a token. This means, of course, that token-level causation does not coincide with type-level causation between relevant types.

Still in the way of an introduction to the two distinct problems of type and token probabilistic causation, and to introduce some of the difficulties involved in characterizing token causation probabilistically, I sketch here an example intended to show that while some token causes (or their types) can raise the probability of their token effects (or the effect types), others can lower the probability of their effects (or the effect types). Again, this kind of case is problematic because it shows that token-level probabilistic causation does not coincide with type-level causation between "corresponding" types.

The example is an elaboration of an example of Dretske and Snyder's (1972). There is a box containing a randomizing device that, once activated, gives the box a physically irreducible 10 percent chance of immediately entering state S (and then remaining in this state) and a 90 percent chance of immediately entering state S' (and then remaining there); before the device is activated, the box is neither in S nor in S'. (The design of the device may be based upon principles from quantum mechanics; and the box may contain a Geiger counter, for example.) Attached to the box is a loaded revolver that fires exactly when the box enters state S. I shall now divide the example into two cases.

In case 1, we take the box and place it next to a healthy cat that we have just purchased from a pet store. We point the revolver at this cat and activate the device inside the box. If the revolver fires, the cat is dead; if the revolver does not fire, we take the cat back to the store. Improbably enough (probability 0.1), the box enters state S: the revolver fires and the cat is killed. I concur with Dretske and Snyder that what we have done (call it token event c) is the token cause of the premature death of this cat (call this token event e). In this case, it seems plausible to say that the probability-increase theory of probabilistic *type* causation is in a kind of natural agreement with the *token*-level causal fact about what we have done. If C is the type of event described in the way I described what we did to the cat, and E is the type of event of a cat of the kind described dying prematurely, then we have both that c token caused e, and that type C is a probability-increasing, type-level cause of type E: C raises the probability of E from some relatively safe, low value to 0.1. And types C and E would seem naturally to correspond to tokens c and e (but see below).

In case 2, we do the same thing, but to a cat of a different kind, a healthy but mean cat that we have taken from the pound and that most likely otherwise would have been "put to sleep" there because of its mean disposition. Before we took the cat away for our "experiment," the probability of its premature death was, say, about 0.8 – thus allowing, say, a probability of 0.191 for the possibilities of its escaping or being reclaimed by its owner, and the rest (0.009) for the possibility of our "rescuing" it and its surviving our experiment (see Chapter 6 for more details). Improbably enough, we do get the cat, we place it in front of the loaded revolver attached to our box, and we activate the device. If the revolver does not fire, we keep the cat alive for observation until it dies of natural causes; and if the revolver does fire, the cat is killed, dying prematurely. We have thus actually reduced the probability of this cat's dying prematurely from 0.8 to 0.1. But, improbably enough, the box lands in state S: The revolver fires, and the cat dies prematurely.

In this case, just as in case 1, it is clearly the token event of what we have done (call this token event c') that is the actual token cause of this cat's dying prematurely (call this cat's premature death token event e'). However, in the circumstances of the example, the *type* of thing we have done (rescue the cat, put it in front of the revolver, and activate the device – call this event type C') *lowers the probability* of the *type* of thing we have token caused (premature death of the cat – call this event type E') from 0.8 to 0.1. In this case, C' lowers the probability of, and is thus a type-level *negative* causal factor for, E'; yet c' is the token cause of e'. And types C' and E' seem naturally to correspond to tokens c' and e' (but see below). In case 2, we have, to put it roughly, a token cause lowering the probability of its token effect.

Examples of this kind will be examined in more detail in Chapter 6, and the claim that there actually are cases of "probability-decrease token causation" will be explained and defended there. Various subtle issues arise. For example, in the cases just described, is the token event of the cat's premature death the same as the token event of the cat's premature death by gunshot wound? Can the idea that token causation coincides with probability increase be rescued by considering different types in such examples? What if, in the previous example, we let type C' be premature death by gunshot wound instead of just premature death? Are some types more "appropriate" than others? The examples just described are simply intended to suggest further reasons here, to be examined in more detail in Chapter 6, for the distinctness and autonomy of probability-increase *type*-level causation from causation at the *token* level, and to suggest that we should not automatically expect a very straightforward explication (if any) of token-level probabilistic causation in terms of type-level probabilistic causation. In Chapter 6, I will de-

19

scribe a kind of probability change that is different from the kind appropriate for the type-level theory. And I will argue that, on this different understanding of probability change, token causation does coincide, in a way, with direction of probability change.

In the first five chapters of this book, I clarify, elaborate, and refine one common current understanding of probabilistic causality, an understanding that I think is appropriate for type-level probabilistic causation, but not for token-level causation. Chapter 1 explains an understanding of probability appropriate for the theory of probabilistic causation, and argues for the importance of relativizing the relation of probabilistic causation to a *token population,* considered to be of a given (appropriate) *kind.* Chapter 2 explains how the common current understanding of probabilistic causation, as probability increase, deals with the well-known phenomenon of "spurious correlation," which arises, most famously, in cases of joint effects of a common cause. This phenomenon, as well as the phenomenon of "causal interaction" discussed in Chapter 3, is why the idea of causal contexts is so central in the theory.

Chapter 4 explores complex questions surrounding the issues of transitivity of type-level probabilistic causation and the role of intermediate, and other kinds of (what I will call) "subsequent," causal factors. In Chapter 5, I argue that the theory needs to require explicitly, in a separate requirement, that causes precede their effects in time. Formulating this requirement is not so straightforward, however, since type-level causes and effects are abstract entities (they are properties, or "factors"), so that it is a somewhat subtle and difficult problem to explain how one such entity can precede another such entity in time. Finally, at the end of Chapter 5, I summarize the theory of type-level probabilistic causation, with all of the qualifications on the basic probability-increase idea that will have been uncovered along the way; and I briefly com-

20

pare the theory elaborated in the first five chapters with other approaches.

The theory of token-level probabilistic causation presented in Chapter 6 bears some interesting formal analogies to the type-level theory. However, the basic probability-increase idea is quite different, focusing on the actual evolution of the relevant probabilities over time. Roughly, the relation explicated is this: Things' being such-and-such at one time and place is a token cause of (or has some other kind of token causal significance for) things' being so-and-so at another time and place. And, roughly, a later event can occur *because of, despite,* or *independently of,* an earlier event, depending on how the probability of the later event actually evolves across the time of the earlier event and between the times of the earlier and later events.

1

Populations and probability

Type-level probabilistic causation is sometimes called "population-level" probabilistic causation and sometimes "property-level" probabilistic causation. And the items that enter into type-level probabilistic causal relations are called "factors," or "properties," or "event types." The basic idea in the theory of type-level probabilistic causation is that causes raise the probabilities of their effects. A factor C is a property-level probabilistic cause of – or a positive causal factor for – a factor E, if the probability of E is higher in the presence of C than it is in the absence of C. And C is causally negative or causally neutral for E if the presence of C lowers or leaves unchanged the probability of E, respectively. But this basic idea needs several clarifications and qualifications.

In this chapter, I explain the importance of the idea of a population to type-level probabilistic causal connection. I argue that type-level probabilistic causation is a relation among four things: a cause factor, an effect factor, a token population within which the first is some kind of cause of the second, and, finally, a kind (of population) that is associated with the given token population. Subsequent chapters reveal how important relativity to populations is for the versatility of the probabilistic theory and how it renders the theory immune to a number of criticisms that have recently been advanced.

Of course, some kind of clarification of the idea of probability is in order. Throughout, I assume an objective and physical understanding of probability, as opposed to a subjective ("degree of belief") or a logical ("partial entailment") interpretation, for example. Although I do not advocate any detailed interpretation of probability in this book, later in this chapter I briefly consider several conceptions, assess their appropriateness for use in a probabilistic theory of causation, and suggest how we should go about understanding objective, physical probability. I sketch an interpretation that I think has some nice formal and conceptual properties but is nevertheless subject to a number of difficulties. Although I suggest that our understandings of probability and causation (and some other ideas) are interdependent, the sketch of an interpretation I offer will, I hope, help as a guide to intuitions about probability, understood as objective and physical.

1.1 POPULATIONS

Consider any property-level probabilistic causal claim, "*C* causes *E*," or "*C* is a positive causal factor for *E*." According to the basic idea, this means that *C* (or the presence of the factor *C*) *raises the probability* of *E* (or the presence of the factor *E*). Of course, no matter how we interpret probability (see Section 1.2), there must be some individual, individuals, or population, *within which C* is understood to raise the probability of *E*. This is because, in the first place, there is no such thing as the probability of *E*, or the probability of *E* in the presence of *C*, unless understood as applying to some individual or population. Also, of course, *which* individual or population is in question can make a difference in the *values* of the probabilities of *E* and of *E* in the presence of *C*. For example, the probability of having a heart attack this year is different in the population of 30-year-olds from what it is in the population of 50-year-olds. And the conditional probability of having a heart attack this year, conditional on having been a heavy smoker for 15 years, is different in the population of 30-year-olds from what it is in the population of 50-year-olds.

Furthermore, I doubt that there are very many factors *C* and *E* such that no matter what population or individual is in question, the presence of C in each case *increases* the probability of E. If this is right, then, given the basic probability-increase idea of positive causal factors, we should expect that whether or not a factor C is a positive causal factor for a factor E will vary from population to population, or from individual to individual. This relativity of causal significance, and of probability change, to populations should, of course, already be expected from the considerations above, about relativity of probability values to populations.

As an example of this, consider the surgeon general's claim that cigarette smoking is a positive causal factor for a heart attack. Obviously, what is intended is that smoking is a positive causal factor for heart attacks in the human population. The obvious intent of the claim is perfectly consistent with the fact that smoking is causally neutral for heart attacks in cigarette smoking machines in laboratories (the machines have no hearts). And the intent of the claim is also perfectly consistent with the hypothesis that there are creatures somewhere in the universe for whom smoking cigarettes is an essential part of daily nutrition, important specifically for the prevention of heart attacks. In the population of these creatures, smoking is actually a negative causal factor for heart attacks, and this is consistent with the obvious intent of our surgeon general's claim.

Although the fact is perhaps obvious, it is worth emphasizing at the outset that the causal significance of one factor for another can vary from population to population and from individual to individual, and that we should not expect there to be very many factors C and E such that the probabilistic causal significance of C for E is the same for all populations and all individuals.

Let us now drop reference to individuals, and think of them as singleton populations. It is now clear that one required refinement of the basic probability-increase idea is this: *A property-level probabilistic causal claim must be made relative to a particular, token, population.* Our first clarification of the basic probability-increase idea of causation is that what we should seek an explication of is the relation that can be expressed as follows: "C is a probabilistic cause of E *in population P*," or "C is a positive causal factor for E *in population P*." Property-level probabilistic causality is a relation among (at least) three things: a causal factor, an effect factor, and a population within which the former is a probabilistic cause of the latter. This simple and quite obvious qualification is extremely significant for the versatility of the theory, as subsequent chapters demonstrate.

A perhaps subtler point, mentioned in the introduction, is that in assessing a probabilistic causal claim, we must always think of the relevant token population (relative to which we wish to assess the probabilistic causal claim) as of a certain given *kind*. A token population is always a token of many types, or kinds, and the kind of causal significance a factor C has for a factor E, in a population P, can depend on what *kind*, Q, of population we think of P as exemplifying. This point deserves further elaboration.

It is clear that probability values can depend on what kind we associate with a given population. In the next section, on probability, I explain this dependence and show how to build it into an interpretation of probability: I will sketch an interpretation of objective and physical probability in terms of frequencies in hypothetical populations *of the relevant kind*. Given the basic probability-increase idea of causation, this dependence of probability on kind of population implies that the causal significance of one factor for another in a given token population can depend on what kind we associate with the population. An example illustrating this was already discussed in the introduction. This was the example in which the actual human population exemplified both the kind *human* and the more complicated kind given by a description of everybody's actual "causal fields," all of which happened to be such that they, together with the factor of becoming a smoker, determine that the individual would still not contract lung cancer. Relative to the token population's being of the first kind, smoking is a positive (or a Pareto-positive or a mixed) causal factor for lung cancer in the token human population of the example; and relative to its being of the second kind, smoking is causally neutral (or perhaps negative) for lung cancer in the population.

The example given in the introduction assumed determinism (though this is not essential for illustrating the point in question here), and it was mainly intended to illustrate a quite different point (about the relation between type-level and token-level causal facts). What follows is a simpler, and more numerical, example. This example assumes indeterminism and perhaps even more clearly illustrates the point at hand.

Consider a population, P, of 10 occasions on which a certain coin may be tossed. There is a device, call it device 1, that we will activate 10 times. Each time, the device first determines whether or not the coin will be tossed, and on those occasions on which it determines that the coin will be tossed, the device tosses it. When activated, this device has a 0.5 chance of tossing the coin; and each time the device tosses the coin, the coin is set into motion in such a way that, as long as nothing interferes with it, it has a 0.5 chance of landing heads and a 0.5 chance of landing tails. There is only one thing that can interfere with the coin in motion. It is a second device, device 2. This device is activated whenever device 1 is. When device 2 is activated, there is a 0.1 chance that it will, at that moment, enter state S, and remain there until the device is reset after the trial. When this device is in state S, it interferes with the coin in motion, if the coin is tossed; and when device 2 interferes, it does so in such a way as to confer probability 1 on the coin's landing tails. Suppose that in the particular population P in question (a particular collection of 10 occasions on which the two devices are activated), the coin happens to get tossed 5 times and device 2 happens to enter state S on each of the 10 occasions. Thus, in P, the coin lands tails on each of 5 tosses.

What is the probability of *heads* in population P? What is

26

the probability of heads given that the coin is tossed, in population P? When these questions are answered, we can say whether the coin's being tossed is a positive causal factor for the coin's landing heads, in population P. The answers to the questions about probabilities depend on how we understand the questions – specifically, on how we think about P when we ask the questions, on just what *kind* of population we think of P as exemplifying. On the one hand, we can think of it as a population of a kind, Q, in which (improbably enough) device 2 always enters state S; this is one type that our token population exemplifies. Relative to P's being of kind Q, the probability of heads and the probability of heads given a toss are both 0; and, of course, the probability of heads given no toss is 0. Thus, relative to P's being of kind Q, we should say that the coin's being tossed is causally neutral for heads: Given that device 2 always enters state S, a toss cannot result in, or affect the chances of, the coin's landing heads.

On the other hand, we can think of the population simply as an instance of the kind, Q', in which the two devices are each activated 10 times; the token population also exemplifies this type. When we think of the population as of kind Q', it seems that we should do the calculations and say that the probability of *heads* in the population is $(0.5)(0.5)(0.9) = 0.225$, and that the probability of *heads, given a toss,* is $(0.5)(0.9) = 0.45$; of course, the probability of heads, given no toss, is 0. And relative to population P's being of kind Q', we should say that the coin's being tossed is a probability-increasing cause of heads.[1]

Both probability values and questions and causal significance, therefore, can depend not only on what *population* is in question, but also on what *kind* we associate with the popula-

[1] There are parallels between this example and the example discussed in the Introduction and alluded to just above. That device 2, improbably enough, always enters state S in the actual token population of this example is analogous to the part of the earlier example according to which, improbably enough, everybody's causal field happens to such that it, together with the factor of being a smoker, determines that the individual will not contract lung cancer.

tion in question. Relative to different token populations, and relative to any given token population's being of different kinds (each of which it exemplifies), the probabilistic facts, and the causal facts, can differ.

However, a population-level causal claim, or a population-level probabilistic claim, cannot be relativized to just *any* kind that the token population in question exemplifies: Some kinds are not appropriate. In assessing the probabilistic or causal role of a factor C for a factor E in a token population P, we must not consider P to be of a kind Q, the fact of whose exemplification *logically implies* the value of the frequency of E, or the value of the frequency of E given C, or the value of the frequency of E given not-C. Nor should the fact of the exemplification of Q logically imply what the numerical relation (equality or direction of inequality) is between any two of these frequencies. Allowing such kinds as appropriate would allow *some* probabilistic causal claims, about logically independent factors, to count as logical truths, which is clearly inappropriate. In addition, evaluating probabilistic causal relations relative to such kinds can even deliver a clearly wrong answer about the causal facts in a population. A simple example clearly illustrates these points.[2]

Suppose I have two coins that I toss simultaneously many times, one coin with my left hand and the other with my right hand. We thus have a large token population, P, of pairs of tosses. All the tosses are "fair," and the results on each side are physically independent of the results on the other. This population of tosses is of kind Q, given by the way I have just described the population. Let C be the factor of the coin on the left landing heads, and let E be the factor of the coin on

---

[2] Note, incidentally, that the previous example is not like this. The fact of the exemplification of Q', of course, does not logically imply the values of the relevant actual frequencies, or the numerical relation between them. Nor does the fact of the exemplification of Q. That device 2 always enters state S only confers probability 1 on the proposition that the relevant actual frequencies will both be 0, without logically implying this proposition. So Q and Q' are not ruled out as inappropriate, for this example, by the requirement just stated. The same goes for the two kinds in the example discussed in the Introduction.

the right landing heads. We know that C is not a causal factor for E: C is causally neutral for E in P, considered to be of kind Q. Now consider any particular large subpopulation of P in which exactly half of the results on the right are heads and in which the results on the two sides always match; call this subpopulation P'. Like P, P' is a token of kind Q. And considering P' to be of kind Q will give us the same answer that C is causally neutral for E in P'; on the understanding of probability given in the next section (in terms of frequencies in hypothetical populations of the relevant kind), C should leave the probability of E unchanged, in P' considered to be of kind Q.

Population P' also is of kind Q': population, of kind Q, in which half the results on the right are heads and in which the results on the two sides always match. In all populations (actual or hypothetical) of *this* kind, the frequency of E is 0.5 and the frequency of E given C is 1 (and the frequency of E given not-C is 0). The fact of the exemplification of Q' logically implies that these are the frequencies; and it is a logical truth that C raises the probability of E in populations of kind Q'. And if we allowed kinds like Q' in the explication of causality, and in the explication of probability, then we would get the *wrong answer* in this example that heads on the left is a probability-increasing cause of heads on the right, in P' considered to be of kind Q'. By hypothesis, the result on the right is *physically independent* of the result on the left.

In this example, the fact of the exemplification of Q' logically implies the values of *each* of the following *three* frequencies: the frequency of E, the frequency of E given C, and the frequency of E given not-C. However, even if we allowed the fact of the exemplification of a kind to logically imply the value or values of just one or two of these frequencies, then the probabilistic theory of causality, when it comes to characterizing the magnitudes of causal impact, could come to an incorrect verdict. And this is true even if we disallowed logical implication of an extreme frequency of 0 or 1. In Chapter 2, I suggest a measure of the magnitude of probabilistic causal impact. But since this measure of magnitude of impact cannot be described at this point, I will not now offer an example illustrating the danger of allowing a kind to fix the frequency of even one or two of the relevant frequencies mentioned above. In any case, I hope it is fairly clear at this point, at least on qualitative or intuitive grounds, that none of the relevant three frequencies should be allowed to be fixed by the fact of the exemplification of an appropriate kind.

The example just described can be used to illustrate a related point. Suppose that in the token population P, the actual frequency of E given C is slightly greater than the overall actual frequency of E.[3] Then P is of kind Q'': population, of kind Q, in which the actual frequency of E given C is greater than the overall actual frequency of E. Although the fact of the exemplification of Q'' does not logically imply any particular values for the relevant actual frequencies, it does logically imply what the numerical relation is between the frequencies. In any actual or hypothetical population of kind Q'', the frequency of E given C will be greater than the frequency of E (and greater than the frequency of E given not-C). Again, if we allowed this kind, Q'', as appropriate, then, given the way probability will be understood in terms of frequencies in hypothetical populations of the relevant kind, we will get the wrong answer that C is a probability-increasing cause of E, in population P considered as of kind Q'' – where by hypothesis, E is physically independent of C.

The way out of these difficulties is simply to require, for the appropriateness of kinds Q, that the values of the relevant frequencies, and the numerical relation between these frequencies, are *logically independent* of the fact of the exemplification of Q. These are two (independent) restrictions: If Q is to be an appropriate kind of population for assessing probabilistic

[3]In the example as described, nothing is more probable than this, and the least likely possibility is that we have strict equality of these two frequencies in the large population. Alternatively, we could consider a large *subpopulation* in which this inequality holds. Also, the same point could be made by considering the reverse of the inequality.

or causal relations between factors C and E in a population P that exemplifies Q, then (1) the values of the frequencies of E, of E given C, and of E given not-C, in a population of kind Q, must *each* be logically independent of the population's being of kind Q, and (2) the numerical relation that holds between any two of these frequencies, in a population of kind Q, must be logically independent of the population's being of kind Q.[4] That is, the proposition that a population is of kind Q must not logically imply propositions stating the values of any of the three relevant frequencies in the population, or propositions asserting what the numerical relation is between any two of these frequencies in the population.[5]

There is one further restriction that we must place on kinds Q. The leading idea in the probabilistic theory of causation is that the causal significance of a factor C for a factor E corresponds to the *difference* that the presence of C makes for E. Roughly, the idea is to compare the probability of E *in the presence of the factor C* with the *overall* probability of E – or, alternatively, with the probability of E *in the absence of C*. If this is to be possible in a way that can reveal a difference that C makes for E, then the kind Q must allow both for the possibility of C's being exemplified in populations of kind Q, as well as for the possibility of not-C's being exemplified in populations of kind Q. Otherwise, we cannot make the relevant *comparisons*. As an extreme case, it is possible for a factor

[4]The two requirements, (1) and (2), are independent of each other because (1) would be violated even if the fact of the exemplification of Q implied the value of *just one* of the relevant frequencies. One value does not determine equality or direction of inequality between any of the relevant frequencies. So failure of (1) does not imply failure of (2). Obviously, failure of (2) does not imply failure of (1), as the last example discussed illustrates. (Of course, all this implies that, also, satisfaction of one of (1) and (2) does not imply satisfaction of the other.) So (1) and (2) are independent of each other.

[5]Chapter 2 reveals that population-level causal claims need to be assessed in terms of probability comparisons *within "causal background contexts"* K_i. When there is more than one such context, I should say that the frequencies, in P, of E given K_i, of E given K_i and C, and of E given K_i and not-C, and the numerical relations between pairs of such actual frequencies, need to be logically independent of P's being of kind Q, for each context K_i – if Q is to be an appropriate kind to associate with P.

31

C to have a causal role for a factor E in a token population P consisting only of C's. But if we consider the population P to be of the kind Q *in which all members are C's*, then all the relevant hypothetical populations, used to assess the relevant probability relations, will consist only of C's – so that the *difference* that C makes cannot be revealed.

This third restriction, then, is this. If Q is to be an appropriate kind of population for evaluating the causal role of a factor C for a factor E in a population P of kind Q, then (3) the fact of a population's being of kind Q must not logically imply that all members of the population have C, nor may it imply that all members of the population lack C. Note that this restriction is consistent with the possibility that the fact of the exemplification of Q implies that *either* all members of the population are C *or* all members are not-C – where Q does not decide between the two, so that some hypothetical Q-populations are all C's while others are all non-C's. And this third restriction is also consistent with the exemplification of Q's implying that the frequency of C is some particular value, or falls within some interval of values – as long as it is not implied either that the value of this frequency is 0 or that it is 1.

As far as I can tell, the three restrictions I have just described – in addition, of course, to the requirement that the relevant token population P must *exemplify* Q – are the only restrictions that we need to place on appropriate kinds Q.

Besides these subtleties involving populations, having to do with their kinds, there are also subtleties involving the individuals in a population, having to do with the *nature and structure* of these individuals. I think we may most naturally or usually think of such individuals as physically discrete, physically compact entities, such as people, rats, coins, or dice. But some causal and probabilistic claims require a more intricate understanding of the relevant "individuals" of a population. Consider, for example, the causal and probabilistic claims, "Careless smoking in forests is a positive causal factor for forest fires" and "Careless smoking in forests increases the probability of a forest fire." For these claims, it would seem that we need a population consisting of "individuals" that are *situations* in which, say, a person is in the proximity of a forest. In addition to such spatial relations among the constituents of an individual in a population, temporal relations may also play an important role. In Chapter 5, I discuss temporal issues in the theory of property-level probabilistic causation. I postpone a detailed discussion of the nature and structure of individuals in a population until then.

I have argued in this section that a probabilistic causal claim at the type level must be made relative to a given token population, and relative to a given kind that the token population exemplifies. It seems that for *any triple* that we may assemble of a candidate causal factor C, a candidate effect factor E, and a token population P, it would be appropriate to inquire about the causal significance of C for E in P. We have seen, however, that we must associate some kind, Q, with P. And I have argued that the appropriateness of the kind Q depends on *all three of* C, E, and P. Of course, P must exemplify Q. There are three more restrictions: First, the fact of a population's being of kind Q must not, by logic alone, fix value of the frequency of E, the frequency of E given C, or the frequency of E given not-C. Second, the fact of the exemplification of Q must not logically imply what the numerical relation is between the values of the frequency of E, the frequency of E given C, and the frequency of E given not-C. And third, Q must logically allow both for the possibility of there being C's among members of populations of kind Q and for the possibility of there being non-C among members of populations of kind Q.

In the next section, on probability, we will see further the rationale for the appropriateness restrictions on kinds, for the explication of physical and objective probability. We will also see how to incorporate relativity to kinds into an interpretation of probability. Following that, in the examples in chapters to come, I will for the most part not explicitly mention the *kind* of the population relative to which we shall be evaluating probability values or causal relations. The relevant kind will be implicit in the descriptions given of the populations, when the examples are described – and, in fact, it will always be the kinds that are more important than the tokens.

1.2 PROBABILITY

There are two main kinds of theories of probability. In the abstract and formal calculus of probability, a probability is any function (defined on a Boolean algebra of entities of any kind) that obeys a set of probability axioms – I shall assume that the reader is familiar with the easy rudiments of the (standard) formal, mathematical theory of probability.[6] Of course, any abstract, formal calculus is subject to multiple interpretations. Thus, there are multiple theories about how the abstract calculus of probability can or should be interpreted, for the various applications to which the idea of probability has been put. Probability has applications in the areas of the theory of confirmation, decision and game theory, the theory of explanation, logic, and the theory of causation, for example. And an interpretation of probability that may be appropriate in one of these areas may not be appropriate in others.

The purpose of this section is to clarify several points about the kind of interpretation of probability I assume for the explication of probabilistic causation, though I do not advocate or try to develop in detail any particular interpretation. The interpretation I sketch and any proposal along the same rough lines is not without its serious difficulties.

First of all, the kind of conception of probability I assume understands probability to be an objective and a physical rela-

---

[6]Otherwise, see Appendix 2, or if necessary, Appendixes 1 and 2. There I present the formal structure (including the idea of a Boolean algebra, in Appendix 1), the axioms, some standard definitions, and several theorems.

34

tion between event types. To say that it is an objective relation means that the probabilistic facts are the way they are, independently of how we conceive the phenomena and independently of what we think the facts are. To say that it is a physical relation means, roughly, that the probabilistic facts are the way they are in virtue of the physical, causal significances that the relevant factors (that is, their presence) have for one another. And the items related will be understood to be event types, since, of course, it is these that enter into type-level causal relations, the level of causation that is the topic of the first five chapters of this book. This kind of understanding of probability is to be contrasted with subjective and logical interpretations, for example.[7]

Subjective interpretations make probability a measure of an individual's (rational) *degrees of belief,* which can differ from individual to individual, depending on an individual's "prior" (rational) degrees of belief and on the evidence (rationally) absorbed by the individual.[8] The kind of thing that a person can be taken to believe (fully or to some degree) is usually understood to be a *proposition* (or a *statement* or a *sentence*), and not an *event type* (or a *property,* or a *factor*). Thus, the kind of things that enter into subjective probabilistic relations are not the kind of things that enter into type-level probabilistic causal relations. Of course, the propositions believed fully or partially may assert that some *token* event oc-

[7]The most prominent kinds of interpretations of probability are called "classical," "subjective," "logical," "actual frequency," "hypothetical frequency," and "propensity" interpretations. All of these except for the classical interpretation will be touched on below. The classical interpretation, associated with LaPlace (1819), takes the probability of an outcome to be, roughly, the ratio of "equiprobable" (or "equipossible") cases favorable to the outcome to the total number of such cases. Although this interpretation is intended to be objective and physical, and as relating event types, it is now widely agreed that there are insuperable difficulties having to do with the applicability of the interpretation, as there often are no "equiprobable equipossibilities" to be found when we wish to apply the interpretation. See Salmon (1966) for a good introduction to interpretations of probability.
[8]For classics on this kind of understanding of probability, see, for example, Ramsey (1926), DeFinetti (1937), and Savage (1954). For more recent appraisals and advancements, see, for example, Hacking (1967), Kyburg (1978), Skyrms (1980, 1984a), Eells (1982), Jeffrey (1983), and the references therein.

35

curs (at a particular time and place), so that we may think of these propositions as corresponding in a natural way to events. But it is not obvious that there are propositions that stand in the same relation to event *types* as propositions asserting the occurrence of a token event stand to those token events.

However, if the idea of an individual's degrees of belief can somehow accommodate the idea of degree of belief *in an event type* (or in the proposition that the type will be exemplified at some time and place), then a subjective interpretation of probability may be suitable for an explication of the idea of our "causal beliefs." I will not pursue this natural idea here.[9] Instead, I will be concerned directly with the explication of the idea of causation itself.

Logical interpretations of probability make probability a measure of the degree to which one proposition "partially entails" another: The probability of a proposition A, conditional on a proposition B, is the "degree to which B logically entails A."[10] This kind of interpretation will not do for the explication of type-level probabilistic causality, for at least three reasons.

First, on a logical interpretation, probability statements (statements of the form, "the probability of A conditional on B is r") are either logically true or logically false, just like statements to the effect that such and such an argument is deductively valid. (In application, the conditioning proposition B is usually taken to include background information satisfying a total evidence condition, or the background information is implicit.) On logical interpretations of probability,

[9] On this, see Skyrms (1980, 1984a), and Eells (1982, especially pp. 153–4, on the distinction between "type-A" and "type-B" beliefs). Also, some of the kinds of event types described in Chapter 5 of this book – the ones that include a specification of a time and a place – may naturally correspond, in the way described above, to propositions in which one may have a subjective probability.

[10] For the classics on the logical interpretation of probability, see Carnap (1945, 1950, 1952). For more recent appraisals and advancements, see, for example, Carnap and Jeffrey (1971), Jeffrey (1980), and Benenson (1984), and the references cited therein.

probability statements assert that a certain logical relation (of "partial entailment") obtains between propositions; they are not "factual" statements. So, on such an interpretation, the probability statements in terms of which a statement of probabilistic causal connection is explicated would be logically true or logically false. In this case, whether the relevant probability increases or decreases obtain would be a matter of logic, so that the question of the truth of a probabilistic causal claim would be a matter of logic.[11] But, of course, questions of probabilistic causal connection are factual, empirical questions, and should not turn on logical relations between the cause and the effect types – no matter how much evidential background information is included in the analysis.

Second, the logical interpretation is usually taken to be the kind of interpretation of probability appropriate for a theory of *confirmation,* for expressing evidential connections relative to background information that is either implicit or included as part of the conditioning proposition in a probabilistic relation. Indeed, logical probability (relative to given background information) is often identified with a rational degree of belief, "credence," or "credibility" function (see Carnap 1952). For this reason, this interpretation may be appropriate for a theory of "rational causal belief," but not for one about objective causation. Not only may different individuals have different systems of rational credence due to the fact that they have different totalities of background information, but also, relative to a given totality of background information, it is hardly plausible to think that there can be a *unique* rational credence function.[12]

Finally, as in the case of subjective interpretations, the

[11] As explained in Chapter 2, the relevant probability comparisons are within background contexts. Once the appropriate contexts are found, the question of whether there is "unanimous" probability increase, decrease, or no change, within the contexts, would be a matter of logic – on the logical interpretation of probability.

[12] Carnap (1952) characterizes, for simple languages, an infinite family of logical probability assignments, or c-functions, in terms of a parameter *lambda* that corresponds, roughly, to how responsive the function is to evidence.

items that enter into probabilistic relations, on a logical interpretation, are propositions (or sentences, or statements), and not event types. Again, while these propositions may assert the occurrence of token events, it is not clear that there are propositions standing in the same relation to event types as propositions asserting the occurrence of token events stand to the token events. So, since it is event *types,* and not event *tokens* or *propositions,* that enter into probabilistic causal connections, it is hard to see how a logical interpretation could work.[13]

Carnap (1945) distinguished two concepts of probability, which he called probability$_1$ and probability$_2$. Probability$_1$ is logical probability, or degree of confirmation, or degree of rational credence; it is a relation between propositions (or between sentences, or statements, or between particular events); and probability statements are, under this interpretation, either logically true or logically false. Probability$_2$, on the other hand, is *empirical* probability, also called "statistical," "objective," or "*frequency*" (sometimes "*relative frequency*") probability; it is a relation between event types; and probability statements are, under this interpretation, factual and empirical. Carnap cites the work of von Mises (1928, see also 1964) and Reichenbach (1935) as examples of developments of the probability$_2$-conception of probability.[14] There are various ways, some more or less adequate than others for our purposes, of developing in detail this probability$_2$, or frequency, kind of understanding of probability; but, at least among the approaches described so far, it is this empirical, frequency, conception that best fits our needs for the theory of probabilistic causality.

Roughly, the frequency conception interprets the probability of one factor A given another B, written $Pr(A/B)$, as the

[13] Compare Carnap (1945), and again Eells (1982, pp. 153–4).
[14] Salmon (1966) cites Venn (1866) as the classic presentation of the frequency interpretation, citing also von Mises (1928) and Reichenbach (1935) as two of the most important twentieth-century expositions.

ratio of the number of occurrences of the conjunctive factor $A\&B$ to the number of occurrences of the factor B. But this rough understanding has to be refined in several ways. First, we want a theory of type-level causation that captures the causal facts, and not mere coincidences of association or correlation. This means that an *actual frequency* interpretation of probability must be ruled out. In interpreting any probability, $Pr(A/B)$, the actual frequency interpretation pays attention just to occurrences of the factors $A\&B$ and B in the course of the world's actual history.[15] But the actual world contains, or may contain, "mere accidental coincidences" that can give rise to actual frequencies and correlations that misrepresent the causal facts, and that therefore do not allow probability, understood as actual frequency, to reflect the causal facts in a natural way (that is, roughly, in a way in which causation coincides with probability increase).

Here is a simple example of how this can happen. If we have a standard cubical die and we round off the corners of the side with just one dot on it, then we would expect that a detailed physical theory would predict that this should result in the one coming up more frequently on rolls of the die than if we had not so altered the die. On an appropriate objective and physical understanding of probability, we expect that shaving the corners around the one-dot side of a die would *increase the probability* of ones coming up on rolls of the die. Also, we would say that shaving the corners around the one-dot side of a die is a *positive causal factor* for ones coming up on rolls of the die.

However, if the outcomes of rolls of dice, shaved as described or not, is a purely probabilistic, or physically indeterministic, matter, then it is entirely consistent with this probability increase, and positive causal factorhood, that in fact the frequency of ones happens to be smaller among the

[15] For further discussion of the actual frequency interpretation, see, for example, Russell (1948), Salmon (1966), Sklar (1970), Suppes (1974), Fetzer (1977), van Fraassen (1977a), Hacking (1980), and Eells (1983).

(perhaps very small number of) rolls of dice that have been shaved in the way described than it is on rolls of standard dice. Obviously, there is a positive probability, though low, that such a pattern of events should arise, in the course of the world's history. So if we let actual frequencies play the role of probability in the theory of probabilistic causality, then, applying our "basic idea" of positive, negative, and neutral causal factorhood, we would get the wrong answer in this case that shaving the corners off dice in the way described is a *negative* causal factor for ones coming up on rolls of dice.

If, on an appropriate objective and physical understanding of probability, the probabilities of the factors relevant in a given problem are all intermediate (that is, not 0 and not 1), then the actual frequencies could take on any values whatsoever (subject only to the probability axioms, of course). If the situation is indeterministic, then which particular frequencies actually arise will *always* be a "mere coincidence," or accident. Thus, if we used actual frequency as probability in the theory of probabilistic causality, then the answers given to questions about how one factor is causally relevant to another, in a given population, will also turn out to be mere accidents, and thus fail to reflect the nonaccidental character of the causal facts in a given population (of a given kind). Since the actual frequencies in a population could, "by accident," be *anything,* we cannot expect them to be of help in probabilistically representing the one way things are with respect to causal relations in a population.

This is not to say that probabilistic correlations (that is, the presence of one factor affecting the probability of another) can never be misleading, if we somehow develop and use an appropriate understanding of probability. However, we must insist on an understanding of probability that is sensitive to the causal facts in such a way that misleading correlations can be detected and untangled in an appropriate way. Here is an example of what I have in mind, an example of a kind that will be discussed in detail in Chapter 2, on spurious correlation. Falling barometers and rain are joint effects of a common cause (approaching cold fronts), and neither of the two joint effects is a cause of the other. Joint effects of a common cause are often correlated: a falling barometer, for example, increases the probability of a rainy day. But this probability increase does not coincide with causation: Falling barometers do not cause rain. So we may call this kind of correlation a misleading correlation, even if it does not arise by mere coincidence, or accident.

As mentioned above, an appropriate understanding of probability must be sensitive to the causal facts in a way that allows us to detect and untangle misleading correlations. In Chapter 2, the solution to the problem, for this example, will be to consider *separately* occasions on which a cold front is approaching and occasions on which no cold front is approaching. This is called "holding fixed" the common cause, positively and then negatively. We expect that in both cases – that is, both when the common cause is held fixed positively and when it is held fixed negatively – the correlation between falling barometers and rain will disappear. In neither case should falling barometers increase the probability of rain. If so, then when the theory tells us to consider the two cases separately (as it will), the theory will tell us, correctly, that falling barometers do not cause rain.

However, suppose that the relevant causal relations are all indeterministic, and suppose that, aside from whether or not a cold front is approaching, there is no other factor that is causally relevant to whether or not barometers fall or to whether or not it rains. This means that the factors of a cold front's approaching, and of no cold front's approaching, should confer *intermediate probabilities* on the factors of falling barometers and rainy days. We saw above that the probability of rain should be the same given an approaching cold front with falling barometers as it is given an approaching cold front without falling barometers. But, because the probabilities are all intermediate, this is consistent with the possibility

41

that, by mere coincidence or accident, rainy days happen to be more frequent when there is an approaching cold front and the barometers are falling than they are given just that there is an approaching cold front. And the same may also be true when there is no cold front approaching, again by mere coincidence or accident.

In this case again, if we used actual frequencies in the representation of causal facts as probability relations, we would still get the wrong answer, that falling barometers cause rain, even when we hold fixed the factor of whether or not a cold front is approaching. The shaved die example showed that frequency correlations can mislead, being in a different direction from that of the correct probabilistic correlations, and can in this way suggest the wrong kind of causal significance. And the falling barometers example shows that even in cases in which we should expect correlation without causation, the actual frequency interpretation is incapable of always untangling the misleading correlation.

These kinds of considerations have led some philosophers to develop "modal" versions of the frequency interpretation of probability that are intended to correspond more closely to what I have been vaguely calling an "appropriate objective and physical" understanding of probability, an understanding on which probability is supposed to be sensitive to the causal facts and not to mere coincidences. Some philosophers (for example, Popper 1959 and Kyburg 1974) have suggested that probability be explicated in terms of limiting frequencies in *hypothetical infinite populations* of an appropriate kind, while others have sought to incorporate the modal element by reference to the idea of *physical propensity* (Popper 1959; Fetzer 1971, 1981; Giere 1973). In addition, Skyrms (1980, 1984a) has investigated the possibility of representing *causal beliefs,* and *the way we reason about objective probabilities,* in terms of ("resilient") *subjective* probabilities that are conditional on appropriate constellations of background factors.

Serious difficulties confront hypothetical infinite frequency and propensity accounts.[16] I believe that an adequate understanding of objective probability should be sought by examining the role that the probability concept plays in what we take to be, for example, theoretically adequate explanations of single events and of regularities, theoretically adequate understandings of events and objects themselves, as well as our understanding, such as it is, of causal facts and laws, including the proper role of the probability concept in understanding the character of rational decision when causal and evidential considerations conflict (see Skyrms 1980, 1984a, and Eells 1982, 1983). In that case, our understandings of probability, causation, explanation, events, and objects, for example, are interdependent. I believe that such an approach is fully consistent with Skyrms's approach, which emphasizes the role of our concept of objective probability in the theory of learning from experience – that is, in confirmation theory, or the theory of proper assimilation of evidence.

To guide intuitions about probability in this book, the interpretation of probability that I sketch below, intended to be an objective and physical interpretation, is a mixture of a propensity kind of account and a hypothetical frequency approach. Although I think there are serious difficulties with such theories, they at least have the advantage of avoiding the problems discussed above (as well as other problems) that confront actual frequency interpretations. I am not *proposing* the theory that I sketch below. After explaining the interpretation, I will mention two difficulties with it – one of which I think confronts, in a conclusive way, all hypothetical frequency accounts of probability, and the other of which I think seriously confronts all propensity accounts of probability. In addition, I make no claim of noncircularity for the combination of the account of *probability* to be sketched here with the theory of probabilistic *causation* that will be developed in the chapters that follow. Note, however, that some

[16] See, for example, Sklar (1970) and Eells (1983).

sort of circularity should be expected anyway, if our understandings of probability and of causation (as well as of other ideas) are interdependent, as suggested above.

We begin with an actual population, say P. We are interested in conditional and unconditional probabilities involving just certain factors, say X, Y, Z, X_1, and so on. I suppose that population P is finite, with N members. One problem with the actual (and finite) frequency conception of probability is that on this account, probabilities $Pr(X)$ and $Pr(X/Y)$ can only take on the values 0, $1/N$, $2/N$, . . . , $(N - 1)/N$, and 1. For example, if the actual population in question consisted of just three tosses of a certain "fair" coin, then the actual frequency of heads would have to be either 0, ⅓, ⅔, or 1 – while the truth may be that the probability of heads is ½. A second problem with this interpretation, noted above, is that the actual frequencies may be mere coincidences that, if taken as probabilities, can misrepresent the causal facts. Hypothetical frequency and propensity accounts of probability are intended (among other things) to deal with these two problems.

Let us think of the way the actual population P is as the result of one "experiment," an experiment that could, theoretically, be repeated. The experiment goes like this: We invest a set of N individuals with a certain "distribution of initial conditions" (as Sklar 1970 expresses the idea). The initial conditions are factors that are causally relevant to the way individuals in P are, with respect to having or lacking the factors X, Y, Z, X_1, and so on; except that the factors X, Y, Z, X_1, and so on, are not among these *initial* conditions. If the initial conditions are I_1, . . . , I_n, then a distribution of initial conditions just specifies the frequencies of the (at most 2^n) possible combinations of I_i's and $\sim I_i$'s. A distribution of initial conditions gives rise to what has been called an "experimental set-up," which gives the individuals in the population "propensities" of various strengths to have or lack the factors X, Y, Z, X_1, and so on. These propensities will differ from

44

individual to individual, depending on which of the initial conditions are present and absent in the individuals.

The *result* of the experiment is the resulting conditional and unconditional frequencies, involving the factors X, Y, Z, X_1, and so on, in P. The initial conditions were distributed among the N members of the actual population P in one certain way. So to repeat the experiment, we distribute these conditions in exactly the same way, in a (possibly) different population of individuals. And we can imagine this being done again and again. If the causal relation between the initial conditions and the factors X, Y, Z, X_1, and so on, is not deterministic, then we should expect the resulting frequencies to differ from one experiment, or hypothetical population, to another.

For now, let us suppose that the experiment is repeated on populations *of the same size, N,* as the actual population, P. As we shall see, this is tantamount to the assumption that the *kind, Q,* that we are associating with P requires, for its exemplification, an *N-membered* population. Later, this account will be generalized to accommodate kinds whose exemplifications *do not* fix the sizes of populations.

Let P_0 ($= P$), P_1, P_2, . . . be the infinite sequence of hypothetical populations that would result from conducting this experiment infinitely many times.[17] And let Fr_0, Fr_1, Fr_2, . . . be functions giving the values of frequencies – involving the factors X, Y, Z, X_1, and so on – in the populations P_0, P_1, P_2, . . . , respectively. Then, according to this sketch of a hypothetical frequency/propensity account of probability, we may define, for any of the factors X of interest,

(1)
$$Pr(X) = \lim_n \frac{\Sigma_{i=0}^n Fr_i(X)}{n} .$$

This is the limit, as n approaches infinity, of the average of the frequencies of X in the first n (or, strictly speaking, $n + 1$) of

[17] But see below for difficulties with this stipulation.

the P_i's.[18] This limit (if it exists – see below) is, intuitively, the analogue, for the infinite case, of the *average* of finitely many frequencies. If the P_i's are all set theoretically disjoint, and # is the cardinality (number of elements) function, then (given that the P_i's all have the same number of elements) this definition comes to the same thing as

$$(2) \quad Pr(X) = \lim_n \frac{\#\{\, a \,:\, a \text{ in } X, \quad \text{and in some } P_i, \, i < n \,\}}{\#\{\, a \,:\, a \text{ in some } P_i, \, i < n \,\}}.$$

Conditional probability, $Pr(X/Y)$, is defined in the usual way in terms of unconditional probability as $Pr(X\&Y)/Pr(Y)$.

It is worth explaining why we consider the different populations P_i one by one, and then average their Fr_i's, rather than combining the P_i's into one overall population and then looking for limits of frequencies in the combined population. First, the value of a limit of frequencies can depend on the *order* in which the individuals in the population are taken. So if we combined all the P_i's into one large population, we would have to decide on some way of ordering the population into a *sequence*. But there is no principled way of doing this.[19] Second, it is possible for the frequency, and for the probability, of some factors to depend on the *size* of the population: The size of a population may itself be an initial condition causally relevant to certain factors. This is why we first identified the Fr_i's, all defined on populations *of size N,* separately, and then "averaged" them, rather than first com-

---

[18] Malcolm Forster and Dan Hausman point out that, on hypothetical limiting-frequency conceptions of probability, the actual frequencies in the actual population P_0 "wash out." The frequencies in the actual finite population have no effect on what the limiting relative frequency is. Thus, on this kind of understanding of probability, both probability and type-level probabilistic causation may be understood to be relations among, in addition to the relevant factor or factors, *just a kind* Q that the actual population exemplifies: other than by way of one of the *kinds* Q that the actual population is considered to exemplify, the *actual* population P_0 plays no role in the analysis. This is, of course, consistent with the fact that it is actual populations (that exemplify many kinds) that are the point of departure for scientific investigations into causes.

[19] There is a related problem for the present approach: namely, how to order the populations P_i. See below.

bining all the N-membered populations into an *infinite* population and then looking for limiting frequencies in this infinite population.

This sketch of a theory avoids the two problems mentioned above for the finite frequency account. First, since a probability will be, roughly, an "average" of *infinitely many* finite frequencies, there is no limit to the precision of probability values, and probability values are not restricted to finitely many possibilities. Any limit of averages of increasing numbers of finite frequencies – that is, any real number between 0 and 1 inclusive – will qualify as a possible probability, under this interpretation. And second, since infinitely many "experiments" must be (hypothetically) conducted, the effect of "accidental coincidences" that inevitably arise in single populations should be dampened. At this point, some philosophers have invoked considerations involving the statistical law of large numbers, claiming that in the limit the effect of accidental coincidences should, with probability 1, be completely eliminated.

In terms of a model like this, we can see how probabilities in a given *token* population are relative to a given *kind* that the token population exemplifies, an idea discussed in the previous section. A given token population, P, will exhibit a given distribution, call it D, of initial conditions $I_1, \ldots, I_n$, as explained above. But $I_1, \ldots, I_n$ may just be a *more or less complete* list of all the factors that are distributed in some way in token population P, and that may be considered as initial conditions. If $J_1, \ldots, J_m$ is a *different* list of initial conditions – maybe a more complete list, maybe a less complete list, or maybe different in some other way – then *these* factors will *also* be distributed in some particular way, call it D', in the token population P. Then it is natural to say that D and D' are two *kinds* or *types* that population P exemplifies.

Relative to P's being of kind D, we hypothetically repeat the experiment P using experimental set-up D, in order to find probabilities on the model just sketched. And relative to

47

P's being of kind D', we repeat the experiment P using experimental set-up D'. Of course, depending on *which* sequence of experiments we look at, the limit of finite averages of the relevant finite frequencies can be different. This is how this interpretation of probability makes probability, in a *token population P*, relative to a given *kind* that token P exemplifies. It is easy to see how this relativity to a *kind* of a token population, in this interpretation of probability, applies in the examples discussed in the previous section and in the introduction. (Henceforth, as before, I will use "Q," "Q'," and so on, to denote the *kinds* that token populations exemplify.)

So far, I have only discussed the case in which the *kind* that we associate with P *fixes the size of the population*. All the hypothetical populations considered (populations of the same kind as P) were required to be of the same size, N, as P. Concerning this, there are three issues to be addressed. First, there surely are kinds that we may wish to associate with a token population that *do not* fix the size of populations that exemplify them. Actual population size is a factor, like those factors in examples discussed in the previous section, that we may not want to hold fixed in the assessment of probabilistic and causal claims. Second, there is the fact that, when we let population size to vary from one hypothetical population P_i to another, the two characterizations of probability given above, (1) and (2), do not always come to the same thing. Examples of this will be given below. So, third, assuming that one of (1) and (2) above is correct in such cases, the question arises of which one it is – since it cannot be both.

Here is a simple example that shows that (1) and (2) can be in conflict, when the exemplification of a given kind is compatible with multiple population sizes.[20] Consider this kind. A population is of kind Q if and only if: By a 50-50 chance process it is determined whether the population shall consist of four or of eight coin tosses; if four tosses, then a fair coin

---

[20] I thank Marty Barrett for helpful discussions about this conflict, and about which of (1) and (2), if either, is always correct.

will be tossed; and if eight tosses, then a coin biased 3:1 for heads will be tossed. Let X be the factor of a coin's landing heads, and let us see what value we should expect for $Pr(X)$ given the two characterizations, (1) and (2) above, of Pr.

Without being very rigorous, we can reason as follows. For (1), we take the average of the frequencies of X in the four- and eight-membered hypothetical populations that result from repeating the experiment. For the four-membered populations, we expect the frequency of heads (X) to average 0.5, since in these populations a fair coin is tossed. In the eight-membered populations, we expect the frequency of heads to average 0.75, since in these populations a coin biased 3:1 for heads is tossed. Since we expect nearly exactly as many four-membered populations as eight-membered populations (this because it is a 50-50 chance process that determines whether a population will be four- or eight-membered), we should expect the overall average of the frequency of X in all the populations to be the straight average of 0.5 and 0.75, or 0.625 (⅝).

For characterization (2) of Pr, we combine the populations of tosses into one series of tosses, where the order of the tosses that occur within one population is not important, and the important thing is that, where $i < j$, tosses from P_i come before tosses from P_j. Again not being very rigorous, we can say that twice as many come from eight-membered populations as come from four-membered populations. Thus, as a preliminary calculation, we have:

$$Pr(X) = (⅓)(\text{frequency of heads among tosses from four-membered populations})$$
$$+ (⅔)(\text{frequency of heads among tosses from eight-membered populations}).$$

And this we should expect to equal (⅓) (½) + (⅔)(¾), or ⅔.

Thus, the method embodied in characterization (1) of Pr gives $Pr(X) = ⅝$; and the method of characterization (2) gives $Pr(X) = ⅔$. Now, of course, the question is: What is the probability of X in a population P considered to be of kind Q?

Is characterization (1) above correct, so that the answer is ⅝? Or is it characterization (2) that is correct, so that the answer is ⅔? An intuitive way of putting the question is this: In the hypothetical populations in terms of which (1) and (2) are formulated, should we, in assessing the probability of X, weight the *populations* equally, or should we weight all the *individuals,* of which the populations are composed, equally? I believe that the correct approach is (1), that the populations should be weighted equally, and the correct probability value in the example above is ⅝.

To motivate these answers, I can only offer another example, one in which the discrepancy between (1) and (2) is more extreme and in which it is (1) that seems clearly to give the right answer. Let us say that a population P is of kind Q if and only if: P consists of entrants in a lottery in which there are exactly 100 losers, where the size of P may range from 101 to 10,000, and the process that generates P gives each of these 9,900 population sizes an equal chance of being realized. Let X be the factor of being a winner. The extremes cases here are tokens of Q in which there are 101 entrants and 1 winner (frequency of winners about 1 percent) and tokens of Q in which there are 10,000 entrants and 9,900 winners (frequency of winners 99 percent). Without going through the calculations here, I simply note that, for the probability of X in a population of kind Q, characterization (1) gives (about) 0.5 and characterization (2) gives (about) 0.98.

Considering a token population P simply to be of kind Q, each of the 9,900 possible population sizes (101, 102, . . . , 10,000) is equally likely for P, as is each of the possible frequencies of winners (1/101, 2/102, . . . , $(1 + i)/(101 + i)$, . . . , 9,900/10,000, for $i = 0$, . . . , 9,899). Thus, the expected frequency of X in P, considering just that P is of kind Q, is 0.5. This is the answer given by (1) for the probability of X in P relative to P's being of kind Q, and it is the answer given by weighting hypothetical populations equally – that is, by weighting equally all the relevant equally probable ways P could be,

50

given that it is of kind Q. This seems to be the appropriate way to proceed.

Method (2), on the other hand, proceeds by weighting equally all the (hypothetical) individuals in all the hypothetical populations. This seems inappropriate for assessing the probability of X *in P*. Method (2) proceeds as if there is an enormous population of hypothetical lottery entrants that have chances of becoming a member of a 101-membered population, 102-membered population, . . . , a 9,999-membered population, and a 10,000-membered population. Of course, these individuals would have better chances of becoming members of the larger populations than members of smaller populations, and more of them end up as members of the larger populations than members of the smaller populations. Thus, by weighing these "hypothetical individuals" equally, method (2) biases the assessment of $Pr(X)$ by weighing the larger populations more heavily than the smaller ones, where by hypothesis each of the 9,900 population sizes is equally likely. Method (2) would be delivering the correct answer for the probability of X in the enormous population of hypothetical lottery entrants (if we wished to consider such a population), but this is quite different from the probability of X *in a token population P relative just to P's being of kind Q*.

Thus, probability should go by (1), and not (2), on this sketch of a hypothetical frequency/propensity interpretation of probability.

I do not wish to defend this kind of interpretation of probability against all kinds of criticisms that may be raised; I only intend this model to help convey some general ideas about the kind of objective and physical conception of probability that I will assume in the chapters that follow. And for this purpose, it seems useful to appeal to the more or less intuitive and natural ideas of propensity and hypothetical limiting frequency, which a number of philosophers have used in this connection – though I do not know of any philosopher's having proposed just the kind of model sketched above, taking

51

probability as an infinite "average" of frequencies in *finite* hypothetical populations. As I have already mentioned, I think there are serious problems with accounts of this kind (Eells 1983). Here I briefly described two difficulties, one pertaining to the appeal to the idea of hypothetical limiting frequencies, and the other to basing the explication of probability on the idea of propensity.

In the theory sketched above, it was stipulated that some sequence, P_0, P_1, P_2, . . . , is *the* sequence of populations that *would* result from performing, infinitely many times, the "experiment" described. But, of course, it is hardly plausible to assume that there is a definite, unique possible outcome of hypothetically replicating the experiment infinitely many times – that is, to assume that there is just one possible such infinite outcome sequence of hypothetical populations. And, of course, the value of the relevant limit of averages of frequencies will vary from outcome sequence to outcome sequence. Furthermore, this value can depend on the *order* in which the P_i's are taken, so there is the problem of formulating and justifying a principle that tells us how the hypothetical populations should be ordered. Philosophers who have advanced hypothetical limiting frequency accounts of probability have recognized and have tried to deal with these problems.

Typically some such more complicated idea as the following is tried. It is suggested that "$Pr(X) = r$" be understood as meaning that for "all or almost all" infinite sequences of populations that might result from replicating the experiment infinitely many times, the limit of the average frequency is r.[21] But then it is natural to ask: "Which is it, *all* or just *almost all?*" Surely it is implausible to insist on "all," for among the presumably *nondenumerably many possible sequences* of populations, there will be two that yield different limits. For example, suppose population P consists of one toss of a fair coin

[21] This is very similar in spirit to what Kyburg (1974) has proposed in this connection, but otherwise his hypothetical frequency theory is quite different in details and basic structure from the account sketched above.

(and the relevant kind is given by this description of P). Then, as Brian Skyrms (1980) puts basically the same point,

On the hypothesis that the coin has a propensity of one-half to come up heads on a trial and that the trials are independent, *each infinite sequence of outcomes is equally possible. If we look at all physically possible worlds* [or sequences of populations that consist of single tosses of the coin], *we will find them all,* including the outcome sequence composed of all heads. (p. 32)

There will also be the outcome sequence composed of all tails. And for *any* real number r, between 0 and 1 inclusive, there will be possible outcome sequences having r as the limit of the frequency of heads. So we cannot insist on "all."

So it must be "almost all." But what does this mean? We cannot take it to mean "all but a finite number." In the coin-tossing example, there will be, among all the possible outcome sequences, some in which the limit of the frequency is $\frac{1}{2}$, and others in which the limit is, say, $\frac{1}{3}$. But if there is at least one of each of these two kinds, then there will be infinitely many of *each* kind. (It is a simple fact that, if the limit of the frequency of an outcome in an infinite sequence of trials is, say, r, then rearranging the outcomes in any finite initial segment of the sequence leaves the limit at r; and there are infinitely many ways of rearranging outcomes on finite initial segments of an infinite sequence.) In general, assuming that the relevant physical system is indeterministic, then for each r strictly between 0 and 1, there will be an outcome sequence in which the limit of the relevant frequency is r; and if there is at least one, there will be infinitely many; so, for each r strictly between 0 and 1, there will be infinitely many outcome sequences for which the limit of the frequency is r.

In the face of this difficulty, the following kind of approach has been suggested.[22] Where S_0, S_1, S_2, . . . is an infinite

[22]This is similar in spirit to the proposal of Fetzer and Nute (1979, 1980); see also Fetzer (1981). Again, however, their proposed hypothetical limiting frequency account is quite different in general structure and details from the account sketched above.

sequence of outcome sequences (an infinite sequence of infinite sequences of the kind of sequence, P_0, P_1, P_2, . . . , described above, where, say, $S_i = P_{i0}$, P_{i1}, P_{i2}, . . .), $Pr(X) = r$ if and only if the limit of the frequency of S_i's, in which the limit of the frequency of X is r, is 1. That is, where LFr_i's are the *limiting average frequency* functions for sequences S_i:

$$Pr(X) = r \text{ if and only if:}$$
$$\lim{}_n \frac{\#\{\, S_i \ : \ i < n \text{ and } LFr_i(X) = r \,\}}{n} = 1.$$

Again, however, there are fatal flaws in this kind of proposal. For one thing, there are, presumably, *nondenumerably many outcomes sequences*, so that this approach requires that we somehow select, from these nondenumerably many, just *denumerably* many, to be the S_i's. The limit, above, as n approaches infinity, must, of course, be a limit of just denumerably many ratios. But if the relevant physical system is indeterministic, then there should be, *for each value of r between 0 and 1*, denumerably many outcome sequences *all of* which have the limit of the relevant frequencies equaling r. So there is the problem of choosing some "appropriate" denumerable set of outcome sequences, from the nondenumerably many that are available.

Furthermore, even if this can be done, there is the further problem of deciding just how to order these denumerably many outcome sequences into a sequence of outcome sequences; and of course the values of limits can depend on the order in which relevant items (the S_i's in this case) are taken. And there seems not to be any natural or principled order in this case.

As a final thorn, not every infinite sequence of averages of frequencies (of any of the kinds described above) need even *have* a mathematical limit.

So the hypothetical limiting frequency idea is not without its difficulties, once we try to be formally precise about the idea (see further Eells 1983). The propensity idea can be criticized somewhat differently. One can either take the propensity idea as a primitive, in which case it is not very illuminating (we might as well take probability itself as a primitive), or we could try to explicate it in terms of hypothetical frequencies, in which case, of course, the idea confronts the same kinds of problems just described above.[23]

Although I have criticized the interpretation of probability sketched above, and while I would advocate a different kind of approach to understanding probability as objective and physical, my hope is just that the discussion has conveyed a rough conception of objecive and physical probability that may help to guide intuitions about the idea of probability in the chapters that follow.

[23] Again, see Eells (1983) for more details.

2

Spurious correlation and probability increase

The first main qualification of the basic probability-increase idea of probabilistic causation, explained in Chapter 1, is the relativity of the causal relation to a given *token population,* considered to be of a given (appropriate) *kind* that the population exemplifies.[1] The second main qualification of the basic probability-increase idea, to be explored in this chapter, involves the possibility of what has been called "spurious correlation." Of course, what is meant by saying that a factor X raises the probability of a factor Y is that $Pr(Y/X) > Pr(Y)$ – equivalently, $Pr(Y/X) > Pr(Y/\sim X)$.[2] Another way of expressing this relation is to say that Y is positively probabilistically *correlated* with X. It is famous that "correlation is no proof of causation," and it is also true that causation does imply correlation. The possibility of spurious correlation is one reason why.

In this book, I will actually explore in detail three general

[1] As noted in Chapter 1 (note 18), the actual token population washes out, or disappears, so to speak, when probability is analyzed in terms of hypothetical relative frequencies involving infinitely many populations or individuals. The actual frequencies in a finite token population have no mathematical effect on a hypothetical infinite limit. All that is relevant, on this kind of interpretation of probability, is the *kind* that the actual token population is considered to exemplify. But, of course, there are other ways of understanding probability than hypothetical limiting-frequency approaches. In any case, in what follows, I will often include (perhaps needlessly) the actual *token population* (as well as its *kind*) as a relatum in the relations of probability and of population level probabilistic causation.

[2] $Pr(Y) = Pr(X)Pr(Y/X) + Pr(\sim X)Pr(Y/\sim X)$. So, assuming that $Pr(X) \neq 0$ and $Pr(\sim X) \neq 0$, $Pr(Y)$ is an average of $Pr(Y/X)$ and $Pr(Y/\sim X)$, and must therefore lie strictly between these two values. Henceforth, I will for the most part use the letters "X," "Y," "F," "G," and so on to refer to factors, rather than "C" and "E," which already suggest "Cause" and "Effect," which can perhaps be misleading in some cases.

ways in which probability increase may fail to coincide with causation, and I will show how the probability-increase idea of causation should be adjusted to accommodate these three possibilities. After briefly describing the three possibilities below, this chapter will concentrate on one of them, the one called "spurious correlation." The other two will be dealt with in subsequent chapters.

One simple way to see that probability increase does not imply causation is to notice that the relation of positive correlation is symmetric. If X raises the probability of Y, then Y raises the probability of X.[3] So if probability increase implied causation, then causation would be symmetric as well, at least for probability-increasing causes. But clearly the relation of causation is not symmetric, and we do not want our theory to imply that Y is a cause of X whenever X is identified as a cause of Y. In fact, most plausibly, the relation of causation is *asymmetric*. If so, then if X causes Y, then Y *does not* cause X. So if X is a probability-increasing cause of Y, then Y is a *probability-increasing noncause* of X.

A natural approach to this kind of probability increase without causation would be to include in the theory of probabilistic causation, along with the probability increase idea, a condition requiring that a cause precede its effect in time. This would handle this kind of probability increase without causation, because temporal precedence is not symmetric. In Chapter 5, I argue that the temporal priority idea *must* be *explicitly* incorporated into the probabilistic theory, in order to handle what we may call this "problem of temporal priority of causes to effects." Until Chapter 5, let us adopt the convention that the factor denoted by the letter "X" is temporally prior to the factor denoted by the letter "Y." In most cases this will be obvious. But the idea of one *factor* (or *property* or *type*, an *abstract* thing) preceding another in time is somewhat

[3] If $Pr(Y/X) > Pr(Y)$, then, by the standard definition of conditional probability, $Pr(Y\&X)/Pr(X) > Pr(Y)$. It follows that $Pr(Y\&X)/Pr(Y) > Pr(X)$, so that again by the definition of conditional probability, $Pr(X/Y) > Pr(X)$.

subtle and puzzling. In Chapter 5, I suggest a way of understanding this idea that is very natural in light of the relativity of probability and probabilistic causation to populations.

A second kind of probability increase without causation involves the possibility of what has been called "causal interaction." This kind of possibility, and its relevance to the probability-increase theory of type-level causation, is easy enough to explain. But in order to motivate the solution to the problem adequately, it is necessary first to lay down some of the rudiments of the theory, the relevant parts of which will be given in this chapter. The following may provide a general idea of the problem, which is addressed in detail in Chapter 3.

It is possible for a causal factor X to *interact* with a factor F, relative to the production of a factor Y, in the sense that the causal significance of X for Y is different when F is present from what it is when F is absent. To use an example of Cartwright's (1979), ingesting an acid poison (X) is causally positive for death (Y) when no alkali poison has been ingested ($\sim F$), but when an alkali poison has been ingested (F), the ingestion of an acid poison is causally negative for death. I will argue that in a case like this it is best to deny that X is a positive causal factor for Y, even if, overall (for the population as a whole), the probability of death when an acid poison has been ingested is greater than the probability of death when no acid poison has been ingested (that is, even if $Pr(Y/X) > Pr(Y/\sim X)$). I will argue that it is best in this case to say that X is causally *mixed* for Y, and despite the *overall* or *average* probability increase, X is nevertheless not a positive causal factor for Y in the population as a whole.

Chapter 3 shows that another problem that arises in thinking about causation, the problem of disjunctive causal factors, is an instance of this kind of problem. First, however, we must deal with a third kind of probability increase without causation, which has been called "spurious correlation." The resolution of this problem will give us the framework, and

58

part of the motivation, for the resolution of the problem of probabilistic causal interaction.

2.1 SPURIOUS CORRELATION

On what seems to be the usual understanding of the term, two factors are *spuriously correlated* when (roughly) neither causes the other and the correlation disappears when a third variable is introduced and "held fixed" – that is, the correlation disappears both in the presence and in the absence of the third factor.[4] This does not quite capture the kind of situation I will explore in this chapter. As Simon (1954) is careful to point out, if a correlation between factors X and Y disappears both in the presence and in the absence of a third factor Z, then the explanation may be *either* that the correlation results from the joint causal effect of Z on X and Y (Z is a common cause of X and Y) *or* that Z is an intermediate causal factor between X and Y (X operates on Y through Z or Y operates on X through Z). We shall not count the second possibility as a case of spurious correlation. In the second case, one of X and Y may in fact be a genuine positive causal factor for the other (of course, given our convention that the factor represented by the letter "X" temporarily precedes the factor represented by the letter "Y," it cannot be that Y causes X). This kind of case will be discussed in Chapter 4, on causal intermediaries and transitivity of causal chains.

So let us for now understand there to be a spurious correlation between two factors X and Y if neither causes the other and they are correlated effects of a common cause Z, where the correlation of (the later) Y with (the earlier) X disappears when Z is held fixed. Because we are excluding the case in which Z is causally intermediate between X and Y, I sometimes refer to the common cause Z as a "separate cause" of factor Y – that is, a cause of Y that is separate from X's causal

---

[4]See, for example, Simon (1954), Suppes (1970), and Skyrms (1980).

role, if any, for Y. This understanding of spurious correlation will have to be generalized in several ways below, but first some explanation of the definition and some examples to illustrate the idea.

Suppose a factor Z is a cause of both X and Y. See Figure 2.1. (In this figure, and in others that follow, the solid lines with arrows represent causal connections, the broken lines represent correlations, and the "+"'s and "−"'s indicate whether the causal impact or the correlation is positive or negative.) In the simplest kind of common cause case (others will be considered later), the following relations hold:

(1) $Pr(X/Z) > Pr(X/\sim Z)$,
(2) $Pr(Y/Z) > Pr(Y/\sim Z)$,
(3) $Pr(Y/Z\&X) = Pr(Y/Z\&\sim X)$,
(4) $Pr(Y/\sim Z\&X) = Pr(Y/\sim Z\&\sim X)$.

Propositions (1)–(4) imply $Pr(Y/X) > Pr(Y/\sim X)$.[5] (1) and (2) correspond to the assumption that Z is a common cause of X and Y, on the probability increase idea. (3) and (4) say what it means for the correlation between X and Y to disappear when Z is held fixed (positively and negatively); and they correspond roughly to the assumption that Z is the *only* factor involved that has any causal influence on any of the others. And the derivation of $Pr(Y/X) > Pr(Y/\sim X)$ from (1)–(4) is supposed to *explain* (in the simple kinds of cases I have in mind now) the correlation of Y with X in terms of the "screening off" common cause Z.[6]

[5] $Pr(Y/X) = Pr(Z/X)Pr(Y/Z\&X) + Pr(\sim Z/X)Pr(Y/\sim Z\&X)$. So, by (3) and (4),
$$Pr(Y/X) = Pr(Z/X)Pr(Y/Z) + Pr(\sim Z/X)Pr(Y/\sim Z).$$
Also by (3) and (4),
$$Pr(Y/\sim X) = Pr(Z/\sim X)Pr(Y/Z) + Pr(\sim Z/\sim X)Pr(Y/\sim Z).$$
Let $a = Pr(Z/\sim X)$ and $b = Pr(Y/\sim Z)$. Then by (1) and (2), and symmetry of correlation, there are positive numbers u and v such that $Pr(Y/X) = (a + u)(b + v) + (1 - a - u)b = av + uv + b$, and $Pr(Y/\sim X) = a(b + v) + (1 - a)b = av + b$. Since $uv > 0$, $Pr(Y/X) > Pr(Y/\sim X)$.
[6] This is the idea articulated by Reichenbach (1956) in his "principle of the common cause," according to which correlated factors can be explained in terms of a common cause. (3) and (4) are what it means to say that Z and $\sim Z$ each screen off Y from X – the same as what it means to say that the correlation between X and Y disap-

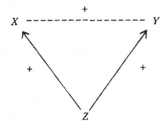

Figure 2.1

Here is a concrete example that is often used to illustrate this: The British statistician Ronald Fisher (1959) once considered the possibility that lung cancer (Y) is positively correlated with smoking (X) not because smoking causes cancer, but because there is a genetic common cause (Z) of the two. In this hypothetical example, X increases the probability of Y, even though X in no way causes Y. The probability increase is due to smoking's increasing the probability of its cause, the "smoking gene," which in turn increases the probability of the gene's effect, lung cancer. But if we hold fixed whether or not the gene is present, smoking will not increase the probability of lung cancer. A different example of this was discussed in Chapter 1. Rain (Y) is correlated with falling barometers (X) that precede the rain. But falling barometers do not cause rain. Again, the two factors are effects of a common cause: approaching cold fronts (Z).

As I mentioned above, our understanding of spurious correlation must be generalized; the characterization given above is too narrow. For one thing, there is the possibility of a spurious correlation arising from the operation of *multiple* separate causes of a factor Y, where these causes are causally indepen-

pears when Z is held fixed. Salmon (1978) calls the kind of structure described by (1)–(4) a "conjunctive fork," which he distinguishes from "interactive forks," in which one or both of (3) and (4) are false. I discuss interactive forks in Section 2.4. For criticisms of the principle of the common cause, and assessments of the purposes to which it has been put, see also van Fraassen (1977b, 1980), Salmon (1984), Sober (1984b) and (1987a), Torretti (1987), and Forster (1988).

61

dent of each other and of the factor X whose causal role for Y is in question. The analysis of this kind of case will be somewhat different from that of the single separate cause (Z) case. I will turn to the multiple separate causes case at the end of this section, and will for now restrict the analysis to the case of the single separate cause.

The more general phenomenon of spurious correlation that I characterize below can arise in a number of ways, with various possibilities for the true causal relation between X and Y, including but not limited to causal neutrality. The basic idea behind the more general understanding of spurious correlation can be expressed, intuitively, like this: Because of the operation of a factor commonly causally relevant to X and Y, the *magnitude* of correlation of Y with X is different from the magnitude of X's causal significance for Y. The idea is to include as cases of spurious correlation, not only cases in which the *direction* of inequality between $Pr(Y/X)$ and $Pr(Y/\sim X)$ fails to coincide in the natural way with the kind of causal significance X has for Y, but also cases in which the magnitude of the difference between $Pr(Y/X)$ and $Pr(Y/\sim X)$ fails to coincide with the degree of causal significance of X for Y. Before being more precise about this (in particular, about how we should understand these magnitudes), I will illustrate the idea with the help of a well-known example, and variations on it.

Nancy Cartwright (1979) cites a study by Bickel, Hammel, and O'Connell (1977) on graduate admissions at Berkeley. It was found that, in the population of all applicants, getting admitted was positively correlated with being male. The frequency of admission among male applicants was higher than the frequency of admission among female applicants. This naturally suggested discrimination against women, and (as Cartwright puts it) "thus rais[es] the question '*Does being a woman cause one to be rejected at Berkeley?*' " (Equivalently: "Is being male a positive causal factor for getting admitted at Berkeley?") However, admissions decisions were made within the academic departments to which one applied. And when the admissions histories of the departments were investigated separately, one by one, it was found that there was no department within which there was a correlation between gender and getting admitted. This is consistent with the fact that, *on average,* the frequency of admission was lower among women than it was among men. The women applicants tended to apply to departments into which it was harder to gain admission.

The table below gives an example of how this can happen; all the entries are the number of accepted applicants over the number applying.

	Department 1	Department 2	Total
Male	81/90	2/10	83/100
Female	9/10	18/90	27/100

In this example, Department 1 accepts 90 percent of all male applicants as well as 90 percent of all female applicants, while Department 2 accepts 20 percent of all male applicants as well as 20 percent of all female applicants. Within each department, there is no correlation between gender and admission. Overall, however, the probability of getting admitted is more than three times greater for male applicants than it is for female applicants. The department-by-department analysis of admissions records was taken as exonerating the Berkeley graduate school from the charge of discrimination: Being male, it now seems, was not after all a positive causal factor for getting admitted, in the population of all applicants to the Berkeley graduate school.

If the question of discrimination against women in the example is equivalent to the question "Is being male a positive causal factor for getting admitted?," and if the more careful look at the data in fact shows that Berkeley is not guilty of discrimination, then we have to conclude that being male is not, after all, a cause of getting admitted. However, there is presumably *no common cause* of being a male and getting admitted. So this seems to be a case of a factor Y (getting admitted) being positively correlated with a factor X (being a male) where X does not cause Y (and of course Y does not cause X), *yet there is no common cause Z of X and Y.* Is this a new kind of correlation without causation, not of any of the kinds described above?

If we look more carefully at the example, it turns out that it really is of the common cause kind. First, let us ask what is responsible for the correlation between *being male* and *applying to a department that is relatively easy to get into.* Of course, it would be implausible to suppose that the latter causes the former or that there is a common cause of the two. Most plausibly, being male somehow causes one to apply to the departments that are relatively easy to get into (possibly because of the way males tend to be brought up, getting them interested in the subjects taught in the larger, better funded departments, for example).

Second, I want to question the assumed equivalence between *there being discrimination against women* and *being male's being a positive causal factor for admission.* Although this is perhaps a fine point, it actually does make a crucial difference in the analysis of this example. If an institution *is* guilty of discrimination against women, then it is not, strictly speaking, *being* male that is necessarily a positive causal factor for admission, but rather *the institution's believing of an individual that he (or she) is male* that is the positive causal factor.

Suppose an institution in fact *does* have a policy of discriminating against women. Suppose also, as a thought experiment, that one year all the men were persuaded to check "female" on their applications, and all the women "male." Then we would expect that *being* male could be a negative causal factor for admission. Nevertheless, the charge of discrimination against women holds up, because the institution's *believing* of an individual that he or she is male *is* a positive causal factor for admission. As another thought experiment, suppose that the institution does have a policy of discrimination against women and that one year all the applicants decided in some random way whether to check "male" or "female" on their applications. Then it could be that actual gender is causally neutral for admission, despite the policy of discrimination against women.

In the Berkeley example, I think that being male *is* a positive causal factor for admission and that the graduate school is *not* guilty of discrimination. Being male is causally positive for admission; being male causes one to apply to departments that are relatively lenient in their admissions policies, which in turn is causally positive for admission – and transitivity of causation is plausible in this case. But there is no discrimination: There is no correlation, within any department, between admission and the department's believing of an applicant that he (or she) is male. And it is *this*, strictly speaking, that the department-by-department analysis of admissions history must have turned up, in order for it to be correct to conclude from the study that there is no discrimination against women applicants at Berkeley. Also, strictly speaking, the rows in the table earlier that describes the example should correspond to *being believed to be* male and *being believed to be* female.

Now let X be the factor of the Berkeley graduate school (or a department) believing of an applicant that he or she is male, let Y be the factor of getting admitted, Z the factor of being male, and W the factor of applying to one of the departments that are relatively easy to get into. Then the rows in the table should be relabeled X and $\sim X$, and the causal structure of the example is as diagrammed in Figure 2.2. Of course, Z is causally positive for X. Now if the only way in which Z can affect Y is by way of its influence on W (which seems plausible given the description of the example), then (as will be shown in Section 4.3) probabilistic type causation from Z to

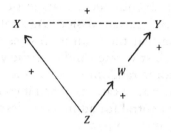

Figure 2.2

W to *Y* will be *transitive:* *Z* causes *Y*. Thus, Figure 2.2 depicts a special case of the common cause structure shown in Figure 2.1. Here, being male really is a cause of getting admitted, but getting admitted is only spuriously correlated with the school's believing of an applicant that he or she is male: *X* is causally neutral for *Y*.

The idea of *Y*'s being spuriously correlated with *X* should be consistent with the causal significance of *X* for *Y* being other than neutrality. For example, *X* could be *causally negative for Y*, consistent with the more general idea of spurious correlation mentioned above and formulated more precisely below. To see this, consider this slight modification of the example just described. Suppose that in each department there is a lower frequency of admission among applicants believed to be male than there is among those believed to be female: suppose there is a certain amount of "reverse discrimination" in each department, so that *X* is causally negative for *Y*. Still, if the tendency of women to apply to departments that are harder to get into is sufficiently strong, then there will remain a positive correlation between being believed to be male and getting admitted. For an example, simply change the entries in the top row of the table above that describes the original Berkeley example to read, "45/90, 1/10, 46/100." In this example, being believed to be male *increases the probability,* overall, of getting admitted, yet being believed to be male

66

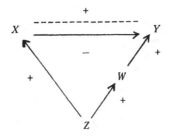

Figure 2.3

is a *negative causal factor* for getting admitted. This situation is diagrammed in Figure 2.3.

Brian Skyrms (1980) has described another example with basically this same feature. Suppose that air pollution in the cities got so bad that city-dwellers tended to refrain from smoking, so as not to put their lungs in double jeopardy. And suppose that people who lived in the country, where there is little pollution, generally felt safe enough to indulge. If the air in the cities is bad enough, and if the ratio of smokers in the cities to smokers in the country is low enough, then the frequency of lung disease could turn out to be lower among the smokers than it is among the nonsmokers. This is because the smokers tend to live in the country, where the air is clean, and the nonsmokers tend to live in the cities, where they are exposed to severe air pollution.

Nevertheless, of course, smoking (as well as exposure to air pollution) is causally positive for lung disease. In this case, a factor X (smoking) is a cause of a second factor Y (lung disease) even though the former lowers the probability of the latter, on average. This is a case of spurious correlation as characterized above, since pulmonary health ($\sim Y$) is positively correlated with, though not caused by, smoking. Also, if we identify living in the country – call this factor Z – as a common cause of pulmonary health and smoking, then we have the common cause structure shown in Figure 2.4.

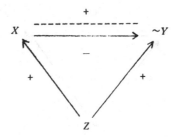

Figure 2.4

Another possibility consistent with spurious correlation is that X is in fact causally *positive* for Y. Here, it is not the *fact* of positive correlation that is spurious (since X *is* a cause of Y), but rather the *magnitude* of the correlation. By the magnitude of Y's correlation with X, I mean simply the difference between $Pr(Y/X)$ and $Pr(Y/{\sim}X)$. Below, I will be more precise about "spurious magnitudes" of correlation, but first, a couple of examples will illustrate the possibility intuitively, and show further the need to generalize the idea of spurious correlation in a way that takes into account magnitudes of correlation and causal significance.

Consider this modification of the Berkeley admissions example. Suppose there is some discrimination against women in each department, but only very little; within each department there is a small correlation between X and Y. Still there could be a large correlation between X and Y overall. For an example, change the bottom row in the table above used to describe the original Berkeley example to read, "8/10, 16/90, 24/100." In this example, there is a slight tendency of X to cause Y, but a large correlation between X and Y that the causal significance of X for Y does not explain. Most of the correlation is explained not by discrimination, but, again, by the tendency of women to apply to the more stringent departments. See Figure 2.5.

All the examples so far are cases that show that correlation

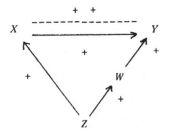

Figure 2.5

is no proof of causation. I mentioned above that it is also true that causation does not imply correlation. This can be seen easily enough by another modification of the Berkeley admissions example. If the males still tended to apply to the easy departments more frequently than the females did, but both departments discriminated, to just the right degree, against applicants believed to be male, then it could turn out that there is no overall correlation between getting admitted and being believed to be male. For an example of this, change the top row of the table above describing the original Berkeley example to read, "26/90, 1/10, 27/100." In this case, there is discrimination against those believed to be male, but exactly 27 percent each of males and females get admitted. So X is causally negative for Y (positive for $\sim Y$), yet there is no correlation, overall, between X and Y (or between X and $\sim Y$). See Figure 2.6 (the 0 above the broken line between X and Y represents probabilistic independence, *no* correlation of Y with X). This we may call a case of "spurious independence."

The examples we have seen show that we need a more general understanding of the idea of spurious correlation than the one first given. One obvious generalization, to accommodate cases of spurious independence and some of the other cases of spurious correlation in which X *is* causally relevant to Y, is to say that Y is spuriously correlated with X if, because of the action of a separate cause of Y, it is not true that X is

69

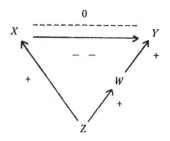

Figure 2.6

causally positive, negative, or neutral for Y, according to whether $Pr(Y/X)$ is greater than, less than, or equal to, $Pr(Y/\sim X)$, respectively. Note that on this understanding, spurious *correlation* includes, somewhat awkwardly, the possibility of probabilistic *independence* combined with some kind of causal relevance (spurious independence).

Note also that even though the relation of correlation itself is symmetric, I have used conditional probabilities only in one direction (from X and from $\sim X$ to Y, and not from Y or $\sim Y$ to X) in this characterization of spurious correlation. That is because the problem of spurious correlation can only be of real interest in one direction. Since at most one of any X and Y can precede the other in time, it follows from the requirement of temporal priority of causes (alluded to at the beginning of this chapter and to be clarified in Chapter 5) that, as far as the causal relation between X and Y is concerned, there is just one question of interest, namely, "What is the causal role of the earlier factor for the later?" And we have already adopted the convention that X precedes Y. Because causes precede their effects, it seems needless to add to the reason why the later Y is not causally relevant to the earlier X any idea that X is only spuriously correlated with Y.

This characterization of spurious correlation is still not fully satisfactory. We have seen a case, intuitively a case of spurious correlation, that does not fit this description. In that

70

example (Figure 2.5), we may say that there are two "components" of the correlation of Y with X: a small component due to X's positive causal significance for Y, and a large component due to the existence of a common cause of X and Y. The fact of the second component makes the correlation of Y with X "largely spurious," or largely unrepresentative of the causal significance of X for Y. In this example, Y is strongly positively correlated with X, yet X is only weakly causally positive for Y, so that the *degree* of correlation of Y with X does not match, intuitively, the "degree of causal significance of X for Y." Although we can at this point be precise about the meaning of "degree of Y's correlation with X" (just the difference, $Pr(Y/X) - Pr(Y/{\sim}X)$), we cannot yet be precise about "degree of causal significance of X for Y." Yet I wish henceforth to understand spurious correlation more generally as follows: Y is spuriously correlated with X if, because of the fact of a separate cause of Y, the degree to which Y is correlated with Y does not equal the degree to which X is causally significant for Y.

Not being precise at this point about degree of causal significance is on a par with not having been precise about what causation is in the characterization earlier of the narrower idea of spurious correlation (as simply common cause correlation without causation). Just as the simpler idea of spurious correlation was characterized in terms of causation without being precise about causation, so also the more general idea is characterized in terms of degree of causal significance without being precise about degree of causal significance. After property-level probabilistic causation itself has been sufficiently clarified in the pages that follow, we will be in a position to be more precise about the meaning of degree of property level causal significance.[7]

The new, more general understanding of spurious correla-

[7]This will be done in the next section. It is perhaps worth noting that, as discussed more fully in that section, there will be a kind of circularity in the definitions, given there, of the various kinds of causal significance. The different kinds of causal significance a factor X can have for a factor Y will be defined in terms of *other* causes of Y. However, the definitions will *not* rely on the idea of *degree* of causal significance.

71

tion includes cases of spurious independence of Y of X, cases in which only part of the correlation between X and Y can be explained by X's causal relevance to Y, as well as cases in which negative causal relevance is accompanied by positive correlation, cases in which positive causal relevance is accompanied by negative correlation, and cases in which positive or negative correlation is accompanied by causal neutrality.

Note, incidentally, that the possibilities of positive correlation with causal neutrality, and of positive correlation with negative causal relevance, show that positive correlation is *not sufficient* for positive causal relevance; and the possibilities of positive causal relevance with probabilistic independence, and of positive causal relevance with negative correlation, show that positive correlation is *not necessary* for positive causal relevance. All four of these possibilities are illustrated either by an example given above, or by the result of changing the Y of an example given above to $\sim Y$, or vice versa. The fact of these possibilities (or, more formally, the fact that any kind of correlation can be reversed or made to disappear in subpopulations), is known as "Simpson's paradox," named for E. H. Simpson (1951).[8] Each of the possibilities of positive correlation, negative correlation, and probabilistic independence can be consistently combined with each of the possibilities of positive causal factorhood, negative causal factorhood, causal neutrality, and (what I will describe in the next section) mixed causal relevance.

So much for examples of (single separate cause) spurious correlation for now. It is time to see how they may be explained in general, so that the possibility of spurious correlation may be appropriately accommodated in the theory of probabilistic causation. Consider first the simple kind of spurious correlation in which Y is positively correlated with X and X is completely causally irrelevant to Y (such as in the

[8]Cartwright (1979) mentions that this is sometimes known as the Cohen–Nagel–Simpson paradox, since it is presented as an exercise in Cohen and Nagel (1934). She also says that Nagel suspects he learned about it from Yule's (1911).

Fisher smoking hypothesis example, the falling barometers and rainy days example, and the first version of the Berkeley admissions case discussed above). The crucial feature of this kind of spurious correlation, a feature that fully explains this kind of correlation, is that a genuine probability-increasing cause (Z) of Y is correlated with the noncause (X) of Y. Whether the correlation between the genuine cause and the noncause is spurious or not is irrelevant to whether or not the correlation between X and Y is spurious. The point is that when the noncause (X) occurs, *the genuine cause (Z) is simply more likely to occur,* thus increasing the probability of Y.

It is because X is correlated with a genuine, probability-increasing cause of Y, that X increases the probability of Y. And X's correlation with a genuine, probability-increasing cause of Y will result in a spurious correlation between X and Y whether or not the correlation between X and the genuine cause is spurious. And it is an entirely different question whether or not there will always be, in cases in which X is correlated with a genuine cause of Y, a genuine cause of Y that is also a genuine cause of X.

In the first, simple common cause examples discussed above, the genetic condition is a genuine cause of lung cancer and it is correlated with (because it causes) the noncause, smoking; and the passing of a cold front is a genuine cause of rain and it is correlated with (because it causes) falling barometers. In the Berkeley admissions example, the spurious correlation between being believed to be male and getting admitted is explained by the correlation between applying to a lenient department (a genuine cause of admission) and being believed to be male (the noncause) – even though this latter correlation is spurious. Of course, the spurious correlation is also explained by the correlation between being believed to be male and *being male,* where the latter factor is a genuine cause of admission as well as of being believed to be a male.

Consider now other kinds of spurious correlation between factors X and Y, cases in which X may be genuinely causally

73

relevant to Y but in which the degree of (overall) correlation between X and Y does not appropriately reflect X's true causal significance for Y. In these cases, it is incorrect to refer to X as a "noncause" of Y, as in the diagnosis given above of the simpler kind of spurious correlation. But the same diagnosis applies. The fact that X is not a noncause of Y does not affect the fact that the spurious correlation (the inequality between the correlation of Y with X and the degree of causal significance of X for Y) is explained by the existence of a factor Z that is correlated with X and is a genuine probability-increasing cause of Y. It is easy to see that this is the explanation for the spurious correlations in the other versions of the Berkeley admissions case considered above and in Skyrms's example involving smoking in the cities and in the country. In all these cases, there is a "component" of the correlation of Y with X that is explained not by X's causal significance for Y, but rather by the correlation, with X, of a separate genuine, probability-increasing cause of Y.

It is important to note, however, that not all cases in which a factor X is correlated with a genuine cause Z of a factor Y are cases of spurious correlation of Y with X. For example, in some cases of transitive causal chains from X to Z to Y, X will be correlated with Z. If X is a genuine cause of Z and Z is a genuine cause of Y, then it can happen that Z, a genuine cause of Y, is correlated with X. Yet Y's correlation with X need not be spurious in such a case: if causation is *transitive* in this case, then X will be a genuine cause of Y.[9] And in such a case, the degree to which Y is correlated with X may exactly equal the degree of X's causal significance for Y, so that the correlation is not spurious. This, of course, is just a reiteration of the reason, given earlier, for explicitly excluding, from cases of spurious correlation, cases in which the reason why a third factor Z screens off a correlation between factors X and Y is that Z is causally intermediate between X and Y.

[9]Transivity of causal chains will be discussed in detail in Section 4.3.

What allows for the possibility of a spurious correlation between two factors X and Y is the existence of a third factor Z that is causally relevant to Y *independently of X's causal role, if any, for Y*. In all of our examples, the spurious correlation of Y with X is explained by the existence of a factor Z such that (i) Z is genuinely causally relevant to Y, (ii) Z is correlated with X, and (iii) X is causally irrelevant to Z. It is exactly this that made it possible, in the examples given above, for the degree of correlation of Y with X to fail to reflect just the causal significance of X for Y. In all the examples above, the factors Z (and, where applicable, both Z and W) satisfy the three conditions just laid down.

In cases of spurious correlation in which there is *just one* separate cause of the later factor Y (or in which other causes W trace back to a single separate cause Z), it seems that it is exactly (i), (ii), and (iii) above that explain the spurious correlation. These we may call cases of "single separate cause spurious correlation." When there are *multiple* separate causes, however, this diagnosis is not quite on the mark. It is possible for a factor Y to be spuriously correlated with a factor X and for this to be explained by the operation of, for example, *two* separate causes, F and G, of Y, *where neither F nor G is correlated, overall, with X*. In such a case, (i) and (iii) above hold of each of F and G, but (ii) fails of each of F and G. However, as we shall see, a condition very much like (ii), but involving *conditional* correlations will hold in such cases.

Examples of this are more complex and harder to grasp intuitively than examples of single separate cause spurious correlations.[10] Figure 2.7 depicts a numerical example of this kind; Figure 2.7a gives the probabilistic relations and Figure 2.7b gives the causal structure of the example. Although I will not describe a "real life" example corresponding to Figure 2.7, I would encourage the reader, after finishing this

[10] I note that Fisher and Patil (1974) give an example of this, and diagnosis it in terms of conditional correlations. Compare also Miettinen (1970) and (1974).

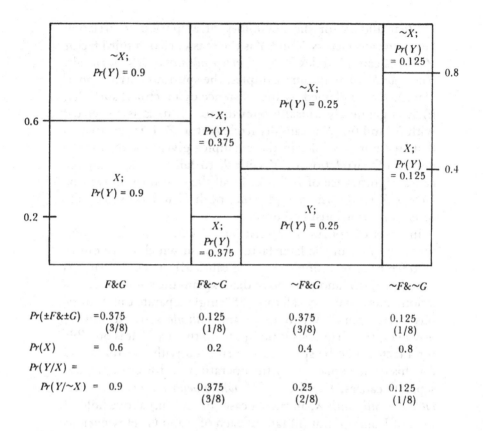

	F&G	F&~G	~F&G	~F&~G
$Pr(\pm F \& \pm G)$	=0.375 (3/8)	0.125 (1/8)	0.375 (3/8)	0.125 (1/8)
$Pr(X)$	= 0.6	0.2	0.4	0.8
$Pr(Y/X)$ =				
$Pr(Y/\sim X)$ =	0.9	0.375 (3/8)	0.25 (2/8)	0.125 (1/8)

OVERALL: $Pr(F/X) = Pr(F/\sim X) = 0.5; Pr(G/X) = Pr(G/\sim X) = 0.75$; so that *each of F and G is uncorrelated with X* - and each of F and G fail to satisfy (ii) above, though of course they *can still satisfy (iii)* (see Figure 7b below).

OVERALL: $Pr(Y/F) = 0.70625 > Pr(Y/\sim F) = 0.10625$, and $Pr(Y/G) = 0.575 > Pr(Y/\sim G) = 0.25$ (and the same *inequalities* hold when G and F, respectively, are held fixed, positively and negatively), so that *each of F and G is a cause of Y* - each of F and G do satisfy (i) above.

OVERALL: $Pr(Y/X) = 0.52375 > Pr(Y) = 0.49375 > Pr(Y/\sim X) = 0.46375$; so that there is a *(spurious)* correlation of Y with X.

Figure 2.7a

76

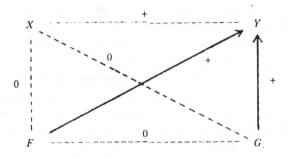

Figure 2.7b

section, to try to think of one. (I should add that there would not be a great loss of continuity if the reader were to skip to the end of this section at this point.)

Here, what explains the spurious correlation of Y with X is *not* the existence of any independent (of X) cause of Y that is, overall, *correlated with* X.[11] One way of explaining the correlation of Y with X in this example would be to note that, when the cause G is operating (which has probability 0.75), there *is* a correlation of the *stronger* cause F with X: A little calculation (or contemplation of Figure 2.7a) shows that $Pr(F/G\&X) = 0.6 > Pr(F/G\&\sim X) = 0.4$. There is this *conditional correlation* of F with X. When G is present, which is more probable than its being absent, X is correlated with the stronger cause, F, of Y. The *overall independence* of F from X is then explained by the fact that, when G is absent, there is a stronger negative correlation of F with X: $Pr(F/\sim G\&X) = 0.2 << Pr(F/\sim G\&\sim X) = 0.8$. These two conditional correlations, in opposite directions, are consistent with X's raising the probability of Y, overall, because G, within which there is a positive correlation between X and F, is so much more probable than $\sim G$, within which there

[11] The idea that there may be a *conjunctive* factor, such as $F\&G$ or $\sim F\&\sim G$, that is an independent cause of Y, and indeed correlated with X, will be discussed below.

is strong negative correlation between X and F, and because F is a stronger cause of Y than G is.

This suggests that spurious correlations should be explained, in general, in terms not only of *correlated independent causes,* but also of *conditionally correlated independent causes,* conditional on other causes – where, as we have seen, these conditional correlations are consistent with overall independence. That is, we could explain a spurious correlation of a factor Y with a factor X as arising from the existence of factors $F_1, \ldots, F_n$, such that (i) the F_i's are causes (positive or negative) of Y, (ii) each F_i is either correlated (positively or negatively) with X *overall* or correlated (positively or negatively) with X *conditionally on some way of holding fixed other F_j's,* and, of course, (iii) the F_i's are causally independent of X. Without trying to prove this diagnosis rigorously here, I conjecture that this is the form of the most general kind of explanation of spurious correlation, where there may be multiple independent causes of Y.[12]

Before closing this section, there are two points I should make, both of which relate what has been discussed in this section to ideas that will be addressed later. First, about the example depicted in Figure 2.7, it might be suggested that, if we are clever enough, we *could* characterize, in terms of F and G, items that are causes of Y, independent of X, and that *are correlated with* X. For example, it might be suggested that some of the conjunctive factors, $F\&G$, $F\&\sim G$, $\sim F\&G$, and $\sim F\&\sim G$, may be independent causes of Y that are correlated with X. Or it might be suggested that we adopt a more general framework in which we consider *partitions* of factors (such as the *four-element partition* of consistent conjunctions

---

[12] However, it is easy to demonstrate this: If (1) there is no overall or conditional correlation between X and independent causes of Y (conditional on ways of holding other such causes fixed), and (2) maximal ways of holding fixed the independent causes screen off Y from X (which we would expect if X is not a cause of Y), then there can be no correlation of Y with X.

of F, G and their negations),[13] rather than simple "On/Off" (or *two-valued*) factors (such as F and G), as the relevant items that enter into causal relations. Note that the conditional probabilities of $F\&G$, $F\&\sim G$, $\sim F\&G$, and $\sim F\&\sim G$, all conditional on X, are all different in the example above, so that there is a kind of "correlation of the partition with X." These are approaches to spurious correlation that I would like to avoid.

As to the first suggestion, the *absence* of a conjunctive factor is disjunctive; for example, the negation of $F\&G$ is $(F\&\sim G) \lor (\sim F\&G) \lor (\sim F\&\sim G)$. And often the different disjuncts of a disjunctive causal factor confer different causally significant probabilities on the effect factor in question. So the question arises of how to average or otherwise combine these probabilities to come up with a *single* probability of the effect factor in the absence of the conjunctive factor, to compare with the probability of the effect factor in the presence of the conjunctive factor. Disjunctive causal factors are discussed in Chapter 3; the question just posed will be answered there in a way that makes the first suggestion ill suited as an approach to spurious correlation.[14]

As to the second suggestion, involving partitions, I think this would be just fine as an alternative approach to structuring our understanding of probabilistic causation. This makes the *form* of probabilistic causal claims quite different from the

[13] A *partition* is a set of factors (or propositions or sentences) that are "mutually exclusive" and "collectively exhaustive." That is, the conjunction of any two of its elements is impossible and the disjunction of them all is necessary. Appendix 1 explains the idea of a partition.

[14] Saying this, however, may help at this point. It may be argued that, in the example, $F\&G$ and $\sim F\&\sim G$ are both correlated with X and causes (positive and negative, respectively) of Y. They are causes of Y because no matter how we average the conditional probabilities of Y, conditional on the disjuncts comprising the absence of these conjunctive factors, the inequality between the probability of Y in the presence of one of the conjunctions and the probability of Y in its absence cannot change in direction. The problem, to be addressed in Chapter 3, is that the same is not true of $F\&\sim G$ and $\sim F\&G$. This is because these factors confer neither the highest nor the lowest probability on Y.

way I have been supposing we may fruitfully understand it. I have been supposing that we may understand the form of such claims to be (roughly), "X is a causal factor for Y in population P (considered to be of a kind Q)," where X and Y are "On/Off" (or two-valued) variables, or factors. I see no compelling reason to abandon this natural approach, and it is the approach I will continue to pursue.

Finally, it is worth emphasizing that the topic of this chapter is just spurious correlation, a phenomenon that involves *separate causes* of the candidate effect factor in question. As mentioned at the beginning of this chapter, there are other reasons, besides spurious correlation (that is, besides the confounding effects of separate causes), for why (degree of) correlation may not coincide with (degree of) causation. These involve *symmetry of correlation and asymmetry of causation* (to be discussed in Chapter 5) and *causal interaction* (to be discussed in Chapter 3). It should be emphasized that the discussion in this chapter is not intended to handle *all* the reasons why probability change does not coincide with causation, but only the reason of separate causes.

Based on the understanding of spurious correlation given above, the second main qualification of the basic probability-increase idea of probabilistic causation will be intended to deal with this phenomenon. We must "control" for correlations (both conditional and unconditional) that may exist between a factor X, whose causal role we want to characterize, and other causes, Z, of the effect factor Y in question. This is the topic of the next section.

2.2 CAUSAL BACKGROUND CONTEXTS

Recall the example about falling barometers (X) and rainy days (Y). Falling barometers do not cause rainy days, though rainy days are correlated with falling barometers: $Pr(Y/X) > Pr(Y/\sim X)$. Also, the factor of an approaching cold front (call it factor F here) is causally positive for rain, and it is correlated with (since it is a cause of) falling barometers. As we saw in the previous section, it is the existence of such factors as F here that explains the spurious correlation of Y with X.

Let us suppose that we know that F *is the only factor that is causally relevant to* Y, so that we have a case of what I called, in the previous section, "single separate cause spurious correlation." (This means, of course, that we know also that X *is causally neutral for* Y.) If we observe the probabilistic impact of X on Y first in the presence of F, and then separately in the absence of F, then we should expect, in each case, that Y will be probabilistically independent of X:

$$Pr(Y/F\&X) = Pr(Y/F\&\sim X)$$

and

$$Pr(Y/\sim F\&X) = Pr(Y/\sim F\&\sim X).$$

This is because we know that the *only reason* for a correlation of Y with X could be a correlation of F, a genuine cause of Y, with X; we know that X itself is causally neutral for Y.[15] And in the above two probability comparisons, F is "held fixed," positively and then negatively; and given each way of holding F fixed, the correlation of F with X disappears. When F is held fixed positively, the probability of F is 1, both conditional on X and conditional on $\sim X$; and when F is held fixed negatively, the probability of F is 0, both conditional on X and conditional on $\sim X$. Since we know that the *only reason* for a correlation of Y with X is the correlation of F, a genuine cause of Y, with X, we should expect any correlation of Y with X to disappear when the correlation of F with X disappears.

Suppose, on the other hand, that all we know, besides the *correlation of* Y *with* X, were the *two probabilistic equalities displayed above*, and that *aside from the possibility of* X, F *is the only factor causally relevant to* Y. Then I claim we should be in a position to conclude that X is causally neutral for Y. We

[15] I am, as always, following the convention that X precedes Y in time, as explained earlier.

should be able to conclude that Y's positive overall correlation with X is *explained entirely by a correlation of X with F*, the "single separate cause" of Y. In fact, in this example, we shall be in a position to apply Reichenbach's principle of the common cause involving conjunctive forks, as discussed at the beginning of the previous section. The problem is to explain the correlation of Y with X. The equalities above, which I have just assumed we know to hold, are parts (3) and (4) of a conjunctive fork (substituting "F" for "Z" in the description of conjunctive forks earlier). Explaining why we would observe inequalities (1) $(Pr(X/F) > Pr(X/{\sim}F))$ and (2) $(Pr(Y/F) > Pr(Y/{\sim}F))$ in this example is a little more intricate.

First, why should we observe inequality (2) to hold? I am assuming that it is true that (and that we know that) aside from the possibility of X, F is the *only* factor that is causally relevant to Y. Now *in fact, X is not* causally relevant to Y (though this is what we will be reasoning *to,* and not *from*). Given this, and the fact that aside from the possibility of X, F is the only cause of Y, it follows that F in fact *is the only cause of Y* (though we cannot reason to this conclusion, just from the equalities above and the rest of what I have assumed, two paragraphs back, that we know). And this implies, I assume, that in fact, F raises the probability of Y overall. F cannot be correlated with *negative* causes of Y, resulting in a *spurious independence* of Y from F, or a *spurious negative correlation* between Y and F, for example: given *what we know to be true,* and given *the fact that X* actually is not a cause of Y, there cannot be any such negative causes of Y. That is why we should *observe* inequality (2) to hold (though we cannot *reason* to it just from what I have assumed, two paragraphs back, that we know). As to inequality (1), it *mathematically follows from* (2), (3), (4), and the overall correlation of Y with X.[16]

---

[16] From the two equalities above, it follows that:
$$Pr(Y/X) = Pr(F/X)Pr(Y/F) + Pr({\sim}F/X)Pr(Y/{\sim}F).$$
and,
$$Pr(Y/{\sim}X) = Pr(F/{\sim}X)Pr(Y/F) + Pr({\sim}F/{\sim}X)Pr(Y/{\sim}F).$$
Let $x = Pr(F/X)$ and $y = Pr(F/{\sim}X)$. We want to show that $x > y$. Let $a = Pr(Y/$

Thus, F, X, and Y form a conjunctive fork. This means that the correlation of Y with X is explained entirely by the existence of the genuine cause, F, of Y, and by F's correlation with X (and by the fact that there are no other separate causes of Y). Thus, we should conclude that the correlation of Y with X does not correspond to causal relevance, and in fact the disappearance of the correlation, when the single correlated cause is held fixed, indicates causal neutrality. From the equalities, and the assumption that, aside from the possibility of X, F is the only factor causally relevant to Y, we conclude that none of X's positive probabilistic relevance to Y could be explained by any *causal* relevance of X to Y. That is, given the assumption that we have held fixed *all* the causes of Y (aside from the possibility of X), we can conclude from *X's probabilistic neutrality for Y* that *X is causally neutral* for Y.

It is worth noting that all that is crucial in the reasoning, above, first, from X's causal neutrality for Y to the two displayed probabilistic equalities, and second, from the two equalities to causal neutrality, is this: The assumption that we know that, *aside from the possibility of X, F is the only factor that is causally relevant to Y*. It is not crucial that we know in addition that there is a correlation of Y with X. This is already more or less explicit for the first line of reasoning, from causal neutrality to the two equalities. For the second line of reasoning, from the two equalities to causal neutrality, it is easy to see that the reasoning can easily be modified so that the crucial assumpiton gives this conclusion: If the two equalities hold, then any correlation that there may be between X and Y would be fully explained by a correlation of F with X. That is, no component of any correlation between X and Y is explainable by any causal relevance of X to Y. Given the basic probability-

$\sim F$). By the assumption that F is the only cause of Y, and hence that Y is positively correlated with F, there is a positive number e such that $Pr(Y/F) = a + e$. Then the correlation of Y with X can be expressed as follows:
$$x(a + e) + (1 - x)a > y(a + e) + (1 - y)a.$$
which reduces to $xa + xe + a - xa > ya + ye + a - ya$, and then to $xe > ye$, and then, since $e > 0$, to $x > y$.

increase idea for probabilistic causation, this means that we can conclude, from the inequalities, that X is causally neutral for Y. (Note also that, given the two equalities and Y's correlation with F, Y will be correlated with X if and only if F is correlated with X.)

Thus, the main point is this: In evaluating the causal role of a factor X for a factor Y, if a factor F is the only factor that, aside from the possible causal role of X, is causally relevant to Y, then, when we hold F fixed positively and negatively, probabilistic independence of Y from X coincides with causal neutrality of X for Y – whether or not Y is, overall, correlated with X. After explaining how this idea deals with spurious correlation more generally (including spurious independence and spurious degrees of correlation, as well as cases of multiple separate causes), I will show how the resulting theory deals with some of the other examples discussed in the previous section.

The basic idea behind this refinement of the theory is that in order for positive, negative, and neutral causal relevance to coincide with positive, negative, and neutral probabilistic relevance, we have to control for all factors that are, independently of the candidate causal factor in question, causally relevant to the effect factor in question. In order for X to count as a positive causal factor for Y, for example, X must have a positive probabilistic impact on Y beyond that which is explainable by *other,* independent, causes of Y that may be correlated with X (where, as noted at the end of the previous section, this correlation may be either unconditional or conditional on other causes of Y). The general strategy for controlling for independent causes goes back a long way; Skyrms (1980) reports that the basic idea was already explicitly formulated by F. Y. Edgeworth (1892, 1910), was anticipated half a century earlier by Bravais, and was used by the English statistical school of Edgeworth, K. Pearson (1897), and G. U. Yule (1910). See also, for example, Reichenbach (1956), H. Simon (1957), C. Granger (1969), Suppes (1970, 1984), Salmon

84

(1971), Cartwright (1979), Skyrms (1980, 1984a), Eells and Sober (1983), and, for a review, Skyrms (1988).

Suppose we want to assess the causal significance of a factor X for a factor Y. And suppose there are exactly n factors, $F_1, \ldots, F_n$, distinct from X, that are causally relevant to Y in a way that is independent of X's causal relevance, if any, to Y. That is, the F_i's are all the factors, other than X and effects of X, that are causally relevant to Y.[17] We saw in the previous section that it is the existence of such factors F_i *that are correlated with* X (either overall or conditional on other such factors) that makes possible a spurious correlation between X and Y. However, in the theory described below, causes of Y that are *uncorrelated* with X (either overall or conditional on other such causes) will be controlled for in the same way as causes of Y that *are correlated* with X (either overall or conditional on other such causes). As explained below, first, it *cannot hurt* to do so, in that whether or not we control for uncorrelated causes, we get the same answers about both direction and magnitude of causal influence; and second, this approach will give "causal background contexts" within which it can only be exactly the causal impact *of X* on Y (as opposed to the causal impact of *other causes* as well) that is measured by the probabilistic impact of X on Y (the magnitude of probabilistic impact within contexts will not be skewed in ways in which it could be if we did not control for uncorrelated causes).[18]

Since the F_i's are *all* the factors that are (independently of X) causally relevant to Y, if we "control" for the presence or absence of each F_i, we can observe the probabilistic impact that X makes upon Y beyond that made by other factors that

---

[17] Again, I will postpone a detailed discussion of the important *independence* requirement until Chapter 4. The idea is that the F_i's should not include factors to which X is causally relevant – and that, as explained below, we should not "hold fixed" causally intermediate factors when assessing the causal role of X for Y.

[18] I am hedging here: We will see in Chapter 3 that there are sometimes factors that are *not causes* of the candidate effect factor, but which nevertheless must be controlled for, namely, "interactive" factors.

are causally relevant to Y (independently of X). If we observe that X increases the probability of Y no matter what other independently causally relevant factors are present, then (pending further refinements of the theory that do not involve spurious correlation) we can conclude that X is causally positive for Y.

In assessing X's causal relevance to Y, we have to hold fixed all the other factors, $F_1, \ldots, F_n$, that are causally relevant to Y, independently of X, and then observe the probabilistic impact of X on Y given each of these ways. With n such other factors, there are 2^n ways of holding each fixed, positively or negatively. That is, there are 2^n conjunctions in which each of these n factors occurs exactly once, either positively (unnegated) or negatively (negated). Of these 2^n "maximal conjunctions," let $K_1, \ldots, K_m$ be exactly those that have nonzero probability both in conjunction with X and in conjunction with $\sim X$. (Thus, for $i = 1, \ldots m$, $Pr(K_i \& X) > 0$ and $Pr(K_i \& \sim X) > 0$.) These K_i's are called "causal background contexts," relative to the assessment of X's causal role for Y.[19]

Then we say that X is a *positive causal factor* for Y if and only if, *for each i, $Pr(Y/K_i \& X) > Pr(Y/K_i \& \sim X)$. Negative causal factorhood* and *causal neutrality* are defined by changing the "always raises" ($>$) idea to "always lowers" ($<$) and "always leaves unchanged" ($=$), respectively. The idea that the inequality or equality must hold for *each* of the background contexts K_i is sometimes called the condition of *contextual unanimity*, or *context unanimity*.[20] This condition, and some alternatives to it, are discussed in the next section. Note that these three relations of positive, negative, and neutral causal factorhood

[19] If the F_i's already contain conjunctions of some of the F_i's and their negations, then it could turn out that m is less than n. Whether or not the F_i's do contain conjunctions of F_i's and their negations turns on the questions of the causal roles of disjunctive causal factors, which will be discussed in Chapter 3.

[20] The term "contextual unanimity" was introduced by John Dupré (1984) to distinguish this kind of "unanimity" from what Elliott Sober and I (1983) called "unanimity," by which we meant what Dupré calls "unanimity of intermediaries." I will discuss the latter idea in Chapter 4.

are not exhaustive of the possible causal significance that a factor X can have for a factor Y: There remains the possibility of various kinds of *mixed* causal relevance, corresponding to various ways in which unanimity can fail. This possibility is discussed further in the next section and in Chapter 3.[21]

Note further that the definition is circular in that it characterizes X's causal role for Y in terms of other causally relevant factors. However, the circularity is not as bad as it could be. As Skyrms (1988) has pointed out in connection with this kind of definition of causation, X's causal role for Y is characterized in terms of the causal roles of factors *other than* X for Y. The idea of X's being causally positive for Y, for example, is noncircularly defined in terms of the causal roles of *other* factors for Y. However, the definition of positive causal relevance in general *is* circular. Thus, what we have is a theory about the *relation* between probability and causation.[22]

Before applying these definitions to some of the examples of the previous section, four further points about the definitions are in order here. First, as stressed in Section 1.1, these definitions must be understood as relative to a particular population, as well as to a kind that the token population exemplifies. This relativity to a population and a kind is already explicit in the assumption that we have a definite probability function to work with: as explained in Section 1.1, probabilistic, as well as causal, relations can differ from population to population, and from kind of population to kind of population.

Second, it is worth reiterating here the important point briefly motivated in the previous section, and alluded to from time to time in this section, that we only hold fixed factors

---

[21] In particular, I will show in the next section that, among the factors $F_1, \ldots, F_n$ that are causally relevant to Y independently of X, and that must be held fixed in background contexts, we must include not only all factors that are causally *positive* or *negative* for Y independently of X, but also factors that are causally *mixed* for Y independently of X. *Mixed causal relevance is a kind of causal relevance.*

[22] In Chapter 5, I will consider a natural suggestion for removing the circularity (the idea is to hold fixed all factors simultaneous to or earlier than the causal factor in question). I will give an example that strongly suggests that we should reject this suggestion.

that are causally relevant to Y *independently of X*. As explained in the previous section, these are factors that are causally relevant to Y but to which X is *not* causally relevant. A simple rationale for this independence condition was, in effect, briefly explained in the previous section: X's correlation with factors that are causally relevant to Y, but *intermediate* in a causal chain from X to Y, may fail to make the correlation between X and Y spurious. However, the condition has been the subject of some controversy in recent literature on probabilistic causality. In Chapter 4, I discuss the condition in detail, explain various parts of the controversy, and defend and generalize the condition.

Third, as promised in the previous section, we can now be more precise about the idea of degree of causal significance that was used in characterizing spurious correlation. A natural definition, suggested by the definition above, is this: The *average degree of causal significance* of a factor X for a factor Y is given by

$$ADCS(X, Y) = \Sigma_i \, Pr(K_i)[Pr(Y/K_i \& X) - Pr(Y/K_i \& \sim X)],$$

where the K_i's are the causal background contexts appropriates for assessing X's causal role for Y (in the relevant population considered to be of the relevant kind).[23] The difference between this and magnitude of correlation, $Pr(Y/X) - Pr(Y/\sim X)$, is that $Pr(K_i)$, enters into the formula for average degree of causal significance *unconditional on X*, which is appropriate since X is causally irrelevant to the K_i's, which specify factors causally relevant to Y *independently of X*. Note, of course, that this is an *average,* and that while positive, negative, and neutral causal relevance imply that the average is positive, negative, and zero, respectively, the converse is not true. This is because of the possibility of mixed causal significance, for

[23] I note that I. J. Good (1961–2, 1983,1985) offers a quite different kind of definition of (what he calls) "the tendency of [X] to cause [Y]." This is discussed briefly in Section 5.3 below.

88

which the average could be any value strictly between -1 and $+1$.

Finally, as mentioned above, the F_i's in the definitions above are *all* the factors, other than X, that are, independently of X, causally relevant to Y – and not just those that are *also correlated with* X, either unconditionally or conditional on other such causes. But spurious correlation between factors X and Y was diagnosed in the previous section as resulting from there being factors that are both independently causally relevant to Y and correlated with X, either unconditionally or conditional on other such causes. So the question naturally arises of why it should be required that we hold fixed, in the background contexts, *all* factors independently causally relevant to Y, and not just those of the kind that have been implicated in the possibility of spurious correlation.

The first thing to note is that even if it were not required to hold fixed uncorrelated causes of Y, it cannot hurt to do so. Suppose, for example, that a factor X is a positive cause of Y, and that X raises the probability of Y for each way, J_i, of holding fixed independent causes of Y that *are correlated* with X, either unconditionally or conditional on other causes of Y. Now suppose that Z is an independent cause of Y that is uncorrelated with X, either unconditionally or conditional on other independent causes of Y. I first show that the average degree of causal significance of X for Y is the same, whether or not Z is held fixed in addition to the other independent causes.

Not holding Z fixed, we have

$$ADCS(X,Y) = \Sigma_i \, Pr(J_i)[Pr(Y/J_i\&X) - Pr(Y/J_i\&\sim X)].$$

Factoring in Z, this is the same as

$$ADCS(X,Y) = \Sigma_i \, Pr(J_i)$$
$$\{[Pr(Z/J_i\&X)Pr(Y/J_i\&Z\&X) + Pr(\sim Z/J_i\&X)Pr(Y/J_i\&\sim Z\&X)] - [Pr(Z/J_i\&\sim X)Pr(Y/J_i\&Z\&\sim X) + Pr(\sim Z/J_i\&\sim X)Pr(Y/J_i\&\sim Z\&\sim X)]\}.$$

89

Since Z is uncorrelated with X, conditional on the J_i's, we have

$$ADCS(X, Y) = \Sigma_i \, Pr(J_i)$$
$$\{[Pr(Z/J_i)Pr(Y/J_i\&Z\&X) + Pr(\sim Z/J_i)Pr(Y/J_i\&\sim Z\&X)]$$
$$- [Pr(Z/J_i)Pr(Y/J_i\&Z\&\sim X) + Pr(\sim Z/J_i)Pr(Y/J_i\&\sim Z\&\sim X)]\}.$$

Then rearranging and simplifying, we have

$$ADCS(X, Y) = \Sigma_i \, Pr(J_i)Pr(Z/J_i)[Pr(Y/J_i\&Z\&X)$$
$$- Pr(Y/J_i\&Z\&\sim X)] + \Sigma_i \, Pr(J_i)Pr(\sim Z/J_i)[Pr(Y/$$
$$J_i\&\sim Z\&X) - Pr(Y/J_i\&\sim Z\&\sim X)]$$
$$= \Sigma_i \, Pr(J_i\&Z)[Pr(Y/J_i\&Z\&X) - Pr(Y/$$
$$J_i\&Z\&\sim X)] + \Sigma_i \, Pr(J_i\&\sim Z)[Pr(Y/$$
$$J_i\&\sim Z\&X) - Pr(Y/J_i\&\sim Z\&\sim X)].$$

And this is $ADCS(X, Y)$ calculated in terms of background contexts $J_i\&Z$ and $J_i\&\sim Z$, which are obtained from the J_i's by holding fixed, in addition to the other independent causes of Y, Z as well. Thus, holding fixed factors like Z (causes of Y that are causally independent of X and uncorrelated with X overall and conditional on other independent causes of Y) does not affect the value of $ADCS(X, Y)$.

This agreement about *average degree* of causal significance does not by itself imply agreement on the *qualitative question* of kind of causal significance. However, if it suffices, for getting the right answer, to hold fixed just the causally independent (of X) causes of Y that are conditionally or unconditionally correlated with X (and the main point of previous section is that this does suffice, at least for dealing with spurious correlation), then, I claim, we should expect the same answer when we hold fixed such factors that are conditionally and unconditionally uncorrelated with X. That is, for contexts J_i and factors X, Y, and Z, as above, we should expect

$$Pr(Y/J_i\&X) > Pr(Y/J_i\&\sim X)$$

90

if and only if both

$$Pr(Y/J_i \& Z \& X) > Pr(Y/Y_i \& Z \& \sim X)$$

and

$$Pr(Y/J_i \& \sim Z \& X) > Pr(Y/J_i \& \sim Z \& \sim X);$$

and the same when we substitute "$<$" or "$=$" for "$>$" throughout. In fact, holding fixed *all* independent causes of Y (those held fixed in the J_i's as well as factors like Z above) gives us contexts within which causal impact is approximated by probabilistic impact more closely than if we only held fixed independent causes that are correlated, conditionally or unconditionally, with X, as I will now explain.

To simplify this discussion, let us suppose that there are no causes of Y that are correlated with X, conditionally or unconditionally, and that, aside from the possibility of X, Z is the only cause of Y; so Z is uncorrelated with X. (For cases in which there are causes of Y that are correlated with X, the possibility I describe below could arise within contexts J_i.) And let us suppose that X is a positive causal factor for Y. It is possible that X can make only either a very large difference for Y or a very small difference – and never a moderate, or intermediate, difference. And this feature of the causal significance of X for Y may not show up unless factors like Z are held fixed. For example, the relevant probabilities may be as follows:

$$Pr(Y/Z \& X) = 0.9 > Pr(Y/Z \& \sim X) = 0.2;$$
$$Pr(Y/\sim Z \& X) = 0.4 > Pr(Y/\sim Z \& \sim X) = 0.3;$$
$$Pr(Y/X) = 0.65 > Pr(Y/\sim X) = 0.25.$$

Here, $Pr(Z) = Pr(Z/X) = Pr(Z/\sim X) = 0.5$, so that the probability values in the last line displayed are 50-50 averages of the values in the first two lines.

In this example, X can either make a huge difference in the probability of Y, or only a tiny difference. When Z holds, the difference is huge: $0.9 - 0.2 = 0.7$. When $\sim Z$ holds, the difference is tiny: $0.4 - 0.3 = 0.1$. And there is no kind of individual (or no kind of concrete situation) in which X can make a moderate, or intermediate, difference in the probability of Y. Nevertheless, on average (that is, not holding Z fixed), X does make a moderate, or intermediate, difference in the probability of Y: as the last line displayed above shows, this average difference is $0.65 - 0.25 = 0.4$. However, this *average difference in probabilities* does not correspond to a *degree of causal influence* that X can have on Y for any kind of individual, or in any concrete situation. For in any individual or concrete situation, Z will either be present or absent, so that the causal significance of X for Y will be either very large (0.7) or very small (0.1), and never intermediate (0.4).

By holding fixed *all* independent causes of Y, and not just those causes that are conditionally or unconditionally correlated with X, we can more accurately observe, in terms of probability comparisons, the causal impact that *just X* has on Y. By doing this, we control for the probabilistic significance that causes other than X can have for Y, even though these causes may not be correlated with X. This yields causal background contexts in which the causal significance of *just X* for Y is more precisely isolated. It is for this reason that the definitions above require holding fixed *all* independent causes of Y.

Let us now see how the definitions given above give the right answers in the examples of the previous section diagrammed in Figures 2.2–2.6. In the first, simple, version of the Berkeley admissions example, we have to hold fixed the factors Z (being male) and W (applying to a stringent department) in assessing the causal role of X (being believed by the school to be male) for Y (admission). It was part of the example that in stringent departments, as well as in departments that are not stringent, the frequency of admission is the same among those believed to be male as it is among those not believed to be male. So, holding fixed W (and holding fixed

92

Z as well should not affect this) makes Y probabilistically independent of X. So the definitions tell us that being believed to be male is causally neutral for admission, which is the right answer.

Figure 2.3 depicts the example in which being believed to be male is slightly causally negative for admission despite the positive correlation: Both stringent departments and departments that are not stringent tend to discriminate against applicants believed to be male. Again, the definitions tell us that we must hold fixed both Z and W, since they are causally relevant to Y independently of whatever causal relevance X has for Y. And the description of the example tells us that given each way of holding these factors fixed, the frequency of Y is less given X than it is given $\sim X$. So we get the right answer that being believed to be male is causally negative for admission. The reader is invited to apply the definition of average degree of causal significance to this example.

In Skyrms's example, diagrammed in Figure 2.4, we have to hold fixed the factor of living in the country (Z). And it is part of the example that both among the country dwellers and among the city dwellers, smoking decreases the probability of healthy lungs, so that the definitions give us the right answer that smoking is causally negative for pulmonary health.

Figure 2.5 diagrams the version of the Berkeley admissions example in which there is a little discrimination in favor of males both in the stringent and in the not so stringent departments, but a big correlation between getting admitted and being believed to be male, which is explained mainly by the common cause Z (being a male). Of course again we must hold fixed the factors Z and W, so that, in the relevant probability comparisons, the component of the overall correlation due to the common cause will disappear. That is,

$$Pr(Y/X) >> Pr(Y/\sim X),$$

but

$$Pr(Y/Z\&W\&X) > Pr(Y/Z\&W\&\sim X) \quad \text{(by a little)},$$
$$Pr(Y/Z\&\sim W\&X) > Pr(Y/Z\&\sim W\&\sim X) \quad \text{(by a little)},$$
$$Pr(Y/\sim Z\&W\&X) > Pr(Y/\sim Z\&W\&\sim X) \quad \text{(by a little)},$$

and

$$Pr(Y/\sim Z\&\sim W\&X) > Pr(Y/\sim Z\&\sim W\&\sim X) \quad \text{(by a little)}.$$

The last four probability comparisons are the ones relevant to assessing the degree of X's causal significance for Y; and the differences in these are smaller than in the first comparison.

In the "causation without correlation" example of Figure 2.6, we again must hold fixed Z and W. And again the effect of holding fixed W reveals the policies of the stringent and the not so stringent departments: Within each kind of department, there is negative probabilistic relevance of being believed to be male and getting admitted, and therefore, according to the definitions, negative causal relevance of the former for the latter.

Thus, the definitions of the different kinds of causal factorhood given above provide plausible analyses of the examples of spurious correlation given in the previous section. Of course there are other ways in which we can have correlation without causation, aside from what we have been calling spurious correlation. There is still the problem of temporal asymmetry of causation and the problem of probabilistic causal interaction, both briefly described at the beginning of this chapter. In the next section, in the course of further evaluating the definitions given in this section, the problem of causal interaction will emerge. This will be dealt with by further refinements of the theory in Chapter 3. In Chapter 5, the problem of temporal priority of causes will be handled.

2.3 CONTEXT UNANIMITY

It has been questioned whether a genuine cause really must raise the probability of a genuine effect of it in *every* causal background context. That is, the condition of context unanimity has been questioned. Skyrms (1980), for example, has suggested a weaker condition, which he calls a "Pareto-dominance condition": X raises the probability of Y in a least one causal background context ($Pr(Y/K_i \& X) > Pr(Y/K_i \& \sim X)$ for at least one i) and X lowers the probability of Y in no causal background context ($Pr(Y/K_i \& X) \geq Pr(Y/K_i \& \sim X)$ for every i). This can be called "Pareto-positive causal factorhood." The corresponding definition of negative causal factorhood, which can be called "Pareto-negative causal factorhood," would be parallel (in effect, X is Pareto causally negative for Y if it is Pareto-positive for $\sim Y$); and the definition of causal neutrality would remain the same. John Dupré (1984) has argued for a more radical departure from context unanimity. He proposes that the condition should be dropped altogether and replaced with an idea he calls "statistical correlation in a fair sample."

In this section, I will examine Skyrms' suggestion and Dupré's argument for rejecting context unanimity. I believe that neither revision is necessary, and that there are advantages in *not* revising the definitions in either of these ways. I will also argue that, while Skyrms's suggestion is fairly "harmless," Dupré's more radical departure from context unanimity is a step backward, one that must ultimately either make causation tantamount to mere correlation or, in order to avoid vagueness, involve arbitrary, unmotivated distinctions.

Skyrms offers no rationale for his suggested Pareto weakening of the original definition, simply calling it a "plausible interesting weakening" of the stronger condition. However, Elliott Sober (1984a) offers the following interesting rationale for the suggestion:

Suppose some other physical condition, apart from smoking, *guarantees* the occurrence of a coronary. If an individual has that physical condition, smoking cannot boost the probability of a heart attack any higher that it already is. Yet it would be overly restrictive to conclude that smoking is not a positive causal factor for heart attacks in a population that happens to include some individuals with the condition. To take account of this sort of case, we should relax the requirement in the following way: The causal factor must raise the probability of the effect in at least one background context and must not lower it in any. (pp. 293–4)

Although I can sympathize with the intuitions that motivate the weakening for Sober, and although I think the revision does not lead to any serious difficulties, it seems to me that there is a better way to account for cases like the one Sober describes.[24]

In this example, we can describe, using the non-Pareto (or strict) version of probabilistic causation, all the causal facts in the general population – by considering *subpopulations*. There are three relevant populations involved: (i) the general population, (ii) the subpopulation of individuals who lack that physical condition, and (iii) the small subpopulation of individuals who have that condition. According to the original definitions, smoking has a mixed causal role for heart attacks in the first population, a positive causal role for heart attacks in the second population, and a neutral causal role for heart attacks in the third.

An advantage to this approach is that it allows for a group of statements about causal relevance in particular populations to have greater descriptive power. Even when considering particular subpopulations, the Pareto-revision approach does not allow for expression of the idea that a causal factor is "unanimously positive" for the effect. Of course subpopulations can be found for which the Pareto-revision approach can express the truth that, within them, the causal factor is neutral for the effect factor. But if all we say about the rest of

[24] I should add that an important reason for Sober's using the Pareto formulation has to do with the strategy of his critique of genic and group selectionism in evolutionary theory (1984a). The Pareto version is *weaker* than the strict version in such a way as to give genic and group selectionism, characterized in terms of probabilistic causality, a "better chance" of being true; and if genic and group selectionism are false on the Pareto interpretation, then they must be false on the strict version as well.

the subpopulation is that the causal factor is Pareto-positive for the effect factor, then, consistent with this statement of Pareto-positive causal significance, the possibility remains that, in a subpopulation of *it,* the causal factor is neutral for the effect factor.[25]

In statements of *positive or negative* causal significance, it seems that we should want our concepts of positive, negative, and neutral causal factors to be just as sensitive to the possibility of causal *neutrality* within some contexts or subpopulations as they are to the possibility of the *reverse* kind of causal significance within some contexts or subpopulations. Intuitively, there are three "pure" (non–Pareto and unmixed) possible causal roles one factor can have for another: positive significance, negative significance, and causal neutrality. The Pareto condition allows mixtures of the first and the last to count as positive causal relevance. But a mixture of the first and the last is just as mixed a causal factor as a mixture of any other two of the three kinds of unmixed causal significance. Indeed, as will become clearer in Chapter 3, it seems best to think of cases of (nontrivial) Pareto probabilistic causation as examples of *causal interaction* (briefly explained at the beginning of this chapter), in which, due to the interaction, we have correlation without (strict) positive causal factorhood.

On the other hand, of course, one may carve up all the possibilities however one wants, and if I do it differently from the way you do it, then we simply arrive at *different concepts.* One set of concepts may be more versatile or descriptive than the other for one purpose, and vice versa for another purpose; and each set of concepts may be just as "legitimate" and coherent as the other. I do not think there is anything

---

[25] Of course if one adds the information that the first subpopulation is the *largest* subpopulation of the general population for which the causal factor is neutral for the effect factor, then we have as much causal truth described as is possible using the strict definitions. But a statement providing this kind of information goes beyond statements of the form "*X* is causally positive (or negative or neutral or mixed) for *Y* in populaiton *P*": such a statement *quantifies over* subpopulations.

conceptually wrong or incoherent with the Pareto-revision, of course. However, for the reasons given, I will stick with the strict understanding of positive and negative causal relevance, given in the original definitions of the previous section.

Let us now turn to the reasons advanced by Dupré (1984) for abandoning context unanimity altogether. The main consideration Dupré advances in support of abandoning the requirement is the possibility of the following kind of case:

Suppose that scientists employed by the tobacco industry were to discover some rare physiological condition the beneficiaries of which were less likely to get lung cancer if they smoked than if they didn't. Contrary to what the orthodox [context unanimity] analysis implies, I do not think that they would thereby have discovered that smoking did not, after all, cause lung cancer. . . . If this is correct it seems to suggest that causes should be assessed in terms of average effect not only across different causal routes, but also across varying causal contexts. (p. 72)

There are three points I would like to make about cases of this kind, the first having to do with our understanding of "causal background contexts," the second having to do with how the definitions of the different kinds of probabilistic causal relevance can deal with examples of this kind (given a proper understanding of contexts), and the third with Dupré's suggestion of assessing causes in terms of their "average effect."

First, if we follow the understanding of contexts explained in the previous section, it is not at all clear that the theory requires us to hold fixed in the background contexts the rare physiological condition in Dupré's example. In the explanation of contexts given in the previous section as well as in, for example, Cartwright (1979), we are required to hold fixed all and only those factors (other than the causal factor in question and its effects) that are *themselves causally relevant* to the effect factor in question. But is having that rare physiological condition a cause – positive, negative, or even mixed – of lung cancer? It need not be, given the way Dupré has formulated his example.

98

Consistent with the description of the example, that physiological condition could be positive, negative, mixed, or neutral for lung cancer. (In Chapter 3, I explain in more detail these possibilities for factors like the rare physiological condition in Dupré's example.) The condition could be a strong positive cause of lung cancer; but for those with the condition, smoking helps a little. (The situation could be as depicted in Figure 3.3 of Chapter 3, where X is smoking, F is the physiological condition, and Y is lung cancer.) The condition could also be a strong negative cause of lung cancer, where the combination of the condition with smoking gives the best possible protection. (See Figure 3.2 with X, F, and Y interpreted as just explained.) Also, the condition could be mixed for lung cancer, where among those with the condition, whether or not one smokes makes a big difference, and among those without the condition, smoking makes little difference. In this case, the condition could be negative for lung cancer among smokers, and positive among nonsmokers. (See Figure 3.4 with X, F, and Y again interpreted as explained above.)

Finally, and most importantly, the condition could be *causally neutral* for lung cancer, if it is causally relevant to smoking in just the right way, so that smoking is causally intermediate between the condition and lung cancer. (Figure 3.5, discussed in Chapter 3, shows how this can happen.) In this case, the definitions given above say we *should not* hold fixed that rare physiological condition when assessing smoking's causal role for lung cancer.

Having said this, I nevertheless think that on a proper understanding of causal background contexts, the rare physiological condition in this example *should* be held fixed, in assessing smoking's role for lung cancer. This means that our understanding of contexts has to be revised, since on the current understanding we do not hold fixed any factors that are causally neutral for the effect factor in question. In Chapter 3, on interaction, we will see why we have to hold fixed some factors, like the physiological condition in Dupré's example, that may be causally *neutral* for the effect factor in question. (By the end of this section it will be apparent why it is necessary to hold fixed independent *mixed* causes of the effect factor in question.) Henceforth in this section, let us suppose, with Dupré, for the sake of discussion of his example, that the context unanimity theory *does* require us to hold the rare physiological condition fixed.

My second point about Dupré's example is simply to show how the strict context unanimity theory succeeds in capturing all the causal truth in the example. As in the discussion Skyrms's suggested Pareto weakening of the theory, we exploit the relativity of probabilistic causality to populations. In the subpopulation of individuals without that rare physiological condition, smoking is causally positive for lung cancer. In the subpopulation of individuals with the condition, smoking is a negative causal factor for lung cancer. And in the combined population, smoking has a mixed causal role for lung cancer.

Given the definitions of the previous section, the fact that smoking has *mixed* causal relevance for lung cancer in the combined population implies that, in the combined population, smoking is *not* a *positive* causal factor for lung cancer. (The definitions imply that *positive, negative, neutral,* and *mixed* causal relevance are mutually exclusive, as well as exhaustive, of the kinds of causal significance one factor can have for another in one population.) I agree that the claim that smoking is not a positive causal factor for smoking in the combined population *can* be *misleading* – especially if we put it, as Dupré does, as the claim that "smoking did not, after all, cause lung cancer." But I think it is misleading only to the extent that we lose sight of the population-relativity of probabilistic causation, and perhaps slip back into interpreting the causal claim in terms of the concept of *token* causation. The claim that *smoking is not a positive causal factor for lung cancer in population P* does *not* imply that there are no *subpopulations* of

100

P within which smoking *is* a positive causal factor for lung cancer, nor does it imply that there are no *individuals* in *P* for whom smoking is a token cause, or would be a token cause, of lung cancer.

When it is clearly seen that a claim of positive causal factorhood in a population *P* is quite a strong claim (involving context unanimity), so that the denial of positive causal relevance in the same population is a correspondingly weak claim, and when the question of population-level causal significance is properly untangled from questions about token level causal significance, then the denial of positive causal factorhood should no longer be misleading. Indeed, when all this is borne clearly in mind, it seems best to say that this is another example of the problem of *probabilistic causal interaction,* of probability increase due to interaction rather than to positive causal factorhood, as briefly described at the beginning of this chapter and discussed more fully in Chapter 3.

Perhaps many of us would still not wish to deny that smoking is a positive causal factor for lung cancer in Dupré's example. We may even wish to say that, in this example, smoking is a positive causal factor for lung cancer in the (overall) human population. And some may wish to say this even after the distinctions of the previous paragraph have been thoroughly digested. Perhaps our intuitive concepts are such that "*X* causes *Y* in population *P*" is judged to be true if in a significant subpopulation of *P*, *X* is a (context unanimous) positive causal factor for *Y*. Suppose (just to have an example) that this understanding perfectly matches our intuitions, that it is a perfectly coherent (though vague) concept, and that it is perfectly serviceable in all contexts in which we may ever actually wish to characterize population-level causal relations. This may all be so, but it does not mean that this way of describing the causal facts cannot be improved on. For example, the understanding of population causation in question is vague, involving the idea of a "significant" subpopu-

101

lation of P. In this respect at least, the context unanimity theory is an improvement.

In general, I think there are the following *two* kinds of criteria for the evaluation of philosophical theories – in particular, theories of probabilistic causation. First, a theory should be appropriately sensitive the ways in which we use the words denoting the concepts the theory is about (for example, the words, "positive causal factor," or "is a cause of"). Before the development of a philosophically adequate theory about something can begin, we must first obtain, from common or scientific usage of the relevant terminology or concepts, at least a rough idea or impression of the thing the theory is supposed to be a theory about. This is at least a typical starting point. But second, the theory should be sensitive also to philosophical standards such as: *avoiding vagueness, securing logical consistency, simplicity, non-"ad hoc-ness," expressive power* (the degree to which a variety of possibilities are describable in terms of ideas described and developed in the theory), and so on. My intention is to weigh heavily this second kind of criterion. In any case, in particular, the *vagueness* of the idea, mentioned in the previous paragraph, of a "significant" subpopulation, brings me to my third point about Dupré's argument.

That point is that there is the following problem for those who would reject the requirement of context unanimity, and would say, for example, that a factor X is causally positive for a factor Y when X raises the probability of Y in all but a *rare* causal context or subpopulation within which the probabilistic significance of X for Y may be reversed. In Dupré's example, that physiological condition is supposed to be "rare." Say that in the relevant combined population, 1 percent of all individuals have the condition, and that this counts as rare. But what if the condition were not so rare? What if 5 percent had it – or 15 or 25 percent, or 55, or 95, or 99 percent? If, for such possible populations, we continue to relax the condition of context unanimity, then it is clear that we would be revert-

102

ing to a "mere positive correlation" theory of probabilistic causation, which we have already seen ample reason to reject.

On the other hand, it seems that part of the intuitive rationale for *not* holding that condition fixed in the original example was that the condition was so rare. No reason for not holding fixed such a condition that is *not* rare was given. Indeed, it seems that our intuitions tell us that if that condition were not rare, but rather intermediate in frequency, then smoking would have a *mixed* causal role for lung cancer. So it seems that at some point in the progression of possible populations in which the condition becomes more and more frequent, we must begin to hold the condition fixed, so that smoking then becomes causally mixed for lung cancer. In addition, it seems that Dupré must also say that, if 99 percent of the relevant individuals have that condition, then smoking is causally negative for lung cancer, since this case is entirely symmetrical with the original case. Thus, later on in the progression, we must again relax the requirement of context unanimity and once again not hold the condition fixed. So the problem arises of specifying, and motivating, "cut-off" frequencies for the condition at which points we should begin and then cease to hold that physiological condition fixed. I cannot see how this can be done in a way that would not be arbitrary.[26]

On this kind of approach, the question of whether smoking is a population-level cause of lung cancer will turn on the population frequency of that physiological condition, and in an unacceptable way. Indeed, it seems that in this example, this question should turn not at all on the frequency of individuals with that condition. For example, a person contemplating be-

---

[26] Actually, Dupré does not advocate any such "cut-off" frequency approach, but rather an idea that probabilistic causation is correlation in a "fair sample" of the original population, a sample in which other causal factors are "fairly represented." His approach would seem not even to allow for the category of "mixed" causal significance in a population, which itself seems to be a step backward. Also, his conception of a "fair sample" is vague and problematic. I will not discuss this idea here; see Eells (1987a) for criticisms.

coming a smoker, and trying to assess the health risks, should not be so concerned with the population frequency of that condition, but with whether or not *he* has the condition. That is, the person should be concerned with which *subpopulation* he is a member of, the subpopulation of individuals with the condition (a population in which smoking is causally negative for lung cancer) or the subpopulation of individuals without the condition (a population in which smoking is causally positive for lung cancer). The population frequency of the condition can provide the decision maker with evidence about whether he is in a subpopulation in which smoking is causally positive, or in one in which it is causally negative, for lung cancer, but (except for the extreme frequencies of 0 and 1) it cannot settle the question, and hence cannot be definitive of whether he is in a population in which smoking is positive, or one in which it is negative, for lung cancer.

Even if cut-off frequencies *could* be properly motivated, and even if some other approach could be developed that is both in harmony with Dupré's intuitions and not tantamount to identifying causation with correlation, there still remains a further problem. If, in Dupré's example, a theory says that smoking is simply causally positive for lung cancer in the combined population, then statements of probabilistic causal connection, interpreted on that theory, would in many cases mask a significant causal truth: the fact that there is an "interaction" between the causal factor (smoking, in the example) and *other* causal factors (the rare physiological condition, in the example). Just as for the suggested Pareto weakening discussed before, statements of probabilistic causal connection interpreted on such a theory could not settle the question of whether or not there is such an interaction, of whether or not the causal significance of one factor for another varies from subpopulation to subpopulation. In Chapter 3, we will see how the definitions of the previous section must be revised in order to properly and generally accommodate the possibility of causal interaction.

104

There is another lesson to be learned from Dupré's example. So far, we have only seen examples that show why, when assessing the causal significance of a factor X for a factor Y, we must hold fixed factors that are, independently of X, either *positive* or *negative* causes of Y. Dupré's example suggests another example, one that shows why we must also hold fixed factors that are, independently of X, causally *mixed* for Y.

Suppose things are the way Dupré describes them in his example, and suppose we are interested in the causal significance of *tobacco-stained fingers, X,* for lung cancer, Y, in the general population. Of course the truth is that the factor of tobacco-stained fingers is causally neutral for lung cancer. We saw above that smoking, in the example, is not a positive or negative cause of lung cancer, but rather mixed. And, as pointed out above, that rare physiological condition also need not be a positive or negative cause of lung cancer. But if we hold neither of them fixed, we can expect the probability of lung cancer given stained fingers to be greater than the probability of lung cancer given clean fingers: $Pr(Y/X) > Pr(Y/{\sim}X)$. This is because people who have stained fingers tend to be smokers and because that physiological condition is so rare in the general population. So if the theory told us to hold neither fixed, it would give the wrong answer that tobacco-stained fingers is a positive causal factor for lung cancer.

Now suppose we hold fixed the factor of that rare physiological condition, F (as already mentioned, the theory will be revised in Chapter 3 to require this). Then we can expect that stained fingers would decrease the probability of lung cancer among those with the condition and increase that probability among those without the condition: $Pr(Y/F\&X) < Pr(Y/F\&{\sim}X)$ and $Pr(Y/{\sim}F\&X) > Pr(Y/{\sim}F\&{\sim}X)$. This is because, among those who have the condition, stained fingers increases the probability that one is a smoker, which, among those with the condition, decreases the probability of lung cancer; and among those without the condition, stained fin-

105

gers again raises the probability that one is a smoker, which, among those lacking the condition, increases the probability of lung cancer. So if we hold fixed the physiological condition but not the factor of smoking, we get the wrong answer that stained fingers is causally mixed for lung cancer.

Recall, however, that the definitions given in the previous section say we should hold fixed *all* factors that are, independently of *X*, *causally relevant to Y*. If we interpret this as meaning, "all factors that have either positive, negative, *or mixed,* causal relevance to *Y* independently of *X*," then we must hold fixed the factor of smoking – call this factor *G*. This is because smoking has mixed causal relevance to lung cancer, independently of stained fingers. And, of course, given each of the four ways of holding fixed both the physiological condition *and smoking,* the correlation between stained fingers and lung cancer disappears:

$$Pr(Y/F\&G\&X) = Pr(Y/F\&G\&\sim X),$$
$$Pr(Y/F\&\sim G\&X) = Pr(Y/F\&\sim G\&\sim X),$$
$$Pr(Y/\sim F\&G\&X) = Pr(Y/\sim F\&G\&\sim X),$$

and,

$$Pr(Y/\sim F\&\sim G\&X) = Pr(Y/\sim F\&\sim G\&\sim X).$$

And this, of course, is simply because the only reason for the correlations (overall and conditional on *F* and on *~F*) between stained fingers and lung cancer is the correlation between stained fingers and smoking, and the causal roles of smoking for lung cancer among the *F*'s and among the non-*F*'s; so holding fixed, in addition to that physiological condition, the factor of smoking – which has mixed causal relevance to lung cancer independently of stained fingers – makes the correlation between stained fingers and lung cancer disappear.

Hence, we must interpret "causally relevant" – in the part of the definitions of positive, negative, mixed, and neutral causal factorhood that tells us what to hold fixed – as meaning "causally positive, negative, *or mixed.*" Also, incidentally,

106

the example just described is a case of a kind of spurious correlation we have not yet encountered. In this example, Y is spuriously correlated with X because X is correlated with (because caused by) a factor G that is, independently of X, a *mixed* cause of Y; and G is a genuine, probability increasing, common cause of both X and Y, *positive* for X and *mixed* for Y.

2.4 INTERACTIVE FORKS

According to our simplest understanding of spurious correlation, explained at the beginning of Section 2.1, two factors were spuriously correlated if neither causes the other, they are correlated effects of a common cause, and their correlation disappears when the common cause is held fixed. And recall that the probabilistic structure of such cases is given by

(1) $Pr(X/Z) > Pr(X/\sim Z)$,
(2) $Pr(Y/Z) > Pr(Y/\sim Z)$,
(3) $Pr(Y/Z \& X) = Pr(Y/Z \& \sim X)$,
(4) $Pr(Y/\sim Z \& X) = Pr(Y/\sim Z \& \sim X)$.

Propositions (1)–(4) characterize common causes Z of factors X and Y in Reichenbach's (1956) sense, in which the presence, as well as the absence, of the common cause screens off the correlated effects from each other. Salmon (1978) calls this kind of probabilistic structure a "conjunctive fork."

However, there is another kind of probabilistic structure that has been recognized as a possibility for common cause situations. This is the kind of structure that Salmon (1978, 1984) has called an "interactive fork."[27] It is the same as a conjunctive fork except that one or both of (3) and (4) are changed to inequalities:

(3*) $Pr(Y/Z \& X) > Pr(Y/Z \& \sim X)$,
(4*) $Pr(Y/\sim Z \& X) > Pr(Y/\sim Z \& \sim X)$,

[27] See also van Fraassen (1977b, 1980).

respectively. If a common cause situation has the probabilistic structure of an interactive fork, then holding fixed the common cause, Z, either positively or negatively, will fail to make its joint effects, X and Y, probabilistically independent each other. That is, one or both of Z and $\sim Z$ will fail to screen off Y from X.

Of course, this kind of possibility must be addressed in the theory of probabilistic causation; for according to the theory as developed so far, holding fixed independent (of X) causes Z of Y should render X probabilistically neutral for Y, if X is causally neutral for Y. There are several ways in which situations can exhibit the probabilistic structure of an interactive fork; that is, there are various *causal* patterns consistent with this kind of *probabilistic* structure. In order to properly assess the bearing of the possibility of interactive forks on the theory of probabilistic causation, we must be careful to distinguish among these.

We have actually already encountered several examples in which holding fixed a common cause fails to screen off its joint effects from each other. Figure 2.5, of Section 2.1, depicts one such example, an example that exactly fits conditions (1), (2), (3*), and (4*) for interactive forks. (Let us assume transitivity of the chain from Z to W to Y, so that Z is a probability-increasing cause of Y.) Recall that in the variation of the Berkeley admissions case depicted in Figure 2.5, Z (being male) is a probability-increasing cause of both X (being believed to be male) and Y (admission); and there is some discrimination in admissions policies in favor of males, so that X is, both in the presence and in the absence of Z, a probability-increasing cause of Y. So this example satisfies the conditions for interactive forks. But, of course, we have seen that the theory of probabilistic causation, as laid down so far, handles this case just fine. Since X *is* a cause of Y in the example (even though both are effects of a common cause), X *should*, according to the theory, increase the probability of Y, when we hold fixed the independent cause, Z, of Y.

108

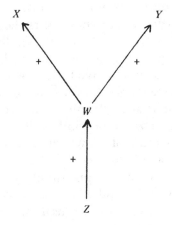

Figure 2.8

Another example having the probabilistic structure of an interactive fork is depicted in Figure 2.8. Here, Z, X and Y will form an interactive fork if these three conditions are met:

(5) W, X, and Y form a conjunctive fork, both conditional on Z and conditional on $\sim Z$,
(6) $Pr(W/Z) > Pr(W/\sim Z)$ (Z causes W),
(7) W screens off each of X and Y from Z, as does $\sim W$.

These three conditions imply that each of (1), (2), (3*), and (4*) will be satisfied by Z, X, and Y.[28]

[28] Essentially this was demonstrated in Eells and Sober (1986). Here is a somewhat different proof. For (1), note first that
$$Pr(X/Z) = Pr(W/Z)Pr(X/Z\&W) + Pr(\sim W/Z)Pr(X/Z\&\sim W)$$
and
$$Pr(X/\sim Z) = Pr(W/\sim Z)Pr(X/\sim Z\&W) + Pr(\sim W/\sim Z)Pr(X/\sim Z\&\sim W).$$
By (6), there are positive numbers a and u such that $a = Pr(W/\sim Z)$ and $a + u = Pr(W/Z)$. By (5), there are positive numbers b and v such that $b = Pr(X/Z\&\sim W)$ and $b + v = Pr(X/Z\&W)$. By (7), $Pr(X/\sim Z\&W) = b + v$ and $Pr(X/\sim Z\&\sim W) = b$. So, $Pr(X/Z) = (a + u)(b + v) + (1 - a - u)b = av + uv + b$, and $Pr(X/\sim Z) = a(b + v) + (1 - a)b = av + b$, from which (1) follows, since $uv > 0$. The proof of (2) is completely parallel. For (3*), note first that
$$Pr(Y/Z\&X) = Pr(W/Z\&X)Pr(Y/Z\&X\&W) + Pr(\sim W/Z\&X)Pr(Y/Z\&X\&\sim W)$$
and

109

Here is an intuitive example of this, a variation of the example given earlier about cold fronts, falling barometers, and rainy days.[29] Let X be barometers falling, Y be rainy days, and W be approaching cold fronts. And suppose that W, X, and Y form a conjunctive fork as before (only now we are using "W," instead of "Z," for approaching cold fronts). Now suppose that a cause of approaching cold fronts, in this part of the world, is westerly winds; call this factor Z. Then Z is a probabilistic cause of W, which in turn is a probabilistic cause of each of X and Y. By transitivity, which is plausible in this case, westerly winds is a probability-increasing cause of both falling barometers and rainy days, with approaching cold fronts as the intermediate causal factor. So (1) and (2) should hold in this example.

Now suppose that there is a westerly wind; that is, let us hold Z fixed positively. This does not necessitate the approach of a cold front. If we now add the information that barometers are falling, this should increase the probability that a cold front is approaching, which in turn increases the probability of rain. And if we instead added the information that the barometers are *not* falling, then this should decrease the probability that a cold front is approaching, which in turn decreases the probability that it will rain. So (3★) should hold. Now suppose that there is no westerly wind; that is, now hold Z fixed negatively. This does not necessitate there being no cold front approaching. So additional information to the effect that the barometers are or are not falling, increases and decreases, respectively, the probability that a cold front is

$$Pr(Y/Z\&\sim X) = Pr(W/Z\&\sim X)Pr(Y/Z\&\sim X\&W) + Pr(\sim W/Z\&\sim X)Pr(Y/Z\&\sim X\&\sim W).$$

By (6) and symmetry of correlation, there are positive numbers a and u such that $a = Pr(W/Z\&\sim X)$ and $a + u = Pr(W/Z\&X)$. Also by (6), $Pr(Y/Z\&X\&W) = Pr(Y/Z\&\sim X\&W) = Pr(Y/Z\&W)$, and $Pr(Y/Z\&X\&\sim W) = Pr(Y/Z\&\sim X\&\sim W) = Pr(Y/Z\&\sim W)$. And by (6) again, there are positive numbers b and v such that $b + v = Pr(Y/Z\&W)$ and $b = Pr(Y/Z\&\sim W)$. So, $Pr(Y/Z\&X) = (a + u)(b + v) + (1 - a - u)b = av + uv + b$, and $Pr(Y/Z\&\sim X) = a(b + v) + (1 - a)b = av + b$, from which (3★) follows. The proof of (4★) is completely parallel.

[29] For another example, see Section 4.2 (the indeterministic version of the example depicted in Figure 4.5).

110

approaching, which in turn increases and decreases, respectively, the probability of rain. And this is (4*).

In this example, Z is a common cause of X and Y, X and Y are causally neutral for each other; yet the common cause Z fails to screen off the correlation between X and Y. However, this kind of situation poses no problem for the theory of probabilistic causation laid down so far. It is not presupposed by the theory that in all cases in which joint effects of a common cause are causally neutral for each other, the common cause must screen off the effects from each other. It is only presupposed, so far, that, in such cases, when *all* independent causes of one of the joint effects are held fixed in one way, the other effect must be probabilistically neutral for the first. The joint effects must be independent conditional on specifications of *all* the causes of one of them (that are causally independent of the other).

In our example, each of W and Z is causally relevant (independently of each of X and Y) to each of X and Y. So, according to the theory, to assess the causal role of X for Y (or of Y for X) we must compare the probability of Y given X to the probability of Y given $\sim X$ (or X given Y and X given $\sim Y$) conditional on each of the four ways of holding fixed both W and Z. And condition (5) of the example implies that X and Y are independent conditional on each of the four ways, $W\&Z$, $W\&\sim Z$, $\sim W\&Z$, and $\sim W\&\sim Z$, of holding fixed W and Z.

This example shows that in at least *some* cases of interactive forks in which the joint effects are causally neutral for each other, when a finer description of the case is made, a conjunctive fork, and a screening off common cause, can be recovered. It is this feature of such cases that allows the probabilistic theory of causation to deliver the correct answers about what causes what in these cases. Also, in the example just discussed, the screening off common cause, W, is a factor that occurs *after* the time of the nonscreening common cause, Z (and, of course, before the time of the joint effects). These two features of this example are shared by examples of inter-active forks that have been discussed recently in the philo-sophical literature, with one important exception that will be discussed below. To illustrate this, and to demonstrate the versatility of the probabilistic theory, I will now turn to some recently discussed examples, and finally to the exception.

Salmon (1984) describes the following example. There are two balls on a pool table, the cue ball and the 8-ball. They are so situated that if a certain novice player attempts to put the 8-ball into a far corner pocket by shooting the cue ball di-rectly at the 8-ball (no banking), and succeeds in doing so, then it is almost certain that the cue ball will fall into the other far corner pocket. Suppose, in fact, that under these circum-stances it is almost certain that either both balls will fall into pockets or neither will. Keeping the distinction between to-ken and population causation clearly in mind, let us consider the population of attempts in which this novice player shoots the cue ball at the 8-ball without first banking. Let X be the event of the 8-ball dropping into one of the far corner pock-ets, Y the event of the cue ball dropping into the other far corner pocket, and Z the event of the cue ball colliding with the 8-ball. Suppose also that the probability of the 8-ball's falling into one of the corner pockets, given the player suc-ceeds in striking the 8-ball with the cue ball, is about 0.5, so that the probability of the cue ball's falling into the other corner pocket, given that the player succeeds in striking the 8-ball with the cue ball, is also about 0.5.

The factors Z, X, and Y in this example clearly form an interactive fork: plausibly, they satisfy, (1), (2), (3*), and (4) of the definition, above, of interactive forks. Most perti-nently, $Pr(Y/Z\&X) \approx 1 > Pr(Y/Z\&\sim X) \approx O$, which is (3*). Also, of course, neither of X and Y is a cause of the other. So we have another example of an interactive fork in which the joint effects are causally neutral for each other.

Let us suppose that, between X and Y, X is the "earlier" factor (that we limit the population to cases in which one of X

112

and $\sim X$ occurs before one of Y and $\sim Y$).[30] Then the issue is the probabilistic theory's verdict concerning the causal role of X for Y. Since X is causally neutral for Y, the problem for the probabilistic theory of causation is to find, and justify, a set of background contexts, for evaluating X's causal role for Y, that screen off Y from X. The question is: Can we find factors causally relevant to Y, independently of X, that, when all are held fixed in any given way, screen off Y from X?

Clearly the answer is yes. For one thing, in macroscopic examples such as this one, classical physics assures us that if we describe the collision of the cue ball with the 8-ball in enough detail – specifying the exact relative positions of the balls, the exact direction of motion and momentum of the cue ball, the exact points of contact, and so on – then we can predict with certainty whether or not the balls will fall into the corner pockets. Let Z_i range over these finer descriptions of the collision. Then, holding fixed, positively, any of the Z_i's will confer probabilities of 0 or 1 on X, Y, and $X\&Y$, and Y is screened off from X. This illustrates the fact that we sometimes have to formulate a common cause *partition* of Z_i's in order to recover a conjunctive forklike structure.[31] This also illustrates what Salmon (1984) calls a "perfect fork," a common cause structure in which the probabilities of the joint effects, conditional on the presence or absence of the common cause, are all either 0 or 1. In perfect forks, the common cause always screens its joint effects off from each other, in that there is no correlation between them conditionally on the presence or on the absence of the common cause.

However, in order for the analysis of this example to apply also to examples in which determinism is false, a different

---

[30] Again, a clarification of the idea of one factor's being earlier in time than another will be given in Chapter 5.

[31] Recall that a *partition* is set of mutually exclusive and collectively exhaustive factors (as noted above and also explained in Appendix 1). The conjunctive fork like structure recovered here is actually a more general kind of conjunctive fork than characterized in (1)–(4) above, where (3) and (4) are replaced by: $Pr(Y/Z_i\&X) = Pr(Y/Z_i\&\sim X)$, for each Z_i in the partition.

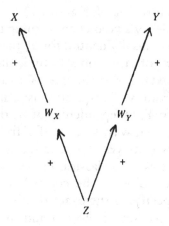

Figure 2.9

kind of approach is needed. Suppose Z occurs, or does not, at a time t_0, and that Y occurs, or does not, at time t_1. Clearly the possible states of the cue ball (including, say, its velocity and direction of motion) *at times between t_0 and t_1* are causally relevant to whether or not Y occurs at t_1 (to whether or not the cue ball falls into a corner pocket at t_1). And these states are causally independent of whether or not X occurs (of whether or not the 8-ball falls into a corner pocket). Let us suppose, for simplicity, that there are just two relevant possible states of the cue ball at a time $t_{1/2}$ between times t_0 and t_1: ball rolling toward the pocket with enough momentum to almost certainly carry it into the pocket (W_Y), and ball rolling in such a way that it is almost certain not to fall into the pocket ($\sim W_Y$).[32] Then Figure 2.9 (ignoring W_X for the moment) de-

[32] A more detailed specification of the intermediate states of the cue ball would specify the possible momenta of the ball, including direction of motion, and so on. In this case, we would have a partition of intermediate factors, or possible intermediate states. Then Z has different kinds of causal significance for the different intermediate factors, and these states have different kinds of causal significances for Y. The simpler analysis given in the text can easily be transformed into such a more detailed analysis.

114

picts the case in more detail than the first description of it above did.

Since W_Y specifies everything that is, at $t_{1/2}$, causally relevant to whether or not Y will occur at t_1, and since the process from W_Y at $t_{1/2}$ to Y (or to $\sim Y$) at t_1 does not interact with (or "intersect") the process from Z to X (or to $\sim X$) – they are now independent processes – W_Y should screen off Y from X. And the same goes for $\sim W_Y$. And since W_Y is causally relevant to Y, and is causally independent of X, it should be held fixed in background contexts in assessing X's causal role for Y. In this example, an appropriate set of background contexts would be the four ways of holding fixed, positively and negatively, Z and W_Y. And within each of these contexts, Y is probabilistically independent of X; so the probabilistic theory gives the right answer that X is causally neutral for Y.

This example shares a feature noted about the example discussed just before. There is a factor *after* the time of the nonscreening common cause that does screen the joint effects off from each other. However, unlike the previous example, this intermediate causal factor does not, in this example, form a conjunctive fork with the joint effects of the nonscreening common cause. Although this is irrelevant to the adequacy of the theory of probabilistic causation to such cases, it is interesting to note that if we define W_X analogously to the way W_Y was defined (W_X is causally intermediate between Z and X as shown in Figure 2.9), and if we set $W = W_X \& W_Y$, then, plausibly, W, X and Y do form a conjunctive fork.[33] In this case, this example has the structure depicted in Figure 2.8, and is analogous in all formal respects.

Salmon (1978, 1984) also gives a microscopic, and indeterministic, example of an interactive fork. The example involves the role of a conservation law in a phenomenon called

---

[33] If the correlation between W_X and W_Y is not perfect, then the partition of the four ways of holding these two factors fixed will, with X and Y, form a conjunctive fork, in the more general way of understanding conjunctive forks, noted above.

Compton scattering. Suppose we can consider a given electron to be at rest, and suppose an energetic photon, with energy E, collides with this electron. Call the event of this collision Z. There is a certain probability that a photon will emerge from the collision with energy E_1; call this event X. And there is a certain probability that an electron will emerge from the collision with energy E_2; call this event Y. Suppose E_1 and E_2 add up to E, as the law of conservation of energy demands. Because of this conservation law, there will be a correlation between events X and Y, conditional on Z. For example (as Salmon illustrates the point), if the probability of X given Z is 0.1 and the probability of Y given Z is 0.1, then the probability of $X \& Y$ given Z is not 0.01, the product of the two probabilities, but rather 0.1. This is because the law of conservation of energy (and the fact that E, E_1, and E_2 are related as the law demands) implies that X will occur if and only if Y does.

So $Pr(X \& Y/Z) > Pr(X/Z)Pr(Y/Z)$, which implies (3*) of the characterization of interactive forks above; plausibly (1) and (2) are also satisfied. And Salmon points out that this example, unlike the one involving billiard balls analyzed above, is not susceptible to analysis as a perfect conjunctive fork. The example, Salmon says, is "irreducibly statistical." No more refined or detailed description of the collision will necessitate the emergence of a photon and electron with given energies. However, if this is the *only* relevant difference between the Compton scattering example and the billiard ball example (which it actually is not, as noted below), then the probabilistic theory of causation can avoid the conclusion that X is a cause of Y, or Y a cause of X, in the same way as explained above for the billiard ball example. Whether or not an interactive fork is analyzable as a *perfect* conjunctive fork does not, by itself, control whether or not the probabilistic theory can correctly analyze the causal relations among the components of an interactive fork.

Let us focus on what the probabilistic theory has to say about the causal role of X for Y; exactly parallel considerations will apply to the question of the role of Y for X. Factor Z is a cause of factors X and Y. So, again, in order for the temporal priority requirement (Chapter 5) to be met, Z must be temporally prior to each of X and Y.[34] Let us say again that Z occurs at t_0 and Y occurs at t_1. Z is the event of the collision at t_0 and Y is the event of the electron having energy E_2 at time t_1. Let W_{Yi} range over states of the electron at a time $t_{1/2}$, between times t_0 and t_1. These will specify the energy of the electron at the intermediate time.[35] Each such state is, of course, causally relevant to Y, independently of X; so they must be held fixed in assessing X's causal role for Y. And, of course, conditional on any of these intermediate states, Y is probabilistically independent of X; so again, the probabilistic theory gives the correct answer that X is causally neutral for Y.

Also, if we let W_{Xj}'s range over intermediate states of the photon, then the partition of $W_{Xj} \& W_{Yi}$'s forms a conjunctive fork like structure with X and Y, in the sense that each element of the partition screens X and Y off from each other. Further, a coarser partition of intermediate states would make this example formally equivalent, in terms of factors and probabilities, to the billiard ball example, as depicted in Figures 2.8 and 2.9. The coarser partition would simply disjoin

---

[34] In discussing this microscopic example involving fast particles, we are coming close to having to take account of relativity of simultaneity and of temporal priority, described in the theory of special relativity. This will be especially pressing when discussing the Einstein–Podolsky–Rosen paradox, below. For now, we may just note that in this example, X and Y each fall in the future light cone of Z, so that each is absolutely future to Z.

[35] So in this example the two particles have definite energies at intermediate times, and I am assuming (falsely) that the example is not of the "Einstein–Podolsky–Rosen" type, which will be discussed below. Whether or not, physically, the Compton scattering phenomenon is of the EPR type is beside the point I want to make here, which is simply that the difference between determinism and indeterminism (or whether or not an interactive fork can be analyzed as a *perfect* conjunctive fork) does not *by itself* control whether or not the probabilistic theory will give the right answers in an interactive fork situation.

those possible intermediate states that are positively causally relevant to each of X and Y into one factor, and disjoin the others into a second factor, the negation of the first.

The billiard ball example and the Compton scattering example are disanalogous in that one is analyzable as a perfect fork and the other is not. However, they are analogous in that they share enough structure for the second analysis given of the billiard ball example to apply also to the Compton scattering example. In each case, the probabilistic theory delivers the correct answers about what causes what.

However, there is another kind of interactive fork possibility for which the analysis is not so clear. These are examples of the "Einstein–Podolsky–Rosen" (EPR) kind. In these examples, roughly, factors Z, X, and Y are described that form an interactive fork, where the joint effects X and Y are spacelike separated, and yet there is no factor (or partition of factors) that describes the state of the system at times before the occurrence of the joint effects and that screens the effects off from each other.[36] This seems especially troublesome for the probabilistic theory of causation, since it seems that we cannot say that either of X or Y is causally relevant to the other, given the "locality" requirement of special relativity theory, understood as meaning that causal processes cannot exceed the speed of light. In the remainder of this section, I will briefly explain some of the issues and some of the bearing of the EPR paradox on the probabilistic theory of causation.

Here is one schematic version of the EPR paradox.[37] Some

---

[36] Two events are *spacelike separated* if they are outside each other's light cones – that is, if they are so spatially and temporally situated that no subluminal, or luminal, process could originate at the time and place of either of the two events and arrive at the time and place of the other. In this case, neither is absolutely future or absolutely past to the other, according to special relativity theory. If, on the other hand, a subluminal process *could* connect two events, then the two events are said to be *timelike separated*, in which case the two events stand in a definite temporal priority order in all reference frames, according to special relativity theory.

[37] For other discussions from a philosophical point of view, see, for example, the following (on which the discussion of the EPR paradox here is mainly based): Skyrms (1980, 1984b), van Fraassen (1982), Jarrett (1984), Salmon (1984).

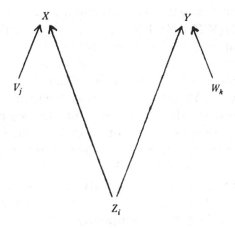

Figure 2.10

event creates two particles that shoot out in opposite directions. At a later time, after the particles are well separated, measurements are made on the two particles. The results of the measurements are called "up" and "down," relative to the spatial orientations of the measuring devices. Let X be that the measuring apparatus on the left gets result "up" ($\sim X$ is "down" on the left), and let Y be that the measuring apparatus on the right gets result "up" at ($\sim Y$ is "down" on the right). Let Z_i range over states of the two-particle system, where the Z_i's can say what the system is like at any time up to, but not including, the time of the two measurements. Finally, let V_j and W_k range over states of the left and right measuring devices, respectively; that is, the V_j's and W_k's range over the different possible spatial orientations of the devices. This much of the causal structure of the example is depicted in Figure 2.10.

Quantum mechanics implies that, given any state of the two-particle system, if the two detectors make measurements in the same direction (that is, with the same spatial orientation), then X and Y will be *perfectly anticorrelated,* and each

119

will have probability 0.5. For all i and j, $Pr(Y/Z_i \& V_j \& W_j \& X)$ $= 0 < 1 = Pr(Y/Z_i \& V_j \& W_j \& \sim X)$, and $Pr(X) = Pr(Y) = 0.5$. Experimental evidence confirms this. The "paradox" of the EPR paradox is that, according to quantum theory, there is this anticorrelation between factors X and Y (or correlation between $\sim X$ and Y), where it seems that neither factor can be causally relevant to the other (since spacelike separated), and where the correlation does not disappear when all factors causally relevant to the events X and Y are specified. Bell (1964, 1971) showed that no theory – no matter what causally relevant factors (or "hidden variables") are admitted into the states Z_i of the system (as long as the hidden variables do not depend on the settings of the detectors) – could *both* give independence of the two measurement results, conditional on the state of the system and detector settings, *and* predict the actual experimental and quantum mechanical statistics. Let me now briefly clarify these ideas.

One kind of "locality" condition, inspired by special relativity theory, is that the result on the right-hand measurement device should be probabilistically independent of both the setting and the result on the left, conditional on a state of the two-particle system and on the state of the measurement device on the right – and the same in the other direction, from right to left. The state of the two-particle system, together with the orientation of the right-hand measurement device, should screen off the result on the right both from the setting on the left and from result on the left, and the same in the other direction, according to this understanding of locality. Roughly following Jon Jarrett (1984), let us call this the condition of *strong locality* (the notation here differs from Jarrett's):

For all i, j, and k,
$$
\begin{aligned}
Pr(Y/Z_i \& W_k) &= Pr(Y/Z_i \& W_k \& V_j \& X) \\
&= Pr(Y/Z_i \& W_k \& V_j \& \sim X) \\
&= Pr(Y/Z_i \& W_k \& V_j) \\
&= Pr(Y/Z_i \& W_k \& X)
\end{aligned}
$$

$$= Pr(Y/Z_i \& W_k \& \sim X), \text{ and}$$
$$Pr(X/Z_i \& V_j) = Pr(X/Z_i \& V_j \& W_k \& Y)$$
$$= Pr(X/Z_i \& V_j \& W_k \& \sim Y)$$
$$= Pr(X/Z_i \& V_j \& W_k)$$
$$= Pr(X/Z_i \& V_j \& Y)$$
$$= Pr(X/Z_i \& V_j \& \sim Y).$$

(Of course, there is some redundancy, on the right-hand sides, in this formulation of strong locality; for example, the first, second, sixth, and seventh lines alone would be sufficient to characterize the idea.) Quantum mechanical predictions violate strong locality. Einstein, Podolsky, and Rosen (1935) concluded that quantum theory must be *incomplete* – that there must be factors that the theory has failed to take account of ("hidden variables") that, if included in the Z_i's, would yield the independence embodied in the strong locality condition.

However, experimental data is in harmony with the predictions of quantum mechanics. And Bell's theorem (1964, 1971), mentioned above, implies that, as long as any new hidden variables do not depend on the V_j's and W_k's, there can be no probability function describing the experiment described above that *both* satisfies strong locality *and* predicts the quantum mechanical and experimental statistics. Given special relativity, it is very plausible that the hidden variables, included in the Z_i's, should not depend on the settings of the measuring devices, the V_j's and W_k's. This is because the devices can be set, each in a *random manner*, and each at "the last moment," at space-time points that are *spacelike separated*, both from each other and from the particles. As Skyrms (1984b) points out, the choices of settings can be made by separate indeterministic quantum mechanical devices whose operations we have every theoretical reason to believe *are* independent of the state of the two particle system.

Strictly speaking, however, what special relativity precludes is superluminal transmission of *information-carrying signals* – that is, the transfer of such signals between spacelike

121

separated space-times. The idea of a "signal" has what has been called a "broad" and a "narrow" connotation. In the broad sense, a signal can be transmitted if there is some kind of *correlation* between the *outcome* on one apparatus and the *outcome* on another. In the narrow sense, a signal is transmitted if there is a *transfer of information* from one apparatus to another. This distinction between the broad and narrow senses of signal has a precise formulation, which I will explain below. Special relativity only precludes superluminal signals in the narrow, information-carrying sense of signal. Because of this, quantum mechanic's violation of strong locality is compatible with special relativity, as I shall now explain.

Suppose two experimenters are stationed at the spacelike separated space-times of the left- and right-hand measurements in the experiment described above, one experimenter on the left and the other on the right. The two experimenters can decide the *orientations* of their respective detectors, but (according to quantum theory) they cannot control the *outcomes* on their detectors. If, given a state Z_i of the two particle system and a state W_k of the right-hand detector, there were a correlation between the *state*, V_j, of the left-hand detector and the *outcome*, Y or $\sim Y$, on the right-hand side, then the experimenter on the left could transmit (at least in a probabilistically reliable way) a superluminal signal about the state of his detector to the experimenter on the right. (The state Z_i of the two-particle system and the state W_k of the right-hand detector could, theoretically, be prearranged.) This would be an information-carrying signal, and thus a signal in the narrow sense of signal. Special relativity rules out this kind of signal.

In addition to ruling out this kind of correlation, between *state* on one side and *outcome* on the other (conditional on any state of the two-particle system and any state of the detector on the other side), strong locality also rules out correlation between *outcome* on one side and *outcome* on the other (conditional on any state of the system and any state of the detector on the other side). While the first kind of correlation is ruled out by special relativity, the second is not. This is because although an experimenter on one side can decide the *state* of his detector, he cannot control the measurement *outcome,* and thus *cannot use the outcome to send a signal to the other experimenter,* in the "narrow" sense of "signal" that means "transfer of information."[38]

In light of special relativity, this suggests a locality requirement that is weaker than strong locality, and, in light of the success of quantum mechanics, more plausible: Given any state Z_i of the two-particle system and any setting W_k of the right-hand measuring device, the measurement result on the right, Y or $\sim Y$, is independent of the *setting* V_j on the left-hand device – and the same in the other direction, from right to left. This means that, conditional on any state of the two-particle system and any state of one of the detectors, the *result* of the measurement on the one detector is independent of the *state* of the other detector, but not necessarily of the *result* of the other measurement.

Again roughly following Jarrett (1984), let us call this the condition of *weak locality:*

> For all i, j, and k,
> $Pr(Y/Z_i \& W_k) = Pr(Y/Z_i \& W_k \& V_j)$, and
> $Pr(X/Z_i \& V_j) = Pr(X/Z_i \& V_j \& W_k)$.

The difference between strong and weak locality is that strong locality requires independence of the outcome on one side from *both settings and outcomes* on the other side, while weak locality only requires independence of the result on one side from the *setting* on the other – all conditional on a state of the two-particle system and a setting on the one side.

The predictions of quantum mechanics are in harmony not only with experiment, but also with weak locality. But as mentioned above, these predictions violate strong locality. Jarrett (1984) shows that strong locality is logically equivalent

[38] Jarrett (1984) reports a proof by Ghirardi, Rimini, and Weber (1980) that no superluminal signal can be produced by quantum mechanical measurements.

to the conjunction of weak locality and a condition he calls "completeness":

For all i, j, and k,
$$Pr(Y/Z_i \& W_k \& V_j) = Pr(Y/Z_i \& W_k \& V_j \& X)$$
$$= Pr(Y/Z_i \& W_k \& V_j \& \sim X), \text{ and}$$
$$Pr(X/Z_i \& V_j \& W_k) = Pr(X/Z_i \& V_j \& W_k \& Y)$$
$$= Pr(X/Z_i \& V_j \& W_k \& \sim Y).$$

(Of course, there is some redundancy, on the right-hand sides, in this formulation of completeness; for example, the first and third lines alone would suffice to characterize the idea.[39]) Quantum mechanical predictions violate completeness. If we hold fixed the state of the two-particle system, and set both detectors in the same direction, then X and Y are perfectly anticorrelated, though this correlation does not make signaling possible.

What is the bearing of this on the probabilistic theory of causation? If we consider a population of experiments in which the state of the two-particle system does not vary, and the orientations of the two detectors are the same and do not vary, then the probabilistic theory tells us that X is a negative causal factor for Y in this population, since in this population $Pr(Y/X) = 0 < 1 = Pr(Y/\sim X)$ (note that this population is homogeneous with respect to all factors causally relevant to Y independently of X).[40] There are several possibilities open

[39] It is easy to see that strong locality is equivalent to the conjunction of weak locality and completeness, given the way in which strong locality and completeness have been redundantly formulated here. *Suppose strong locality.* Weak locality is just the equalities, for i, j, and k, between the first and fourth probabilities and between the fifth and eighth probabilities given in the formulation of strong locality above, which equalities follow by transitivity of equality. And completeness is just the equalities, for all i, j, and k, among the probabilities in the first triple, and among the probabilities in the second triple, of right hand sides in the formulation of strong locality above. *Now suppose weak locality and completeness.* Let i, j, and k be arbitrary, for the proof of strong locality. $Pr(Y/Z_i \& W_k) = Pr(Y/Z_i \& W_k \& V_j)$, by weak locality; and $Pr(Y/Z_i \& W_k \& V_j) = Pr(Y/Z_i \& W_k \& V_j \& X)$, by completeness. This gives us that $Pr(Y/Z_i \& W_k) = Pr(Y/Z_i \& W_k \& V_j \& X)$, the first line of the formulation of strong locality above. The rest of this part of the proof follows the same pattern.
[40] It is perhaps worth noting that, given quantum mechanical statistics, when the independent causally relevant factors (the Z_i's, the V_j's, and the W_k's) are allowed to

as to how assess, or adjust, the probabilistic theory in light of this.

If we wish to deny that X is causally relevant to Y in the example, then we could include in the theory, by fiat, the condition that one event can only be causally relevant to events that lie in the first event's future light cone, a possibility noted by Skyrms (1984b). However, it seems that this approach cannot really get to the heart of the matter, that it cannot make contact with the reasons why one would wish to deny that X is causally relevant to Y in the example. When X occurs, so does $\sim Y$. Now suppose $\sim Y$ is, in a noncontroversial way, positively causally relevant to a third event, U, that *is* within the future light cone of X; and suppose that X is otherwise irrelevant to U (for example, Y and $\sim Y$ each screen off U from X). The requirement that effects must lie within the future light cones of causes will not prevent us from saying that X is a cause of U. But it seems that the idea that X causes U in this example should be just as unsatisfactory as the idea that X is causally relevant to Y, for those who wish to deny that X is causally relevant to Y.

Another possibility is to deny that causation must be local – in the strong sense of locality, of course, whose denial is consistent with weak locality, does not imply the possibility of superluminal, information-carrying signals, and is thus not in conflict with special relativity. In this case, the probabilistic theory would give what may be the *right answer*, namely, that X *is* causally relevant to Y: In the population described above, X *is* causally negative for Y. Note that if we adopt this position, then we are forced to say also that Y is causally negative for X, violating asymmetry of causation. This is because (1) the population is homogeneous with respect to all factors that have to be held fixed in assessing the causal role of either of X or Y for the other, (2) correlation is symmetric,

vary (specifically, when the orientations of the detectors with respect to each other differ from each other in different ways), then X's probabilistic (or "nonlocal causal") role for Y will vary as well.

and (3) neither event is (in special relativity theory) "absolutely" temporally prior to the other, the two events being spacelike separated. This third fact means that, if we agree that X is causally relevant to Y, then we cannot rule out the idea that Y is causally relevant to X on the basis of considerations involving the temporal order of events. This follows unless we say (implausibly, it would seem) that causal relevance is relative to frame of reference.

A denial of strong locality requires a little care in the formulation of a temporal priority requirement. In Chapter 5, the requirement will read roughly like this: *If* X and Y are *timelike* separated (so that they are within each other's light cones), *then* X can be causally relevant to Y only if X is before (and thus absolutely before) Y. This formulation does not rule out symmetry of causation between spacelike separated events, but it does rule out "absolutely backwards" causation.[41] Of course, a denial of (strong) locality does not come with an account of a *mechanism* of nonlocal causation (and the idea of physical mechanisms does not explicitly enter into the theory of type level probabilistic causation anyway). But this would nevertheless seem to be one consistent and coherent way of developing one concept of cause.[42]

Clearly, all this cannot be settled here, since the question of the possibility of *physical mechanisms* behind nonlocal "connections" is so highly relevant, and since only physical theory can address this question.[43] I have nothing more to say about the bearing of EPR phenomena on the probabilistic theory of causation – except that it also seems clear that it would be premature to conclude that the probabilistic theory of causation must exclude EPR phenomena, or to conclude that it simply does not apply to these cases.

[41] It is perhaps worth reiterating here that the delicate idea of ordering event *types* in time will be clarified in one way in Chapter 5.
[42] Compare Skyrms (1984b).
[43] Compare Salmon (1984, pp. 258–9).

3

Causal interaction and probability increase

For the examples of spurious correlation discussed in Chapter 2, it sufficed to hold fixed all (independent) positive, negative, and mixed *causes* of the candidate effect factor, in order for the probability-increase idea to deliver the right answers about what caused what. For these examples, only factors that were *causally relevant* to the candidate effect factor needed to be held fixed. In this chapter, I will argue that other kinds of factors, which may be causally *irrelevant* to (*neutral* for) the effect factor in question, must be held fixed as well, if the probability-increase theory is to deliver the right answers in other kinds of cases.

For example, if the right answer in Dupré's example, discussed in Chapter 2, is that smoking has a *mixed* causal role (not positive, negative, or neutral) for lung cancer, then it will be necessary to hold fixed the factor of that rare physiological condition. Otherwise, causal relevance would go by *average* probabilistic impact of smoking on lung cancer, across the presence and absence of that condition, and this cannot give the correct answer of mixed causal relevance. However, as noted in Chapter 2 and explained more fully in this chapter, that physiological condition need not itself be a positive, negative, or mixed cause of lung cancer.

At the beginning of Chapter 2, the possibility of there being such factors as that physiological condition in Dupré's example was called *the problem of causal interaction*. In Section 3.1, I give a simple formulation of the idea of interaction that characterizes such cases, and I argue that interacting causal factors must be held fixed in assessing causal roles.

In Section 3.2, the idea of causal interaction will be generalized, and the approach to interaction advocated in Section 3.1
will have to be generalized as well. This is to take account of
cases in which a causal factor, or its negation, is, within background contexts as characterized so far, *disjunctive* in character, in a causally significant way. I will show how we can
view such cases as cases of causal interaction. It is possible
that, within causal background contexts as characterized so
far, the disjuncts of a causal factor, or the disjuncts of its
negation, confer *multiple* causally significant probabilities on
the effect factor in question. In such a case, the presence and
absence of the disjunctive factor do not determine the *two
single* causally significant probabilities that we need for comparison within contexts. I call this possibility *the problem of
disjunctive causal factors*. A generalization of the approach
taken in Section 3.1 to the problem of causal interaction will
provide a solution to the problem of disjunctive causal factors
as well.

3.1 CAUSAL INTERACTION AND CONTEXTS

Consider this example, given by Cartwright (1979). She calls
it a case of "interaction," where, according to her understanding of interaction there, "two causal factors are interactive if
in combination they act like a single causal factor whose effects are different from at least one of the two acting separately" (pp. 427–8). The example involves the two factors of
ingesting an acid poison and ingesting an alkali poison (say in
the human population). The presence of one without the
other causes death, but the presence of neither or both does
not. Cartwright gives the example as a case in which a cause
is not "unanimous" for its effect. Dupré's example is also a
cause of interaction, in Cartwright's sense of interaction. And
as in Dupré's example, we can ask the rhetorical question,
"Should we deny that acid poison causes death just because
there is a kind of individual (namely, those who have just

128

ingested alkali poison) for whom ingesting the acid decreases the probability of death?" Of course, my answer here is an emphatic yes.

I will discuss two general ways in which a theory of probabilistic causation may attempt to deal with examples of this sort, two ways that may initially seem to come to the same thing, but which actually do not. For now, I will consider only simple cases in which there are just *two* relevant "On/Off" causal factors and *one* On/Off effect factor – for example, acid, alkali, and death in Cartwright's example, and smoking, the hypothetical physiological condition, and lung cancer in Dupré's example. The general case will be dealt with later in this section and in Section 3.2. I take it to have already been established, in Chapter 2, that in Dupré's example we should not say simply that smoking is causally positive for lung cancer, and, for similar reasons, that in Cartwright's example we should not say simply that ingesting acid poison is causally positive for death. In both cases, the first factor is causally *mixed* for the second. And the same goes for other cases of the same general kind, in which the direction of the probabilistic significance of a causal factor X for a factor Y is different in two subpopulations, or in the presence and absence of a second causal factor F.

According to the first way of dealing with such cases, we may say that neither of the two relevant causal factors (for example, the ingestion of acid poison and the ingestion of alkali poison in Cartwright's example, and smoking and that rare physiological condition in Dupré's example) *separately* has a definite causal role for the effect factor in question, and that *we should focus instead* on the four *combined, conjunctive* factors specified by saying *both* whether or not the first causal factor is present *and* whether or not the second is. In the acid/alkali case, then, neither the ingestion of acid poison nor the ingestion of alkali poison has a definite causal role for death. According to this approach, the causal truth is captured by four causal statements, each to the effect that one of the four combined factors (±acid & ±alkali) causes, or does not cause, death. This approach is adopted by Cartwright (at least for this example), and she says that it "accords" with her explication of probabilistic causation. But she adds that "considerably more has to be said about interactions" (1979, p. 428). Let us call this the "combined-factors approach" to the problem of interaction.

A second approach – the one that I favor and that I have actually already applied to Dupré's example in Chapter 2 – makes use of the fact that probabilistic causality is a relation among (at least) three things: a causal factor, an effect factor, and *a population* within which the former has some kind of causal significance for the latter. (A fourth item in this relation, as we have seen in Chapter 1, is a *kind* that a *token* population exemplifies; this subtlety has some pertinence to the general problem at issue here, as will be explained subsequently). On this approach, the causal truth in Cartwright's example would be expressed as follows: Among individuals who have just ingested an alkali (acid) poison, ingesting an acid (alkali) poison is causally *negative* for death; among individuals who have not just ingested an alkali (acid) poison, ingesting an acid (alkali) poison is causally *positive* for death; and in the combined population, ingesting an acid (or alkali) poison is causally *mixed* for death.

To get the theory of probabilistic causation to deliver the last of these judgments (the one about the combined population), we must require that the factor of whether or not one has just ingested an alkali poison be held fixed in background contexts when assessing the causal role of acid poison for death, and that the factor of whether or not one has just ingested an acid poison be held fixed when assessing the causal role of alkali poison for death. Holding these factors fixed is required by the same condition, laid down later in this section, that also requires holding fixed the presence or absence of that rare physiological condition when assessing the causal role of smoking for lung cancer, in Dupré's example.

130

However, the theory of probabilistic causation, as stated so far, does not include or imply any such condition. The theory only requires that we hold fixed independent positive, negative, and mixed *causes* of the effect factor in question, and the factors of the kind in question (such as ingesting an acid or alkali poison, and that rare physiological condition, in the examples we have discussed) *need not* be positive, negative, or mixed causes of the relevant effect factors. Again, this has already been pointed out in connection with Dupré's example. The physiological condition could, consistent with the description of the example, be causally neutral for lung cancer. Exactly how this can happen will be clarified below (Figure 3.5). In Cartwright's example, ingesting acid (or alkaline) could be neutral for death, if it is a cause of ingesting an alkaline (acid).[1]

Let us call the general approach to interaction just outlined – on which we always hold fixed factors across the presence and absence of which the causal factor in question has different probabilistic significances for the candidate effect factor – the "revised contexts approach." This approach will be developed and defended more fully in this and the next section.

Before evaluating the "combined-factors" and "revised-contexts" approaches to interaction, I should be more precise about how I will understand the idea of interaction. For now, let us continue to assume that there are just three relevant On/Off factors that need to be taken account of in our examples – for example, acid, alkali, and death; or smoking, that rare physiological condition, and lung cancer; in general, X, F, and Y. Then let us say that *a causal factor X interacts with a factor F, with respect to Y as the effect factor,* if X lowers the probability of Y in the presence of F and X raises

---

[1] This is consistent with there being nonzero probabilities for ingesting only acid and for ingesting only alkaline, if we suppose that the combination of acid and alkaline is especially healthy. Again, the issue of the role of intermediate causal factors, to be discussed in Chapter 4, is relevant.

the probability of Y in the absence of F (or vice versa, reversing "raises" and "lowers").[2]

This understanding of interaction is quite different from Cartwright's, quoted above (1979, p. 427–8). For one thing, note that on Cartwright's characterization, interaction between factors X and F, with respect to a factor Y, is *symmetric* in X and F. On the definition just given, however, interaction of X with F, with respect to Y, is *not* symmetric in X and F: Figure 3.2 depicts a schematic example in which symmetry fails. Let "$I_c(X,F,Y)$" mean that X and F are interactive for Y in Cartwright's sense, and "$I_d(X,F,Y)$" mean that X interacts with F in the production of Y in the sense of the definition just given. Then in the Figure 3.2 case we have $I_c(X,F,Y)$ and $I_c(F,X,Y)$, since it is clear that $X\&F$ has a different causal role· for Y, namely negative, from the causal role of $X\&\sim F$ for Y, namely positive. The two factors in combination ($X\&F$) have a different causal role for Y than *at least one of them* (in this case, X) has acting separately ($X\&\sim F$). However, it is also easy to see that $I_d(X,F,Y)$ but *not-$I_d(F,X,Y)$*.

In addition, it is possible for there to be interaction in Cartwright's sense with *no interaction at all* in the sense defined above. For example, suppose that $X\&F$, $\sim X\&F$, $X\&\sim F$, and $\sim X\&\sim F$ are equiprobable and confer, respectively the following probabilities on Y: 0.9, 0.3, 0.2, and 0.1. Then,

$$Pr(Y/X\&F) \quad = 0.9 > Pr(Y) = 0.375 > Pr(Y/\sim(X\&F))$$
$$= 0.2;$$
$$Pr(Y/\sim X\&F) = 0.3 < Pr(Y) = 0.375 < Pr(Y/\sim(\sim X\&F))$$
$$= 0.6,$$

so that X and F in combination have a different causal role for Y (namely, positive) from at least one of them (F) acting alone has for Y ($\sim X\&F$ is negative for Y).[3] However, there is no

---

[2]Later on, this understanding will be generalized to cover cases in which more factors than three (corresponding to X, F, and Y in this definition) are relevant in one way or another, as well as cases in which the presence or absence of a factor F makes a difference as to whether or not X affects the probability of Y at all.

[3]For the assessment of $\sim X\&F$'s causal role for Y, the issue of disjunctive causal factors is pertinent, since the absence of this factor is a disjunction one of whose disjuncts

interaction in the sense defined above, either of X with F or of F with X: X increases the probability of Y both in the presence and in the absence of F, and F increases the probability of Y both in the presence and in the absence of X. Here we have: $I_c(X,F,Y)$, $I_c(F,X,Y)$, but $not\text{-}I_d(X,F,Y)$ and $not\text{-}I_d(F,X,Y)$. However, in all the examples I use in evaluating Cartwright's approach to interaction (the "combined-factors approach"), there will be interaction in both senses of interaction. We have $I_d(X,F,Y)$ as well as $I_c(X,F,Y)$ (or if not certainly $I_c(X,F,Y)$, then certainly $I_c(\sim X, \sim F, Y)$).

Although Cartwright's combined-factors approach to inter-action seems to work alright in the acid/alkali poison case, it will not do in general. I think part of the reason why it seems to work alright in the acid/alkali case is that we picture a kind of symmetry in that example. But there are cases of interaction that lack the kind of symmetry that I think we naturally tend to envision in this example.

Figures 3.1–3.5 depict five rough possibilities for the kind of probabilistic significance that four combined factors, $\pm X \& \pm F$, can have for a fifth, Y, consistent with X's inter-acting with F, with respect to Y, in the way defined above. The figures are intended to represent situations in which factors corresponding to X, F, and Y are *all* the causally relevant factors. In these figures, the "columns" correspond to the four combined factors, $\pm X \& \pm F$; and within a given column, the position of the horizontal line corresponds to the probability of Y conditional on the combined factor to which the column corresponds, the lower the line the smaller the probability of Y given the relevant combined factor. In Figures 3.1–3.4, the spacing of the vertical lines is not meant to indicate the probabilities of the four combined

($X\&F$) confers a *greater* probability on Y than $\sim X\&F$ does, while the other disjuncts each confer a *lower* probability on Y than $\sim X\&F$ does. The issue is not so pressing for the assessment of $X\&F$'s causal role for Y, since *each* disjunct of the negation of $X\&F$ confers a *lower* probability on Y than $X\&F$ does. As mentioned above, this issue will be explored in detail later in this chapter, but nothing there will contradict the possibility of the causal roles of the various factors being just as described.

133

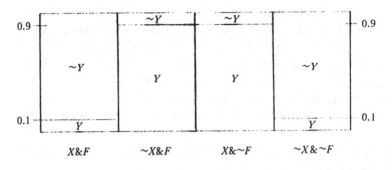

Figure 3.1

factors, so that these diagrams give only an incomplete specification of the probabilistic relations among the three factors X, F, and Y, in the population. In Figure 3.5, however, the spacing of the vertical lines *is* significant, and this will be explained below. For each of Figures 3.1–3.5, the two columns left of center represent the presence of F, with and without X, and the two columns right of center represent the absence of F, with and without X. It is easy to see that for each of Figures 3.1–3.5, X lowers the probability of Y in the presence of F and X raises the probability of Y in the absence of F.

I think Figure 3.1 corresponds most closely to the way we picture the acid/alkali example, with the causal role of acid poison (X) for death (Y) being entirely symmetrical with the causal role of alkali poison (F) for death. Note that for the Figure 3.1 case, no matter what the probabilities of the four combined factors are (as long as they are all nonzero), $\sim X\&F$ and $X\&\sim F$ are both clearly probabilistically positive for Y, and $X\&F$ and $\sim X\&\sim F$ are both clearly probabilistically negative for Y. (It follows, incidentally, that there is interaction in Figure 3.1 type cases in Cartwright's sense of interaction, since X and F in combination have a different causal role for Y than at least one of them acting alone does: $X\&F$ is causally negative for Y, while $X\&\sim F$ and $\sim X\&F$ are causally positive for Y.) So the combined-factors approach will be correct in

134

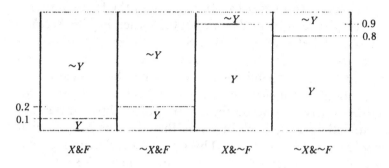

Figure 3.2

saying that $\sim X\&F$ and $X\&\sim F$ are causally positive for Y and that $X\&F$ and $\sim X\&\sim F$ are causally negative for Y. This agrees with our intuitions in the acid/alkali case, and I think the combined-factors approach gives right answers in cases of the Figure 3.1 type.

But now consider the Figure 3.2 possibility of interaction. In this case, $X\&\sim F$ is clearly probabilistically positive for Y, and $X\&F$ is clearly probabilistically negative for Y – regardless of the probabilities of the four combined factors. This is because, among the four combined factors, $X\&\sim F$ uniquely confers the highest probability on Y, and $X\&F$ uniquely confers the lowest probability on X. (From this it follows, incidentally, that there is interaction in Cartwright's sense as well, since X and F in combination ($X\&F$) is clearly causally negative for Y, while one of them (X) alone (that is, $X\&\sim F$) is clearly causally positive for Y.) And the combined-factors approach will say, correctly, that $X\&\sim F$ is causally positive for Y, and $X\&F$ is causally negative for Y. But what about the probabilistic and causal roles of $\sim X\&F$ and $\sim X\&\sim F$ for Y? Their *probabilistic* roles for Y (that is, whether they increase, decrease, and leave unchanged the probability of Y) depend on the probabilities of the four combined factors. For example, if $Pr(X\&F) = Pr(\sim X\&F) = 0.49$ and $Pr(X\&\sim F) = Pr(\sim X\&\sim F) = 0.01$, then $Pr(Y/\sim X\&F) = 0.2$

135

$> Pr(Y/\sim(\sim X\&F)) \approx 0.13.$[4] But if each of the four combined factors has the same probability of 0.25, then $Pr(Y/\sim X\&F) = 0.2 < Pr(Y/\sim(\sim X\&F)) = 0.6$. So, on the combined-factors approach, the *causal* roles of $\sim X\&F$ and $\sim X\&\sim F$ for Y will depend on the probabilities of the four combined factors.

Although the *probabilistic* roles of $\sim X\&F$ and $\sim X\&\sim F$ for Y thus clearly depend on the overall probabilities of the four combined factors, it seems clear that the question of their *causal* roles for Y should *not*. This is because different individuals that are in fact $\sim X\&F$, or that are in fact $\sim X\&\sim F$, may have different propensities to distribute themselves among the other combined factors *were they not* $\sim X\&F$, or *not* $\sim X\&\sim F$. And these propensities, for different individuals, *may not correspond to the overall probabilities of the four combined factors*. And the way in which such individuals may be disposed to distribute themselves among the remaining three combined factors, were they different from the way they in fact are with respect to X and F, may depend on what *kind* we implicitly associate with the token population in question, and Chapter 1 showed that we must always associate a kind with a token population in assessing causal and probabilistic relations in the population. These ideas – the "counterfactual affiliation of individuals with alternative combined factors" and the pertinence of *kind* of population to this – will be explored and developed more fully in Section 3.2.

The situation depicted in Figure 3.3 poses the same problems for the combined-factors approach. The combined factor $\sim X\&F$ is clearly probabilistically positive, and $\sim X\&\sim F$ is clearly probabilistically negative, for Y, no matter what the overall probabilities of the four combined factors are. And the combined-factors approach will say, plausibly, that these factors are causally positive and negative for Y, respectively. But again on this approach, there are factors, in this case $X\&F$ and $X\&\sim F$, whose probabilistic, and hence causal, roles for

[4]Throughout this book, the infrequently used symbol "$\approx$" will mean "is approximately equal to."

136

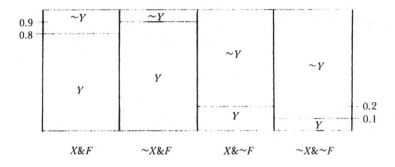

Figure 3.3

Y will depend on the overall probabilities of the other combined factors. Again this is unsatisfactory. (Incidentally, in Figure 3.3 type cases, it seems that X and F may not themselves interact for Y, on Cartwright's characterization of interaction; but $\sim X$ and $\sim F$, as well as $\sim X$ and F, clearly do.)

The cases of Figures 3.2 and 3.3 pose difficulties for the combined-factors approach to interaction, and in general for the assessment of the causal roles of the combined factors for Y. In the next section, I propose a solution to the problem of assessing the causal roles of the combined factors. But the cases of Figures 3.2 and 3.3 pose no problem for the theory of probabilistic causation, as stated so far, for the assessment of the causal role of the *uncombined* factor, X, for Y. First consider the roles of F for Y in these examples. In Figure 3.2, F lowers the probability of Y, no matter what the probabilities of the four combined factors are; and in Figure 3.3, F raises the probability of Y, no matter what the probabilities of the four combined factors are. (And all this is true whether or not we hold X fixed.) So F is a *negative cause* of Y in Figure 3.2, and F is a *positive cause* of Y in Figure 3.3. This means that, according to the theory as stated so far, F *must be held fixed in assessing X's role for Y*, so that the theory will give the correct answer that X is a *mixed causal factor* for Y.

Now consider Figure 3.4. Clearly X interacts with F with

137

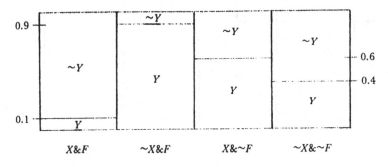

Figure 3.4

respect to Y, since X lowers the probability of Y in the presence of F and X raises the probability of Y in the absence of F. And there is interaction in Cartwright's sense as well, since X and F acting in combination ($X\&F$) has a different causal role for Y (clearly negative) from at least one of them acting alone ($\sim X\&F$ is clearly positive for Y). Also, this kind of case poses all the same problems that arose in the cases of Figures 3.2 and 3.3 for the combined-factors approach, and in general for assessing the causal roles of the combined factors – in this case it is the causal roles of $X\&\sim F$ and $\sim X\&\sim F$ that are problematic. How does the probabilistic theory, as stated so far, fare for the assessment of the role of the uncombined factor, X, for Y, in this example?

In the case of Figure 3.4, we should again regard X as causally mixed for Y: X lowers the probability of Y in the presence of F, and X raises the probability of Y in the absence of F. But the probabilistic theory will give us this correct assessment of mixed causal relevance only if it requires us to hold F fixed. Note that in this case F itself is probabilistically mixed for Y: In the presence of X, F lowers the probability of Y, and in the absence of X, F raises the probability of Y. Plausibly, we should regard F as causally mixed for Y. However, the *theory* will not identify F as mixed for Y unless it tells us to hold X fixed in evaluating the causal role of F for Y:

138

Overall, F is probabilistically neutral for Y, if the four combined factors are equiprobable. On the other hand, the theory will not *correctly* identify X as causally *mixed* for Y, thereby requiring us to hold X fixed in assessing F's role for Y, unless it tells us to hold F fixed in assessing X's role for Y: if the four combined factors are equiprobable, then overall, X is probabilistically *negative* for Y. Neither X nor F can correctly (and for the right reason) he identified as mixed for Y unless the other is correctly identified as mixed for Y (so that the other must be held fixed is assessing the first's causal role for Y).

Although it would be desirable for the theory to avoid this kind of circularity, we have noticed already in Section 2.2 that the theory is circular in that it characterizes the causal role of one factor for a second in terms of the causal roles of *other* factors for the second. In the example of Figure 3.4, F in fact *is* causally mixed for Y, so, whether or not the theory itself identifies F as mixed for Y, the theory tells us to hold it fixed in assessing the causal role of X for Y – and thus the theory *does* deliver the correct answer that X is causally mixed for Y. This nevertheless seems somewhat unsatisfactory; we would like the theory itself to identify factors as mixed, if the theory should tell us to hold them fixed because they *are* mixed. When the theory is revised, below, to incorporate the revised-contexts solution to the problem of interaction, then it will have this feature.

We have seen examples in which the interactive factor F is negative (Figure 3.2), positive (Figure 3.3), and mixed (Figure 3.4) for the effect factor Y in question. In each of these cases, F *is causally relevant* to Y (independently of X, of course), so that the theory tells us to hold F fixed in assessing the causal role of X for Y, and we get the right answer of mixed causal relevance of X to Y. Thus, although these examples pose difficulties for the combined-factors approach, there is no problem in them for the probabilistic theory in correctly identifying the causal role of the uncombined factor X (or of F) for Y. And this is just because, in each of these

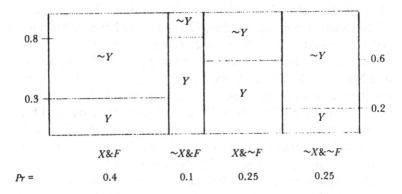

Figure 3.5a

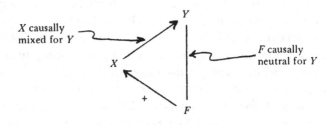

Figure 3.5b

examples, F is (independently of X) causally relevant to Y. However, as I have already noted several times above, there are cases of interaction in which the interactive factor F is causally *neutral* (not positive, negative, or mixed) for the effect factor Y in question, so that F is *not* causally relevant to Y. Now I give a detailed example of this.

Figures 3.5a and 3.5b give the probabilistic and causal structures of such a situation. (Incidentally, it is clear that X interacts with F with respect to Y in this example, that $\sim X$ and $\sim F$ are interactive in Cartwright's sense, and that this example as well has in it all the problems raised above for the combined-factors approach to interaction.) Here, F precedes X in time, and is a positive causal factor for X: $Pr(X/F) = 0.8 > 0.5 = Pr(X/\sim F)$.

140

Although X is causally mixed for, and hence causally relevant to, Y, it is not causally relevant to Y *independently of F; so X must not be held fixed in assessing F's causal role for Y.*[5] And we have

$$Pr(Y/F) = Pr(X/F)Pr(Y/X\&F) + Pr(\sim X/F)Pr(Y/\sim X\&F)$$
$$= (0.8)(0.3) + (0.2)(0.8) = 0.24 + 0.16 = 0.4,$$

and

$$Pr(Y/\sim F) = Pr(X/\sim F)Pr(Y/X\&\sim F) + Pr(\sim X/\sim F)Pr(Y/\sim X\&\sim F)$$
$$= (0.5)(0.6) + (0.5)(0.2) = 0.3 + 0.1 = 0.4.$$

So F is causally neutral for Y.

As a concrete hypothetical example of this, let F be that physiological condition in Dupré's example discussed earlier, let X be smoking, and let Y be lung cancer. (The physiological condition is not so rare in the version of the example depicted in Figure 3.5, but this is beside the point, of course; and obviously the causal structure shown in Figure 3.5b is compatible with a redrawn Figure 3.5a in which the condition is rare.) Smoking prevents cancer when the condition is present and causes cancer when the condition is absent; and the condition causes smoking. The best protection is to lack the condition and not smoke; the worst combination is to have the condition and not smoke. But the combination of smoking with the condition is almost as good as the best protection, and the condition is sufficiently efficacious in producing smoking that its probabilistic impact on lung cancer is the same as that of its absence.

Since F is causally neutral for Y, the theory of probabilistic causation, as stated so far, does not require us to hold F fixed in assessing the causal significance of X for Y. Thus, the theory cannot give us the right answer that X is causally mixed for Y. In fact, $Pr(Y/X) \approx 0.42 > Pr(Y/\sim X) \approx 0.37$, so

[5] X is an intermediate causal factor, between F and Y. Again, the role of causal intermediaries, and why they should not be held fixed in background contexts, will be discussed in detail in Chapter 4.

that the theory, as stated so far, gives the wrong answer that X is causally positive for Y. This example shows that not only the combined-factors approach, but also the theory of probabilistic causation as stated so far, cannot properly accommodate the phenomenon of interaction.

Let me now summarize the arguments and the significance of the examples given in this section. Figures 3.2–3.5 represent examples of interaction that pose difficulties either for the theory of probabilistic causation as stated so far, or for the combined-factors approach to interaction, or for both. In each case, the correct answer is that X is causally mixed for Y. In cases of interaction, the interactive factor F may be (independently of X) causally *negative* (Figure 3.2), *positive* (Figure 3.3), *mixed* (Figure 3.4), *or neutral* (Figure 3.5) for the effect factor Y in question. In cases in which F is (independently) causally negative, positive, or mixed for Y, the theory already tells us to hold F fixed in assessing the role of X for Y, so that the theory already gives the right answer of mixed causal relevance of X for Y. But since causal neutrality of F for Y is a possibility, the theory will not necessarily require us to hold fixed an interactive factor F in assessing the causal role of a factor X for a factor Y, which it must if it is to deliver the correct answer of mixed causal relevance. Although the combined-factors approach seemed to work fine in cases like those depicted in Figure 3.1, we have seen that in cases like those depicted in Figures 3.2–3.5 the approach fails to deliver principled answers. Furthermore, not only does the combined-factors approach fail to give us principled answers to questions about the causal roles of some of the *combined* factors in Figures 3.2–3.5, it also fails to address the question of the causal roles of the *uncombined* factors, X and F, *which is the question with which we begin*. I conclude that we should reject the combined-factors approach to interaction.

The "revised-contexts" approach to interaction, sketched above, is simply to revise the theory of probabilistic causation to require that, in addition to independent *causes* of the effect factor Y in question, we must also hold fixed *independent factors F with which X interacts with respect to Y*, when evaluating the causal significance of X for Y. Given this revision of the theory, we get the right answer of mixed causal relevance in Figure 3.5 type cases. Although F in the example is causally neutral for Y, X interacts with F in the production of Y and hence must be held fixed, on the revised theory. Also, in cases like those depicted in Figure 3.4, the theory need not identify F as a mixed cause of Y in order to insist that it be held fixed in assessing the causal significance of X for Y. It is sufficient that X interact with F with respect to Y. Of course, the revised theory *will* identify F as a mixed cause of Y. Since F interacts with X in the production of Y, X must be held fixed in assessing F's role for Y.

Note the independence condition for interactive factors: Not only should we hold fixed *only causally independent* (of X) *causes* of Y, but also *only causally independent* (of X) factors with which X interacts with respect to Y. (Again, a factor F is causally independent of X if X is neither causally positive, negative, nor mixed for F – that is, if X is causally neutral for, or not causally relevant to, F.) Here is a simple example that shows why.[6] Suppose that smoking (X) is a strong cause of that physiological condition (F) of Dupré's example (so that the condition may not be all that rare). And suppose that the physiological condition is strongly causally negative for lung cancer (Y), which, as we have seen, is consistent with smoking's interacting with the condition with respect to lung cancer. Figure 3.2 represents this situation, except that the vertical lines should be redrawn to indicate that $\sim X \& F$ and $X \& \sim F$ are very rare. In this case, it is clear that smoking is strongly *negative* for lung cancer. However, if we held fixed the intermediate interactive factor of the physiological condition in assessing smoking's role for lung cancer, the probabi-

[6]Again, the independence issue will be examined in more detail in Chapter 4.

listic theory would give us the wrong answer that smoking is causally *mixed* for lung cancer.

Here is a full statement of the revised theory. Let $F_1, \ldots,$ F_n be all and only those factors that are, in population P (of kind Q), either (i) a positive, negative, or mixed cause of Y, independently of X, or (ii) a factor that is causally independent of X and with which X interacts with respect to Y. Let K_i's be those maximal conjunctions of the F_i's that, relative to population P (of kind Q), have nonzero probability both in conjunction with X and in conjunction with $\sim X$. Then X is a positive causal factor for Y in population P if and only if, for each i, $Pr(Y/K_i \& X) > Pr(Y/K_i \& \sim X)$. Negative causal factorhood and causal neutrality are given by changing the ">" to "<" and "=," respectively. And X is causally mixed for Y in P if not causally positive, negative, or neutral.

This approach solves the problem of assessing the causal roles of the *uncombined* causal factors in the examples we have discussed (Figures 3.1–3.5), since in each of the examples F will be held fixed, giving us the correct answer of mixed causal relevance of X for Y. However, *it shares a difficulty that confronts the combined factors approach:* It has not been explained how to evaluate the causal significance of all the *combined* factors. In addition to the uncombined factors, it should also be *allowable to ask* what the causal roles of the combined factors are. But it seems that all the problems raised above for the combined-factors approach to interaction (except for the problem that it does not address the role of the uncombined factors) now apply equally well to the revised theory of the revised-contexts approach. In the next section, I formulate the problem in a very general way, and I offer a solution to the problem that exploits a more general understanding of interaction.

3.2 DISJUNCTIVE CAUSAL FACTORS

According to the theory of probabilistic causation, to evaluate the causal role of a factor X for a factor Y, we must

144

compare the probability of Y in the presence of X to the probability of Y in the absence of X, in each background context. We must have at hand these two *single* probabilities, for each background context. What was problematic in the previous section about combined causal factors was the probability of the effect factor in the *absence* of a *combined*, or *conjunctive*, causal factor. In the examples, there was more than one causally significant probability of the effect factor given the absence of the combined factor, and no principled way of averaging, or otherwise combining, these probabilities into a single probability. The problem arose because the absence of a combined, or conjunctive, causal factor is disjunctive in character. A natural generalization of the problem would include cases in which there are multiple causally significant probabilities of the effect factor both in the *presence* and in the *absence* of the causal factor – that is, cases in which the causal factor and its negation are both disjunctive.

Of course, the negation of a disjunction is a conjunction of the negations of the disjuncts; but it may nevertheless be a nontrivial disjunction of other factors. Suppose, for example, that we are interested in the causal roles of truth-functional compounds of the factors A, B, and C for a factor Y; and consider the disjunctive factor $X = A \lor (B \& C)$. Expressed as a disjunction of maximally specific relevant factors (put into "disjunctive normal form" in terms of A, B, and C), X is equivalent to $(A \& B \& C) \lor (A \& B \& {\sim}C) \lor (A \& {\sim}B \& C) \lor (A \& {\sim}B \& {\sim}C) \lor ({\sim}A \& B \& C)$. And its negation, ${\sim}(A \lor (B \& C))$, is equivalent to ${\sim}A \& {\sim}(B \& C)$, as well as to $({\sim}A \& B \& {\sim}C) \lor ({\sim}A \& {\sim}B \& C) \lor ({\sim}A \& {\sim}B \& {\sim}C)$.

If the probabilities of Y conditional on the various disjuncts of X are not all *all* greater than (or *all* less than or *all* equal to) *each* of the probabilities of Y conditional on the various disjuncts of ${\sim}X$, then whether or not the probability of Y given X is greater than (and whether or not it is less than, and whether or not it is equal to) the probability of Y given ${\sim}X$ will turn on the probabilities of the maximally specific disjuncts of X and of

145

~X. However, the question of the *causal* significance of X for Y should not turn on this, it seems. Suppose, for example, that there is a causal background context K that determines a subpopulation of individuals, some that have X and others that lack X, for each of which it is true that *if they were not X, they would have the factor* ~A&~B&~C (those in K that in fact lack X are of course in fact already ~A&~B&~C). Then for *this* kind K of individual, the relevant probability of Y in the absence of X would be $Pr(Y/\sim A\&\sim B\&\sim C)$ (= $Pr(Y/K\&\sim X)$), and not $Pr(Y/\sim X)$, the latter being an *average* whose value depends on the probabilities of *all* the disjuncts of ~X. This idea will be explained more fully and more generally later in this section.

Of course, the problem of multiple causally significant probabilities arises when we wish to compare probabilities *within background contexts*. The problem is that, even holding fixed everything we have so far learned to hold fixed, it is still possible that, within the background contexts, the presence of a causal factor, or its absence, or both, are "probabilistically inhomogeneous disjunctions," with respect to the effect factor in question. (Let us say that a disjunction is *probabilistically inhomogeneous for a factor* if the disjuncts do not all confer the same probability on the factor in question. Then a disjunction is homogeneous for a factor if all the disjuncts confer the same probability on the factor.) For simplicity in what follows, let us suppose that we are interested in the causal role of a factor X for a factor Y *within some one given background context* that has been identified by the conditions laid down so far. Alternatively, we may suppose that there are no independent causal or interactive factors that must be held fixed, according to the theory as laid down so far, as was the case in the examples depicted in Figures 3.1–3.5. The application of what follows to cases in which this simplifying assumption is not true is obvious: Do it within each of the relevant, previously identified, background contexts.

Here is the problem of disjunctive causal factors at the level of generality at which I will discuss it in this section. Let $Z_1, \ldots, Z_M$ be the consistent, exhaustive, maximal (hence mutually exclusive) conjunctions of all factors under discussion in a given problem that may be truth-functional ingredients of any factor causally relevant to a given factor Y. Thus, each of $Z_1, \ldots, Z_M$ is a factor that is probabilistically homogeneous for Y: There is no further factor under discussion that can be used to divide any of these factors into two or more disjuncts that could confer different probabilities on Y. Then when X is any disjunction, $Z_{i1} \vee \ldots \vee Z_{im}$, of a subset of $\{Z_1, \ldots, Z_M\}$, the question may arise of the causal significance of X for Y. Let the negation of X be $Z_{j1} \vee \ldots \vee Z_{j(M-m)}$ (so that $\{Z_{i1}, \ldots, Z_{im}, Z_{j1}, \ldots, Z_{j(M-m)}\} = \{Z_1, \ldots, Z_M\}$, an M-membered set).

In order to state a solution to the problem succinctly in terms of interaction below, it will be convenient notationally to reformulate the problem as follows. Let the set $\{X_{i1}, \ldots, X_{in}\}$ be the result of disjoining any factors Z_{ik} that confer any given equal probability on Y into a single factor X_{ik}, so that $n \leq m$, $X_{i1} \vee \ldots \vee X_{in}$ is equivalent to $Z_{i1} \vee \ldots \vee Z_{im}$, *no two of the* X_{ik}'s confer the same probability on Y, and no X_{ik} is a disjunction of Z_{ik}'s that confer different probabilities on Y. Similarly, let $X_{j1} \vee \ldots \vee X_{j(N-n)}$ be equivalent to $Z_{j1} \vee \ldots \vee V_{j(M-m)}$ such that $N - n \leq M - m$, *no two of the* X_{jl}'s *confer equal probability on* Y, and no X_{jl} is a disjunction of Z_{jl}'s that confer different probabilities on Y.

Thus, we have X equivalent to the disjunction $X_{i1} \vee \ldots \vee X_{in}$, each of whose disjuncts confers a different probability on Y, and each of which cannot, using factors under discussion in the given context, be decomposed into a disjunction whose disjuncts themselves confer different probabilities on Y. And $\sim X$ is equivalent to the disjunction $X_{j1} \vee \ldots \vee X_{j(N-n)}$, each of whose disjuncts confers a different probability on Y, and each of which cannot, using factors under discussion in the given context, be decomposed into a disjunction whose disjuncts themselves confer different probabilities on Y. So there are

147

multiple causally significant probabilities of Y both in the presence of X (n such probabilities) and in the absence of X ($N - n$ such probabilities). Let us call a disjunction *maximally probabilistically inhomogeneous for a factor* if (i) *all* its disjuncts confer different probabilities on the factor in question and (ii) all of its disjuncts are probabilistically homogeneous for the factor in question. Then each of the disjunctions, $X_{i1} \vee \ldots \vee X_{in}$ and $X_{j1} \vee \ldots \vee X_{j(N-n)}$, is maximally probabilistically inhomogeneous for Y.[7]

The problem of disjunctive causal factors, then, is how to identify a single causally significant probability of Y in the presence of X and a single causally significant probability of Y in the absence of X. Another way of putting the problem is this. How can we appropriately identify causal background contexts within which the presence of X and the absence of X each yield single causally significant probabilities for Y?

I will *first* offer part of a solution to a special case of the problem. This result will be: *If,* in the special case in question, we hold fixed in background contexts factors of a certain kind that I will describe, *then* we can identify the two single probabilities we need within each background context. *Next,* I will explain why we should hold those factors fixed; that will involve a more general conception of interaction than the one advanced above. And *finally,* I will move from the special case to a more general case.

The special case handled first is the one in which for each individual who has X, there is a determinate factor X_{jl} (a disjunct of $\sim X$) that that individual *would* have if it were not X, and for each individual who has $\sim X$, there is a determinate

---

[7] Compare Salmon (1971, 1984). Formally, the X_{ik}'s for X, and the X_{jl}'s for $\sim X$, are analogous to what Salmon calls "broadest homogeneous reference classes," or "maximal classes of maximal specificity." The condition that the X_{ik}'s for X, and the X_{jl}'s for $\sim X$, all confer different probabilities on Y is analogous to what he calls the "multiple homogeneity rule." Conceptually, however, the classes Salmon identifies are more analogous to causal background contexts than they are to disjuncts of a single disjunctive causal factor. However, *this analogy is intended to hold: Just as for Salmon's classes, it is supposed to be exactly the X_{ik}'s, within X, and the X_{jl}'s, within $\sim X$, that confer the multiple causally significant probabilities on Y.*

factor X_{ik} (a disjunct of X) that that individual would have if it were not $\sim X$. (Of course, which X_{jl} an X would be if not X can depend on more than just which of the X_{ik}'s the individual actually has, and which X_{ik} a non-X would have if it were an X can depend on more than just which of the X_{jl}'s the individual actually has; this is explained below.) In the more general case, handled at the end of this section, there may not be a *determinate* X_{jl} that a given X would have if not X, but rather there may be *various* X_{jl}'s that it might have if not X, with different *probabilities;* and similarly for actual non-X's, were they to have X.

Let us begin with the simple kind of case diagrammed in Figure 3.4. (Here the notation will diverge from that used to describe the general case.) Suppose we are interested in the causal significance of $X\&\sim F$ for Y. Recall that $X\&\sim F$ is one of the "problematic" combined factors for the Figure 3.4 case. Given the assumption of the special case just described, there are the following three kinds of individuals in the example: (i) those who would be $X\&F$ if not $X\&\sim F$ (these include all the actual $X\&F$'s and possibly some of the $X\&\sim F$'s); (ii) those who would be $\sim X\&F$ if not $X\&\sim F$ (these include all actual $\sim X\&F$'s and possibly some of the $X\&\sim F$'s); and (iii) those who would be $\sim X\&\sim F$'s if not $X\&\sim F$ (these include all actual $\sim X\&\sim F$'s as well as possibly some of the $X\&\sim F$'s).[8]

[8]Thus, these different kinds of individuals are characterized by which of the relevant *counterfactual conditional schemes* they satisfy. But counterfactual conditionals are vague and subject to various interpretations. What I have in mind is a so-called nonbacktracking, or standard, interpretation, as opposed to a backtracking interpretation. On the intended interpretation, roughly, we evaluate a counterfactual, "if A then C," by seeing whether C is true in a possible world in which A is true, in which things are exactly as they are in the actual world up to a time just before the time of A, and in which the laws of nature are the same as in the actual world. This works when A and C are about particular events. What I have in mind is roughly the analogue of this for when A and C are factors, or properties. The point is just that the main thing that evaluation of a counterfactual conditional "if A then C" should be sensitive to is the *causal* role of just A for C in the relevant circumstances (and not, for example, the evidential role of A for C). See Lewis (1979). And I am not claiming that counterfactual conditionals can be fully interpreted without appeal to causal notions. As mentioned in Chapter 1, I believe that our understandings of a number of concepts – including probability, causation, explanation, and so on – are interdependent. The idea of nonbacktracking counterfactual connection is another such item, and I will not here offer an analysis or explication of this idea.

To make the example, now involving these "counterfactual factors," more concrete, recall that X is the factor of ingesting an acid poison and F is the factor of ingesting an alkali poison. Then the inividuals in $X\&\sim F$ are those who have ingested an acid poison but not an alkali poison. These individuals may differ in the ways they would be different with respect to X and F, were they to have differed from the way they actually are ($X\&\sim F$) with respect to X and F. For example, there may be those $X\&\sim F$'s who immediately seek, and almost obtain, an antidote to $X\&\sim F$ – where the most similar counterfactual state of affairs in which they are different with respect to X and F, from the way they actually are ($X\&\sim F$), is one in which they succeed in finding the antidote F to X. These $X\&\sim F$'s are of the kind (i) described above: Were they not to have consumed the acid poison and *only* that poison, they would have found and consumed the alkali poison as an antidote to the acid poison consumed (they would have been $X\&F$).

There also may be, among the $X\&\sim F$'s, the suicidal types to whom both kinds of poison are available. For such individuals, it is possible, and plausible to assume, that had they not consumed just the acid poison ($X\&\sim F$), they would have consumed just the alkali poison ($\sim X\&F$). Individuals like these are of the kind (ii) described above. Finally, there may the nonsuicidal types who inadvertently consume only the acid poison, where the alkali poison is not even available. These individuals, had they not consumed only the acid poison (that is, were they not $X\&\sim F$), would have consumed neither poison (would be $\sim X\&\sim F$); and these $X\&\sim F$ individuals are of the kind (iii) described above.

Of course, it is an *assumption* in this example that these are the only three kinds of $X\&\sim F$'s that there are. It is possible that some of the $X\&\sim F$'s would distribute themselves over the other three combinations of the presence and absence of X and the presence and absence of F with various *probabilities,* which probabilities may differ from individual to individual;

this case will be considered below. (I also note that, when the disjunctive structure of the formal examples becomes more complicated, later, it may become more difficult to think up a concrete example illustrating the structure.)

Call these three kinds of individuals in this example K_1, K_2, and K_3 – corresponding to (i), (ii), and (iii), described above, respectively. These three kinds of individuals are mutually exclusive and (in our special case) collectively exhaustive: holding each of them fixed positively or negatively gives us back just the three factors K_1, K_2, and K_3 as consistent maximal conjunctions of them and their negations. Holding each of them fixed, then, gives us three background contexts in the example. And we have the following three two-way probability comparisons:

$$Pr(Y/K_1\&(X\&\sim F)) = 0.6 > 0.1 = Pr(Y/K_1\&\sim(X\&\sim F));$$
$$Pr(Y/K_2\&(X\&\sim F)) = 0.6 < 0.9 = Pr(Y/K_2\&\sim(X\&\sim F));$$
$$Pr(Y/K_3\&(X\&\sim F)) = 0.6 > 0.4 = Pr(Y/K_3\&\sim(X\&\sim F)).$$

The assignment of probabilities on the left-hand sides above relies on the plausible assumption that the probability of Y depends just on the way an individual *is* with respect to all factors that are in any way causally significant for Y, and not, beyond that, on the way an individual *would* be if it were different with respect to these factors. We have been supposing all along in this simple example that we have already identified, in X and F, all the factors that have any kind of causal role for Y, so that the K_i's cannot be causally relevant to Y once X and F have each been specified positively or negatively. More precisely, the assumption, for this example, is that $X\&\sim F$ and $(X\&F)\vee(\sim F\&X)\vee(\sim X\&\sim F)$ are each maximally probabilistically homogeneous disjunctions, for Y (here, $X\&\sim F$ is a "degenerate," one-disjunct, disjunction). That is, the K_i's cannot be probabilistically relevant to Y once we have specified X and F in one way, as we have in the left-hand sides in $X\&\sim F$.[9]

[9] If, for some reason, these K_i's *are* causally relevant to Y, then we just have a less simple example, and we must start over with a different set, $\{Z_1, \ldots, Z_M\}$ (in the notation

On the right-hand sides, however, if we ignore the K_i's, then neither X nor F is specified positively or negatively; $\sim(X\&\sim F)$ is consistent with each of X, $\sim X$, F, and $\sim F$. But $\sim(X\&\sim F)$ together with a K_i determines a specification of each of X and F, which gives us a unique causally significant probability for Y.

So, referring to the probability comparisons above, we see that *if we should hold fixed the K_i's*, and if there *are* individuals of all six kinds determined by specifying a K_i and whether or not $X\&\sim F$ (so that all six probabilities are defined), then in this example, $X\&\sim F$ is not unanimous for Y: $X\&\sim F$ has a *mixed* causal role for Y. Of course, as usual, we can only compare the probability of Y given X with the probability of Y given $\sim X$ within causal contexts that have positive probability both in conjunction with X and in conjunction with $\sim X$. Thus, for example, if K_2 had probability 0, then we would have to say that $X\&\sim F$ is a *positive* causal factor for Y.

Note that in this example the probabilities of the K_i's are constrained somewhat, but not completely, by the probabilities of the four combined factors involving X and F. Since $X\&F$ implies K_1, we must have $Pr(X\&F) \leq Pr(K_1)$. Similarly, $Pr(\sim X\&F) \leq Pr(K_2)$ and $Pr(\sim X\&\sim F) \leq Pr(K_3)$. However, the K_i's could distribute themselves in any way whatever among the $X\&\sim F$'s. That is, the only constraint on $Pr(K_1/X\&\sim F)$, $Pr(K_2/X\&\sim F)$, and $Pr(K_3/X\&\sim F)$, is that these three probabilities add up to 1; these values are not at all constrained by the probabilities of the four combined factors involving X and F. And, of course, their values will be determined by the *kind Q* that we associate with the token population in question. The dependence of these probabilities on population kind was already in effect illustrated in the introduction: Recall the example in which the human population was considered to be of such a (counterfactual conditional) kind that all members are such that, were they to become a smoker, they would be

---

used above to describe the general case of the problem), of causally relevant factors, and then generate, in the way described below in the solution to the general case of the problem, *different* contexts K_i, appropriate for such a less simple example.

152

physically determined to die of some cause other than lung cancer before lung disease had a chance to develop.

Note the effect, in the current example, of the assumption that for each individual there is a determinate factor it would exemplify if it were different from the way it is with respect to the causal factor in question: The population is of such a kind that, for each individual, there are only *two possibilities* regarding which combination of X and F it will exemplify. For example, individuals in K_1 *could* only be either $X\&F$ or $X\&\sim F$. This feature of this population, and analogous features of more complicated cases, will disappear later on, when the determinateness is dropped and replaced with the assumption that individuals would distribute themselves over alternate factors with different *probabilities,* were they different from the way they are with respect to the causal factor in question.

This approach also delivers the clearly correct answer for the intuitively less problematic factor, $X\&F$, in the Figure 3.4 example. In this case, the three relevant kinds of individuals, which I would like to treat as background contexts, correspond to the following counterfactual conditionals:

K'_1: If $\sim(X\&F)$, then $\sim X\&F$
(this includes all the $\sim X\&F$'s as well as possibly some $X\&F$'s).
K'_2: If $\sim(X\&F)$, then $X\&\sim F$
(this includes all the $X\&\sim F$'s as well as possibly some $X\&F$'s).
K'_3: If $\sim(X\&F)$, then $\sim X\&\sim F$
(this includes all the $\sim X\&\sim F$'s as well as possibly some $X\&F$'s).

Holding these factors fixed, we have the following three two-way probability comparisons:

$$Pr(Y/K'_1\&(X\&F)) = 0.1 < 0.9 = Pr(Y/K'_1\&\sim(X\&F));$$
$$Pr(Y/K'_2\&(X\&F)) = 0.1 < 0.6 = Pr(Y/K'_2\&\sim(X\&F));$$
$$Pr(Y/K'_3\&(X\&F)) = 0.1 < 0.4 = Pr(Y/K'_3\&\sim(X\&F)).$$

Thus, $X\&F$ is a negative causal factor in the Figure 3.4 case, as expected – as long as enough of the probabilities are defined.

The reader may wish to try out this approach on the other "problematic" and "nonproblematic" factors involved in the examples depicted in Figures 3.1–3.5. Also, the reader is invited to generalize the intuitions that I have tried to convey, in terms of the structures of Figures 3.1–3.5, to cases in which the disjunctive structures of formal examples are more complicated than those depicted in Figures 3.1–3.5. Such more complicated structures will be considered just below.

In any case, the straightforward and intuitive idea behind the more general approach, described below, can be captured by this somewhat complex question, with its parenthetical comments: Do *all* the (possibly disjunctive) *alternatives* (that have nonzero probability) to the (possibly disjunctive) causal factor in question confer a lower (or a higher, or the same) probability on the effect factor in question than (or as) does the *presence* of the (possibly disjunctive) causal factor in question (in which case we will, eventually, conclude that the appropriate *unanimity* requirement is met), or *not* (in which case we will, eventually, be able to conclude that the answer is *mixed* causal relevance)? In terms of the kinds of *figures* we have seen, this question amounts to the same thing as this: whether *all* horizontal lines (representing the probability of an effect factor Y) corresponding to disjuncts of a causal factor X are higher than (lower than, or are at the same level as) *all* the horizontal lines (for the probability of Y) corresponding to disjuncts of $\sim X$. I will not, however, provide figures for such cases that are more "disjunctively complicated" than those depicted in Figures 3.1–3.5.

Let us now consider a somewhat more complicated example than those depicted in the figures above. Turning back to the notation used to describe the general case, and using a previous more complicated example, let

154

$$X_1 = A\&B\&C \qquad X_5 = {\sim}A\&B\&C$$
$$X_2 = A\&B\&{\sim}C \qquad X_6 = {\sim}A\&B\&{\sim}C$$
$$X_3 = A\&{\sim}B\&C \qquad X_7 = {\sim}A\&{\sim}B\&C$$
$$X_4 = A\&{\sim}B\&{\sim}C \qquad X_8 = {\sim}A\&{\sim}B\&{\sim}C$$
$$X = A\bigvee(B\&C) = X_1 \bigvee X_2 \bigvee X_3 \bigvee X_4 \bigvee X_5$$
$${\sim}X = X_6 \bigvee X_7 \bigvee X_8$$

Assume that each of $X_1, \ldots, X_8$ is *causally homogeneous* for Y – that is, that each specifies, positively or negatively, all factors causally relevant to Y. More precisely, again, the assumption is that each of $X_1 \bigvee X_2 \bigvee X_3 \bigvee X_4 \bigvee X_5$ and $X_6 \bigvee X_7 \bigvee X_8$ is a maximally probabilistically inhomogeneous disjunction for Y.

Now there are eight counterfactual factors that have to be held fixed in background contexts. For $i = 1, \ldots, 5$, and $j = 6, 7, 8$, let $K_{i,j}$ be the conjunctive counterfactual factor: If X then X_i, and if ${\sim}X$ then X_j. The $K_{i,j}$'s are 15 mutually exclusive and collectively exhaustive kinds of individuals; and assuming nothing else needs to be held fixed, these are our 15 causal background contexts for this example. More precisely, of course, our causal background contexts will be those of these 15 that have positive probability both in conjunction with X and in conjunction with ${\sim}X$. So to determine the causal significance of X for Y, we must compare $Pr(Y/K_{i,j}\&X)$ with $Pr(Y/K_{i,j}\&{\sim}X)$, for each such $K_{i,j}$. These will all be comparisons involving just two causally significant probabilities, since, for each i and j, $K_{i,j}\&X$ is a subset of X_i and $K_{i,j}\&{\sim}X$ is a subset of X_j: Just note that $K_{i,j}$ is a subset of $X_i\bigvee X_j$, for each i and j. Also, $Pr(Y/K_{i,j}\&X) = Pr(Y/X_i)$ and $Pr(Y/K_{i,j}\&{\sim}X) = Pr(Y/X_j)$, since, again, the X_i's and X_j's are assumed to specify all factors that have any kind of causal role for Y.

As in the previous examples, the probabilities, here, of the X_i's and X_j's constrain somewhat, but not completely, the probabilities of the $K_{i,j}$'s. For example, $Pr(K_{i,j}) \leq Pr(X_i \bigvee X_j)$,

155

for all $i = 1, \ldots, 5$, and $j = 6, 7, 8$. Also, for each $i = 1, \ldots, 5$, $Pr(X_i)$ can be no greater than the sum of the probabilities $Pr(K_{i,j})$, for $j = 6, 7, 8$ (and of course there is a similar constraint for the X_j's). However, let i^* be any of $1, \ldots, 5$. Then there is no constraint at all on the conditional probabilities $Pr(K_{i^* j}/X_{i^*})$, except that they must add up to 1, when summed over the j's (and a similar fact holds of the X_j's, of course). These probabilities will be determined, of course, by the *kind* Q we associate with the token population in question. Note again how the determinateness assumption constrains the kind of population. Each individual is such that there is only one X_i and only one X_j that it *could* exemplify. For each i and j, individuals in $K_{i,j}$ could only be either X_i or X_j. Again, this feature of the population will disappear when the determinateness assumption is dropped below.

In the general case described above, involving the X_{ik}'s and X_{jl}'s, we must consider $n + (N - n)$ counterfactual factors. For $k = 1, \ldots, n$ and $l = 1, \ldots, N - n$, consider the conjunctive counterfactual factors: If X then X_{ik}, and if $\sim X$ then X_{jl}. Let $K_{k,l}$'s be those of these that have positive probability both in conjunction with X and in conjunction with $\sim X$. These $K_{k,l}$'s are (at most) $n(N - n)$ exclusive and exhaustive kinds of individuals; and assuming that nothing else needs to be held fixed, they are the causal background contexts. The probabilities $Pr(Y/K_{k,l}\&X)$ and $Pr(Y/K_{k,l}\&\sim X)$ must be compared, for each causal background context $K_{k,l}$. And these will all be comparisons between two single causally significant probabilities, since $Pr(Y/K_{k,l}\&X) = Pr(Y/X_{ik})$ and $Pr(Y/K_{k,l}\&\sim X) = Pr(Y/X_{jl})$, for each k and l. And recall that the X_{ik}'s and X_{jl}'s are assumed to specify all factors that have any kind of causal role for Y in the general case (except that some of them may be disjunctions of such specifications Z_{ik} and Z_{jl} when more than one confers the same probability on Y).

As in the simpler cases discussed above, the probabilities of the X_{ik}'s and X_{jl}'s impose loose constraints on the probabilities

156

of the $K_{k,l}$'s, where the precise values these probabilities have are determined by the kind Q associated with the token population in question. And the determinateness assumption (about the way individuals would be were they different from the way they are with respect to the causal factor in question) imposes a constraint on the kind Q analogous to those mentioned in the simpler cases.

This procedure has the result that, in the general case, X is causally positive, negative, or neutral for Y, according to whether $Pr(Y/X_{ik})$ is greater than, less than, or equal to, $Pr(Y/X_{jl})$ for all k and l, and otherwise, X is causally mixed for Y. This gives us the intuitively correct answer in cases of the kind depicted in Figure 3.4. And the way causal factors of the form $A\vee(B\&C)$ were handled, and the way the more general case involving the X_{ik}'s and X_{jl}'s was handled, are just straightforward generalizations of the procedure applied to the Figure 3.4 kind of case. But what else can be said in the way of justifying or motivating this way of dealing with the problem of disjunctive causal factors?

As far as justification goes, it seems that on any interpretation of probability suitable for developing probabilistic theories of causality, it cannot hurt to hold fixed factors that may not absolutely need to be held fixed, as long as we do not hold fixed factors, such as *nonindependent* causes of the effect factor in question, that the theory tells us *must not* be held fixed. Once enough has been held fixed to give us the right answer, the theory should give the same right answer if more and more factors are held fixed in the background contexts, as long as holding fixed the additional factors is not strictly prohibited.[10] If this is right, then we cannot be doing anything wrong in holding fixed the counterfactual factors described here.

I will now give positive motivation for holding these fac-

[10] This position is defended in Eells and Sober (1983) against a contrary position argued in Cartwright (1979). Also, it seems that the sketch of an interpretation of probability given in Section 1.2 should support this.

157

tors fixed. This will involve a generalized understanding of probabilistic causal interaction, an understanding on which disjunctive causal factors actually interact, with respect to the effect factor in question, with the factors I have suggested should be held fixed – and we have already seen that it is important to hold fixed interactive factors. I will then turn to the more general case in which there may be no determinate disjunct of the negation of a causal factor to which an individual would belong were it to lack the causal factor in question.

In the previous section, a causal factor X's interacting with a factor F with respect to an effect factor Y was characterized as there being a *reversal,* across F and $\sim F$, of an *inequality* between the probability of Y given X and the probability of Y given $\sim X$. Intuitively, this roughly corresponds to X's being a positive cause of Y in either the presence or absence of F and a negative cause in the other case. If this is the rough intuitive basis for our understanding of interaction, then clearly the characterization of interaction given earlier needs to be broadened. For one thing, there are (besides mixed causal factorhood, which we would like to disappear in specifications of background contexts) three kinds of causal significance that a factor X can have for a factor Y: positive, negative, *and neutral.* So the first natural generalization of probabilistic causal interaction would be to say that X interacts with F with respect to Y if the probabilistic significance – where this could be *positive, negative, or neutral* – of X for Y is just different in the presence of F from what it is in the absence of F (where, of course, X has positive, negative, or neutral "probabilistic significance" for Y according to whether the probability of Y given X is greater than, less than, or equal to, the probability of Y given $\sim X$).

Another natural generalization would allow a causal factor to interact not just with an "On/Off" factor F (contrasted simply with $\sim F$), but with a *three-way partition* $\{I_1, I_2, I_3\}$ of factors (where I_1, I_2, and I_3 are mutually exclusive and collectively exhaustive).[11] Then we can say that X interacts with the partition $\{I_1, I_2, I_3\}$ with respect to Y if the probabilistic significance of X for Y, there being three kinds, is different within each of the three factors of the partition.

It was perhaps natural to pick *three*-way partitions, since we have discussed three kinds of causal significance (aside from mixed, which we would like to disappear within specifications of causal and interactive factors that are causally independent of the causal factor in question). But there is no reason to stop at three. Another natural generalization would be to let a factor X interact with partitions of any size. We could say, for example, that the different possible probabilistic significances of X for Y are positive, negative, and *neutral at any particular value of the probability of Y*. Then the possibilities are $Pr(Y/X) > Pr(Y/\sim X)$, $Pr(Y/X) < Pr(Y/\sim X)$, and $Pr(Y/X) = Pr(Y/\sim X) = r$, for any r, $0 \leq r \leq 1$. In this sense, there are infinitely many different probabilistic significances that X could have for Y. Now we can say that X interacts with a partition $\{I_1, \ldots, I_n\}$, with respect to Y, if, in this broader sense, the probabilistic significance of X for Y is different within each of the I_i's.

This too is a natural understanding of interaction. First note that it is in line with Cartwright's formulation of the idea of interaction, discussed in the previous section. Suppose that X interacts with a partition $\{I_1, \ldots, I_n\}$, with respect to Y, in the sense just described. And suppose that I_i and I_j are such that $Pr(Y/I_i\&X) = Pr(Y/I_i\&\sim X) > Pr(Y/I_j\&X) = Pr(Y/I_j\&\sim X)$, so that I_i and I_j are two of the "new" kinds of factors that could be members of a partition with which X interacts with respect to Y. Then, within the disjunctive context $I_i \vee I_j$, X and I_i have a different role for Y than at least one of them (X) acting alone has: $Pr(Y/I_i\&X) > Pr(Y/\sim I_i\&X)$ $(= Pr(Y/I_j\&X)$, in the disjunctive context $I_i \vee I_j)$. Thus, it is clear that

---

[11] The idea of a *partition*, a set of mutually exclusive and collectively exhaustive factors, is explained in Appendix 1.

which of the I_i's X (or $\sim X$) combines with makes a difference for Y, and for the kind of impact X (or $\sim X$) has on Y.

Parallel to the move from the idea of "leaving the probability of Y unchanged" to the more specific idea of "leaving the probability of Y unchanged at a particular value of the probability of Y," we can also be more specific about the ideas of "raising the probability of Y" and "lowering the probability of Y." For example, a more specific version of "raises the probability of Y" would include the particular values from which and to which X raises the probability of Y. Here is what seems to be the most general possible understanding of X's interacting with a partition $\{I_1, \ldots, I_n\}$ with respect to Y. For each $i = 1, \ldots, n$, let $p_i = Pr(Y/I_i \& X)$ and $q_i = Pr(Y/I_i \& \sim X)$. Then let us say that X interacts with $\{I_1, \ldots, I_n\}$ with respect to Y if the n pairs $<p_i, q_i>$ are all distinct. Again, this is a very natural understanding of probabilistic causal interaction, since again in this sense of interaction, which of the I_i's X (or $\sim X$) combines with clearly makes a difference for Y, and for the kind of impact X (and $\sim X$) has on Y.

Of course, the move to this general conception of interaction blurs the distinction between interactive factors, in the narrower senses, and separate causes of the effect factor Y in question. And perhaps a more general term, like "effect-modification,"[12] would be more appropriate for what I am calling "a generalized conception of interaction." Indeed, I will show at the end of this section that, when the generalized conception of interaction is appropriately incorporated into the theory of probabilistic causality, separate reference to independent causally relevant factors is unnecessary. In any case, because of the smoothness of the move from the narrower senses of interaction to the broader senses, I shall continue to use the term "interaction." (It will nevertheless be useful at times to distinguish the narrower senses of interactive factors from the idea of separate causes.)

[12] I believe that what I am calling a generalized understanding of interaction is at least roughly the same as what Miettinen (1974) calls effect-modification.

Here is the revised theory of probabilistic causality that accommodates the generalized understanding of interaction. It should suffice now just to characterize the background contexts K_i. Let $F_1, \ldots, F_n$ be just those factors that are, in a population P (of a kind Q), *either* (i) a positive, negative, or mixed cause of Y, independently of X, *or* (ii) a member of some partition, whose members are causally independent of X, with which X interacts with respect to Y. Then the K_i's are those maximal conjunctions of the F_i's that, relative to population P (of kind Q), have nonzero probability both in conjunction with X and in conjunction with $\sim X$.

Now consider the partition of $K_{k,l}$'s ($k = 1, \ldots, n$, and $l = 1, \ldots, N - n$) of the solution to the general case of the problem of disjunctive causal factors given above. Call this $n(N - n)$-membered partition R. It is easy to see that all that is required for X to interact with R with respect to Y on the general understanding of interaction is for the following two conditions to hold: ·

(i) the $Pr(Y/X_{ik})$'s are all distinct, $k = 1, \ldots, n$ (recall that $Pr(Y/K_{k,l}\&X) = Pr(Y/X_{ik})$, for each k and l).

and

(ii) the $Pr(Y/X_{jl})$'s are all distinct, $l = 1, \ldots, N - n$ (recall that $Pr(Y/K_{k,l}\&\sim X) = Pr(Y/X_{jl})$, for each k and l).[13]

But the X_{ik}'s and X_{jl}'s were defined above in terms of the Z_{ik}'s and Z_{jl}'s in just such a way as to guarantee (i) and (ii).

We saw in the previous section that it is important to hold fixed causally independent factors with which a causal factor interacts. And we also saw just above that holding fixed the members of R, all of whose members are of course causally

[13] Suppose $K_{k,l}$ and $K_{k',l'}$ are two distinct members of R, and compare the pair $<Pr(Y/K_{k,l}\&X), Pr(Y/K_{k,l}\&\sim X)>$ with the pair $<Pr(Y/K_{k',l'}\&X), Pr(Y/K_{k',l'}\&\sim X)>$. The first pair is the same as $<Pr(Y/X_{ik}), Pr(Y/X_{jl})>$, and the second is $<Pr(Y/X_{ik'}), Pr(Y/X_{jl'})>$. Since we have two distinct members of R, either $k \neq k'$ or $l \neq l'$. In the first case, the two pairs will differ in their first members, by (i); and in the second case, they will differ in their second members, by (ii).

independent of X, provided a solution to the problem of disjunctive causal factors. So we now see that when the idea of interaction is suitably generalized, the solution to the problem of probabilistic causal interaction provides a solution to the problem of disjunctive causal factors as well.

Let us finally turn to the general problem of disjunctive causal factors *without* the assumption that there is a determinate X_{jl} that an X would be if not an X, and a determinate X_{ik} that a non-X would be if it were an X. This means that all the $K_{k,l}$'s (if X then X_{ik}, and if $\sim X$ then X_{jl}) may be *false*, and have probability of *zero*. However, I shall assume that the various individuals that are X would, were they not X, distribute themselves over the X_{jl}'s with different *probabilities* (perhaps different probability distributions for different individuals that are X) – and similarly for the non-X's.

As a first approximation to the solution to this more general case of the general problem, let r range over probabilities such that $r(X_{i1} \lor \ldots \lor X_{in}) = 1$ and s over probabilities such that $s(X_{j1} \lor \ldots \lor X_{j(N-n)}) = 1$. Then for such r and s, consider the factor: If X, then one of $X_{i1}, \ldots, X_{in}$ with probabilities given by r; and if $\sim X$, then one of $X_{j1}, \ldots, X_{j(N-n)}$ with probabilities given by s. Now let $K_{r,s}$'s be those of these that have positive probability both in conjunction with X and in conjunction with $\sim X$.

Will treating the $K_{r,s}$'s as causal background contexts provide a formally adequate and principled solution to the general problem – assuming again that nothing else needs to be held fixed in background contexts (or that all other such factors have already been held fixed in the context of this discussion)? Paralleling the treatment of the deterministic special case, two questions arise. First, of course, is the question: Does each $K_{r,s}$ determine *single causally significant* probabilities for Y in the presence and in the absence of X? And second, can treating the $K_{r,s}$'s as causal background contexts be motivated independently (perhaps in terms of a generalized understanding of interaction)?

162

The answer to the first question is yes, given a plausible assumption similar to one used in the special case above, and one other assumption. Consider first the probability of Y in the presence of X, in the background context $K_{r,s}$:

$$Pr(Y/K_{r,s}\&X) = \Sigma_{k=1}^{n} Pr(X_{ik}/K_{r,s}\&X)Pr(Y/X_{ik}\&K_{r,s}\&X).$$

Of course X_{ik} implies X, so the conjunct X can be eliminated from the conditioning part of the second terms. Also, the $K_{r,s}$'s can be eliminated there as well, since, by assumption, the X_{ik}'s already specify everything causally relevant to Y (this, of course, has the same rationale as the parallel move made in handling the deterministic case above, where it was assumed that it is only the way individuals *are* with respect to factors causally relevant to Y, and not the way they *would* be, if different, that determines the probability of Y). So we have

$$Pr(Y/K_{r,s}\&X) = \Sigma_{k=1}^{n} Pr(X_{ik}/K_{r,s}\&X)Pr(Y/X_{ik}).$$

Also, given the meaning assigned to $K_{r,s}$, it is natural to identify $Pr(X_{ik}/K_{r,s}\&X)$ with $r(X_{ik})$. So we have

$$Pr(Y/K_{r,s}\&X) = \Sigma_{k=1}^{n} r(X_{ik})Pr(Y/X_{ik}).$$

Similarly,

$$Pr(Y/K_{r,s}\&\sim X) = \Sigma_{l=1}^{N-n} s(X_{jl})Pr(Y/X_{jl}).$$

So it is clear that the $K_{r,s}$'s yield single probabilities for Y in the presence and in the absence of X – indeed, single *causally significant* probabilities, since different individuals are characterized by different r's and s's that are supposed to reflect the individuals' causal propensities to distribute themselves over the different kinds of X's (the X_{ik}'s) and non-X's (the X_{jl}'s), where these different kinds of X's and non-X's are already assumed to confer genuinely causally significant probabilities on Y.

The question now is whether the $K_{r,s}$'s *should* be treated as causal background contexts (or as factors that should be held fixed in causal background contexts). This cannot be justified in the way this treatment of the $K_{k,l}$'s was justified in handling the deterministic case above. This is because X does *not* necessarily interact with the partition of $K_{r,s}$'s with respect to Y, in the general sense of interaction described in this section. A $K_{k,l}$ of the deterministic case is the same as some $K_{r,s}$ of the general case in which r and s assign probability 1 to some X_{ik} and to some X_{jl}, respectively. And the distinctness of the probabilities conferred on Y by the X_{ik}'s and the X_{jl}'s guaranteed interaction in the deterministic case. However, multiple given distinct values (if there are more than two) can be *averaged* in many *different* ways so as to yield the *same* average. So the last two expressions displayed above can remain the same in value when calculated in terms of different probabilities r and s. And this just means that X may not interact with the $K_{r,s}$'s with respect to Y (depending on which of the $K_{r,s}$'s get non-zero probability).

However, now that we have the $K_{r,s}$'s to work with, it is easy to construct, from them, a partition with which X *does* interact, with respect to Y, each member of which will still determine a single causally significant probability of Y in the presence and in the absence of X. Where p and q are any numbers between 0 and 1, inclusive, let $K_{p,q}$ be the disjunction of all the $K_{r,s}$'s such that $Pr(Y/K_{r,s}\&X) = p$ and $Pr(Y/K_{r,s}\&\sim X) = q$.

Then the $K_{p,q}$'s are mutually exclusive and collective exhaustive factors, each of which *determines a single causally significant probability* for Y given X and for Y given $\sim X$; and X clearly *interacts with the partition* of $K_{p,q}$'s with respect to Y, in the general sense of interaction explained in this section. We have interaction, of course, because if $K_{p,q}$ and $K_{p',q'}$ are distinct, then the pairs $<p,q>$ and $<p',q'>$ are distinct, so that the two pairs $<Pr(Y/K_{p,q}\&X),Pr(Y/K_{p,q}\&\sim X)>$ $(= <p,q>)$ and $<Pr(Y/K_{p',q'}\&X),Pr(Y/K_{p',q'}\&\sim X)>$ $(= <p',q'>)$ will be distinct.[14] And I say the probabilities determined are *causally*

[14] This technique of disjoining K's could have been used above in handling the deterministic case, thereby avoiding the move from the Z_{ik}'s and Z_{jl}'s to the X_{ik}'s

significant because of the way they are determined by different kinds of individuals' propensities (the r's and s's) to be different kinds of X's or non-X's, were they different, with respect to X and $\sim X$, from the way they are – the values determined are not simply averages that depend on the probabilities of the different kinds of X's and non-X's (that is, they do not depend on the overall probabilities of the disjuncts of X and of $\sim X$).[15]

Thus, in the even more general, nondeterministic case as well, the solution to the problem of interaction provides also a solution to the problem of disjunctive causal factors. Again, the solution to the problem of interaction is that we should hold fixed, in addition to all independent *causes* of the effect factor in question, also all factors that are both causally independent of the effect factor in question and members of a partition with which the causal factor in question *interacts*. This, of course, is the same as holding fixed, as well as all independent *causes* of the effect factor in question, all factors in any partition such that the causal factor in question *interacts with the partition* and all of its members are causally independent of the causal factor in question.[16]

It is worth explaining, finally, how the characterization of causal background contexts given above, accommodating the generalized understanding of interaction described in this section, can be simplified. That is, the revised theory of probabilistic causality can be simplified. Given the way in which the

and X_{ji}'s. For the deterministic case, we had a choice; but in the general case, since there is more than one way to average more than two given distinct values to arrive at another given value, we are forced to disjoin $K_{r,s}$'s that yield given fixed values for the probabilities of X given Y and X given $\sim Y$.

[15] Note, incidentally, that this formulation gives an alternative way of characterizing positive, negative, neutral, and mixed causal relevance. X is causally positive (negative, neutral) for Y if and only if every $K_{p,q}$ is such that $p > q$ ($p < q$, $p = q$); and x is mixed for Y if not positive, negative, or neutral.

[16] A trivially equivalent reformulation of the solution could be stated in terms of a *single* "maximally fine" partition with which X interacts with respect to Y. A partition I is *at least as fine as* a partition J if every member of I is included in (or implies) some member of J. If I is at least as fine as J, then I will have at least as many members as J.

idea of interaction has been generalized and accommodated in the characterization of causal background contexts above, it is no longer necessary to *explicitly* and *separately* take account of factors that are independently causally relevant to the effect factor in question. If a factor F is, independently of X, causally relevant to a factor Y, then its possible confounding role will be taken care of by virtue of the way in which it must figure in a partition with which X interacts, relative to Y, as I shall now show.

Suppose $\{K_i\}_i$ is the set of background contexts that results from holding fixed *both* (i) all factors that are causally independent of X and causally relevant to Y *and* (ii) all factors that are causally independent of X and members of some partition with which X interacts, with respect to Y, in our generalized understanding of interaction. Now let $\{K'_j\}_j$ result from $\{K_i\}_i$ as follows. For each $<p,q>$ pair of possible probability values, disjoin all K_i's such that $Pr(Y/K_i\&X) = p$ and $Pr(Y/K_i\&\sim X) = q$, if any, into a single K'_j, separate from the other K'_j's determined by different $<p,q>$ pairs. It is clear that X interacts, with respect to Y, with the partition $\{K'_j\}_j$, and that the K'_j's are causally independent of X, since the K_i's are. It is also clear that treating $\{K'_j\}_j$ as our set of background contexts will give the same answer about the causal role of X for Y as does using $\{K_i\}_i$ as our set of background contexts.

Now let $\{K''_k\}_k$ be the set of background contexts that results from holding fixed just ((ii), above): all factors that are causally independent of X and members of some partition with which X interacts, with respect to Y, in our generalized understanding of interaction. Obviously, the partition $\{K_i\}_i$ is *at least as fine as* the partition $\{K''_k\}_k$, since the K_i's hold fixed everything the K''_k's do ((ii) above), and perhaps more ((i) above).[17] Also, the partition $\{K''_k\}_k$ is *at least as fine as* the partition $\{K'_j\}_j$, since the K''_k's hold fixed all factors that are

---

[17] Again, one partition I is *at least as fine as* another partition J if every member of I is included in (implies) some member of J; and if I is at least as fine as J, then I will have at least as many members as J.

independent of X and in any partition with which X interacts, with respect to Y, and since, as noted above, X interacts with $\{K'_j\}_j$, with respect to Y and the K'_j's are causally independent of X. From this, of course, it follows (what was obvious anyway) that $\{K_i\}_i$ is at least as fine as $\{K'_j\}_j$. But what is relevant is that $\{K''_k\}_k$ is intermediate in fineness between $\{K_i\}_i$ and $\{K'_j\}_j$ (or the same in fineness to one or both of the latter). From this, and the fact that using $\{K_i\}_i$ and using $\{K'_j\}_j$ as our set of causal background contexts give the same answer about X's causal role for Y, it is clear that using the K''_k's as background contexts also gives the same answer.

It thus follows that, for assessing the causal role of X for Y in a population (of a kind), it suffices to hold fixed just all factors that are causally independent of X and members of a partition with which X interacts, with respect to Y. Of course, this is not to deny the importance of holding fixed independent (of X) causes of Y; it is just that the generalized understanding of causal interaction described in this section already accommodates, as a kind of interaction, the role of such independently causally relevant factors. If F is, independently of X, causally relevant to Y, then it appears that F will be held fixed, positively and negatively, in virtue of being a member of some partition, of factors causally independent of X, with which X interacts, relative to Y. Despite the formal subsumption, here, of separate causes under the definition of generalized interaction, it will nevertheless be useful at times to distinguish the conceptually different roles of interactive factors, in the narrower senses, from separate causes.

So far, we have been concerned with what kinds of factors *must* be held fixed in causal background contexts. The verdict on this question is: All factors that are *both* causally independent of the causal factor in question *and* members of a partition with which the causal factor in question interacts, with respect to the effect factor in question – where interaction is understood in the general way described in this section. In the next chapter, we turn to the question of what kinds of factors

167

must not be held fixed. This issue naturally involves the role of causes of the candidate effect factor that are *not causally independent* of the causal factor in question. This is the question of the role of *intermediate causal factors*. Of course, the question of *transitivity* of property level causality naturally arises in this connection, and this question also will be addressed in detail in the next chapter.

4

Causal intermediaries and transitivity

In Chapter 2, a spurious correlation of a factor Y with a factor X was characterized as a situation in which, because of separate causes of Y, the degree of correlation of Y with X is different from the degree of causal significance of X for Y. The possibility of a spurious correlation of Y with X was diagnosed as arising when there are factors Z that are correlated with X and that are positive, negative, or mixed causes of Y, independently of X, where the correlation in question may be unconditional or conditional on other such factors Z. But as noted in Chapter 2, not all cases in which X is correlated with separate causes of Y give rise to a spurious correlation: The separate causes must be causally independent of X. In Chapter 3, we saw that when a factor X interacts, with respect to a factor Y, with a factor Z that is causally independent of X, then we should say that X is causally mixed for Y. But as noted in Chapter 3, not all cases in which a factor X interacts with a factor Z in the production of a factor Y are cases of mixed causal relevance of X for Y. Again, Z must be causally independent of X.

A factor X may be a genuine cause of Y, and the degree of correlation between X and Y may correctly reflect the strength of causal significance of X for Y, if X causes Y *by way of causing* Z, so that Z causes Y and there is a correlation of Z with X due to X's being a cause of Z. This kind of possibility was the reason, at the beginning of Section 2.1, for explicitly excluding from cases of spurious correlation cases in which the third factor Z screens off a correlation between factors X and Y because Z is *causally intermediate* between X and Y. And in

169

Section 2.2, this was the reason for holding fixed in background contexts, among the causes of Y other than X, only causally independent (of X) causes of Y, when evaluating the causal significance of X for Y. And in Section 3.1, this was the reason for holding fixed, among factors with which X interacts with respect to Y, only those that are causally independent of X.

Suppose Z is a positive, negative, or mixed cause of Y. Then, relative to a factor X, Z is an *independent cause of Y* if X is not a positive, negative, or mixed cause of Z; and if X *is* a positive, negative, or mixed cause of Z, then Z is an *intermediate cause of Y*, or an *intermediate causal factor*, relative to X's causal role for Y. Now suppose that X is some kind of cause of Z – that is, Z is not causally independent of X. Then two conditions that each suffice for Z's being an intermediate causal factor, relative to X's causal role for Y are (i) Z is some kind of cause of Y (as in Chapter 2), and (ii) X interacts with Z with respect to Y (as in Chapter 3). The sufficiency of (i) is obvious, from the preceding characterization of intermediate causal factors. For (ii), just note that if Z is not a positive or negative cause of Y, then Z must interact with X with respect to Y (and, of course, X is causally independent of Z, by asymmetry of causal relevance), so that Z will be causally mixed for Y, which again makes Z an intermediate causal factor, on the above characterization of interaction.[1]

Thus, if Z is not causally independent of X, then the issue of intermediate causal factors arises both when Z is a separate cause of Y (the kind of factor investigated in Chapter 2) and when Z is interactive for Y (the kind of factor investigated in Chapter 3). In the second case, where X interacts with Z with respect to Y, Z in fact is a kind of "separate cause" of Y: a

[1] For example, suppose, where X, Y, and Z are the only pertinent factors, that $Pr(Y/Z\&X) > Pr(Y/Z\&{\sim}X)$ and $Pr(Y/{\sim}Z\&X) < Pr(Y/{\sim}Z\&{\sim}X)$, which is one way in which X can interact with Z with respect to Y. This makes it impossible that both $Pr(Y/Z\&X) = Pr(Y/{\sim}Z\&X)$ and $Pr(Y/Z\&{\sim}X) = Pr(Y/{\sim}Z\&{\sim}X)$, so that Z must be either a positive, a negative, or a mixed causal factor for Y.

mixed causal factor. In this chapter, I will examine the role of intermediate causal factors in more detail, including the independence requirement regarding what factors may be held fixed in causal background contexts.

Here is a concrete example that illustrates the general reason for not holding fixed intermediate causal factors. Your phoning me (X) causes me to lift the receiver on my phone (Y), and it does so by causing my phone to ring (Z). My phone's ringing is an intermediate causal factor between your phoning me and my lifting the receiver on my phone; it is not an independent cause. Although my phone's ringing is a *positive cause* of my lifting the receiver and it is *correlated with* your phoning me, this of cause does not mean that the correlation between your phoning me and my lifting the receiver is spurious. And the reason why we should not hold fixed the factor of my phone's ringing when assessing the causal significance of your phoning me for my lifting the receiver is that if we did so, the positive probabilistic significance of your phoning me for my lifting the receiver would disappear, incorrectly indicating that your phoning me is causally neutral for my lifting the receiver on my phone. On the assumption that my phone rings, as well as on the assumption that it does not, the probability of my lifting the receiver given that you call is the same as that of my lifting the receiver given that you do not call.[2]

In this example, the intermediate causal factor Z is a positive causal factor for the final factor Y of my lifting the receiver: Z is an intermediate *separate (positive) cause* of Y. An example showing that we should not hold fixed causally intermediate *interactive* factors was already given near the end of Section 3.1. In that example, the intermediate causal factor Z

---

[2]This assumes, of course, that it is only by causing my phone to ring that your phoning me can cause me to lift the receiver: the causal chain is "singly connected," as explained in Section 4.3. This telephone example was discussed in Eells and Sober (1983).

171

was causally *mixed* for the final factor Y. Other examples involving intermediate interactive factors are somewhat controversial, and will be discussed below.

If we hold fixed intermediate causal factors in a causal chain from an earlier factor to a later factor, then we do not allow the relevant conditional probabilities to reflect the earlier factor's causal significance for the later factor via the intermediate causal factors in question. If X causes Y *by causing Z* (which in turn causes Y), then, according to the basic probability increase idea for type level causation, X should increase the probability of Y *by increasing the probability of Z* (which in turn raises the probability of Y); but if we hold Z fixed, positively or negatively, then X *cannot raise the probability of Z*. Of course, this rationale for not holding intermediate factors fixed applies whether the earlier factor is positive, negative, or mixed for intermediate factors, and whether the intermediate factors are positive, negative, or mixed for the later one.[3]

Two issues involving the role of intermediate causal factors have especially engaged philosophers recently. First, Nancy Cartwright (1979, 1988a, 1988b, 1989) has argued that it is sometimes *correct* to hold fixed intermediate causal factors, thus contending that the independence requirement is not always correct. She argues that, roughly, if some individuals have or lack an intermediate causal factor, but do so independently of any token causal action of the presence or absence of the earlier factor, then, "for these individuals," the intermediate causal factor should be held fixed. In Section 4.1, I will examine her reasons for this and argue that they are mistaken. Then in Section 4.2, the "problem" of intermediate causal factors and contexts will be generalized, and a solution to the more general problem – to hold fixed not only intermediaries but also (what I shall call) "subsequents" – will be given.

Second, granting that intermediate factors should not be

[3]In Chapter 5, on temporal priority and asymmetry of causation, I clarify the idea of one factor's being temporally prior to another.

172

held fixed, there is still the question of what we should say about the causal role of a factor X for a factor Y when some of the causal chains from X to Y are positive and others are negative – for example, when X is positive both for some positive causes of Y and for some negative causes of Y. This issue will be discussed in detail in Sections 4.3 and 4.4, on transitivity and "unanimity of intermediaries." In Section 4.3, a sufficient condition for transitivity will be given which assumes that, within background contexts, it is the *average probabilistic impact of a causal factor across intermediaries* that reveals the factor's causal role for the effect factor in question. In Section 4.4, I will argue – contrary to Hesslow (1976), Dupré (1984), Otte (1985), and Cartwright (1979, 1988a, 1988b, 1989) – that, within background contexts, it is this *average* probabilistic impact across intermediaries that reveals the causal significance of the earlier factor for the later one.

4.1 CAUSAL INTERMEDIARIES AND CONTEXTS

At one point, Cartwright (1979, p. 423) adopted the condition of never holding fixed intermediate causal factors; this is her condition (iv), on factors that should be held fixed in background contexts. And she explains the rationale for the condition along the lines given above. But she adds:

Unfortunately, it is too strong. For condition (iv) excepts any factor which *may* be caused by C even on those particular occasions when the factor occurs for other reasons. Still, (iv) is the best method I can think of for dealing with this problem, short of introducing singular causal facts, and I let it stand for the nonce. (p. 427)

Cartwright believes that it is appropriate to hold fixed an intermediate causal factor when it occurs for reasons other than the earlier causal factor in question. Note that this idea of the occurrence of an intermediate factor for reasons other than the causal factor in question involves *token,* or *singular,* causation. She refers here to "those particular occasions when

173

the factor occurs for other reasons." More recently (1988a, 1988b, 1989), Cartwright has argued that, because of this, a proper account of property-level causation must invoke singular causal facts: "to pick out the right regularities at the generic level requires reference not only to other *generic* causal facts, *but to singular facts as well*" (1988a, p. 79).[4]

So Cartwright abandons her old condition (iv) and replaces it with her new condition *.

*Test situations [*causal background contexts*] should be subpopulations homogeneous with respect to [intermediate causal factors] K, *except for those individuals where K has been produced by* [*the earlier factor*] C. These individuals belong in the population [background context] where they would otherwise have been *were it not for the action of* C. (1988a, p. 82)

(See also Cartwright 1989.) This is subtle and complicated. I think it can also be expressed as follows. If a factor Z is causally intermediate between factors X and Y (on the property level, of course), then, in evaluating the causal role of X for Y, the causal background contexts K_i should hold fixed the factors Z_1, Z_2, and Z_3 defined as follows:

Z_1a if and only if a has Z and a's having Z is token causally independent of any action of Xa or $\sim Xa$ (or, roughly: a has Z and a would still have had Z had a been different from the way it is with respect to X);

Z_2a if and only if a has $\sim Z$ and a's having $\sim Z$ is token causally independent of any action of Xa or $\sim Xa$ (or, roughly: a has $\sim Z$ and a would still have had $\sim Z$ had a been different from the way it is with respect to X);

and

Z_3a if and only if not Z_1a and not Z_2a.

---

[4]She also concludes that "singular causal facts are not reducible to generic ones" (1988a, p. 79), a thesis we have seen different reasons for earlier, and which will be explored in more detail in Chapter 6.

That is, according to Cartwright's new condition $\star$, each causal background context K_i should hold fixed, positively, exactly one of Z_1, Z_2, and Z_3. Z_1 and Z_2 hold Z fixed positively and negatively, respectively, "for the individuals" that have or lack Z token causally independently of the way they are with respect to X. Z_3, on the other hand, is not necessarily homogeneous with respect to the intermediate causal factor Z, and individuals are in Z_3 if and only if their having or lacking Z is token caused by the way they are with respect to the causal factor X.

It seems that the new condition $\star$ is intended to ensure that, for background contexts K_i, the only differences between individuals in $K_i \& X$ and individuals in $K_i \& \sim X$ are differences for which X *is actually* causally responsible, in the relevant actual population in question. According to $\star$, if an individual happens actually to have (or lack) an intermediate causal factor *token causally independently of the causal factor X in question,* then its having (or lacking) the factor should be held fixed in appropriate K_i's. It seems that the worry is that if we did not hold fixed an intermediate factor for individuals who have (or lack) it *for reasons other than the action of X* (or of $\sim X$), then these individuals may, so to speak, be different in one of $K_i \& X$ or $K_i \& \sim X$ from the way they actually are with respect to the intermediate factor, despite the fact that these differences are not traceable to any actual action of X (or of $\sim X$) – so that in this case probability comparisons would misrepresent actual causal relations in the population in question. To accommodate this intuition, $\star$ makes the nature of the background contexts K_i heavily dependent on what the actual token causal relations happen to be.

Our earlier independence condition, Cartwright's condition (iv), prohibited keeping constant the possession of *any* property level cause of Y that X (or $\sim X$) is causally relevant to at the property level. Cartwright's condition $\star$, on the other hand, is weaker in that it *only prohibits* keeping constant

175

the possession of property level causes of Y that X (or $\sim X$) is actually, token causally responsible for: If some individuals possess a cause of Y token independently of X, then their possession of it must still be kept constant in the background contexts.

While Cartwright says the original independence condition, her condition (iv), is too strong, I think Cartwright's $\star$ is too *weak:* because of the way in which $\star$ makes the nature of the K_i's depend on actual token causal relations, $\star$ *allows too much to be held fixed.* After explaining this, I will explain why I think, contrary to Cartwright, that the original independence condition is *not too strong,* but rather just fine. This will involve an idea that Elliott Sober and I have advanced elsewhere (1983). Finally, I will consider an alleged counterexample of Cartwright's to our account, and I will argue that the counterexample is not genuine; this will involve, in part, an application of the idea that we should hold fixed independent interactive factors, as explained in Chapter 3. And this will lead to the other issues concerning intermediate causal factors alluded to at the beginning of this chapter, some of which will be taken up mainly in Section 4.4, on "unanimity of intermediaries."

Here is a simple example in which condition $\star$ makes a difference, but an unfortunate one. It is just like the telephone example discussed at the beginning of this chapter, except that there is a device attached to my phone that, at any time at which anyone may call me, has a 10 percent chance of interrupting any signal that may be coming to my phone from any other phone and then ringing my phone independently of any such signal. That is, at any time at which anyone may dial my number, there is a 10 percent chance that the following two things will both happen: Any signal that may be coming from another phone will be interrupted and my phone will ring token independently of any such signal. It is important that the device is always operating; it is not incoming signals that activate its operation. Thus, my phone may

176

ring due to the operation of this device even when nobody calls me.

Now consider the causal chain from your calling me to my phone's ringing, to my lifting the receiver. Here, the intermediate factor of my phone's ringing may occur as a result of your calling me, or it may occur because the attached device was activated (10 percent chance of the latter). Now consider an actual population of times at which you may call me. As it happens, improbably enough, the device was activated at each of these times. Improbably enough, at each such time, whether or not you called, any signal that might have been coming from your phone was interrupted and my phone rang token independently of whether or not you called.

Thus, in evaluating the causal impact of your calling me on my lifting the receiver, the intermediate factor of my phone's ringing has to be held fixed, according to *. This is because at each time, my phone happened to ring token independently of whether or not you called. And since my phone's ringing screens off my lifting the receiver from your calling, we get the answer that your calling me is causally neutral for my lifting the receiver. But this is clearly wrong. Your calling in fact has *mixed* causal relevance for my lifting the receiver (or "Pareto-positive" relevance, as explained in Section 2.3). In this example, there are two causal background contexts: Device rings my phone (10 percent chance), and device does not ring my phone (90 percent chance). In the first context, your phoning does not affect the probability of my lifting the receiver, since the phone rings anyway. And in the second context, your calling increases the probability of my lifting the receiver, as in the original example.

It might be objected that since the device is always activated in the token population of the example, the second context has probability zero, so that this context is irrelevant and in any case the conditional probabilities, conditional on this context, are undefined. However, the fact that something never happens does not imply that it has zero probability. In

177

fact, in the example, it was stipulated that it was *extremely improbable that the device would always be activated,* since on each occasion it has only a 10 percent chance of being activated. It was stipulated that the second context, in which the device is *not* activated, has a probability of 0.9. Also, of course, the fact that instances of one factor never in fact token cause any instances of a second does not imply that the first factor is not a positive (or Pareto-positive) causal factor for the second, as explained in the introduction.

As also explained in the introduction, property-level causal and probabilistic facts depend on what *kind* we consider the relevant token population to exemplify.[5] If we consider the token population just described to be of the kind, "population of times at which the device is always activated," then, of course, the correct answer is that your phoning me is causally neutral for my lifting the receiver. But if we consider the population to be simply of the kind, "population of times at which you may call me," then the correct answer is mixed, or Pareto-positive, causal relevance.[6]

This example provides another illustration of the fact, explained in the introduction and more fully below in Chapter 6, that what is true at the property level may have little connection with what actually happens at the token level. In fact, the independence of property-level causal relations from

[5]As I argued there, property-level causation should be understood as a relation between *four* things in all: a *causal factor,* an *effect factor,* a *token population,* and a *kind of population* that we associate with the actual population in question (though, as already noted, the *token* population "washes out" in the formal analyses of probability and property-level probabilistic causation, if we adopt the kind of hypothetical limiting frequency interpretation of probability sketched, though not endorsed, in Chapter 1).

[6]In connection with this, Elliott Sober points out that we can consider a different realization of the set up of this example, one in which my phone does sometimes ring because of your call and not because of the device. There are causal facts that this and the realization described in the text have in common (about the causal significance of your calling me for my lifting the receiver) that Cartwright's condition * cannot capture. Another way of putting this point is that the two realizations of the set up are both token populations that exemplify the kind "population of times at which you may call me"; this is what the two realizations have in common, in virtue of which they share certain population level causal facts.

token-level causal relations already makes it doubtful that *
can be part of a correct explication of causation at the property level. Condition * appeals to token-level causal relations
and makes what we hold fixed depend on what happens at the
token level. For this reason, * does not allow us to consider
the relevant population to be an instance of a *kind* of population in which certain token causal facts are not necessarily just
as they actually are. That is, it does not allow the question of
the property level causal significance of one factor for another
to be relative to a *kind* of population, where a kind would be
allowable even if certain token causal facts in hypothetical
populations that exemplify the kind are different from the
actual token causal facts. Thus, condition * is overly restrictive about the *kinds* that we may legitimately associate with
token populations, and the enforcement of condition * would
make property-level causal claims insensitive to certain property level causal relations.

Of course, the fact that intermediate causal factors can occur for reasons other than the action of the causal factor in
question can have a bearing on the correct assessment of
causal roles. And this is a possibility worthy of clarification.
However, as I shall now explain, I do not believe that this
kind of possibility requires any revision of the independence
requirement, Cartwright's original condition (iv). In particular, the account need not invoke singular causal facts, which
in any case may have very little to do with causal relations at
the property level.

Here is an example that shows how independently occurring intermediate causal factors can bear on the causal relation
between an earlier factor X and a later one Y (see Figure 4.1).
Suppose that X is a very weak positive cause of two intermediate factors Z and W, that Z is strongly causally positive for Y,
and that W is strongly causally negative for Y. Suppose also
that there is a factor F, simultaneous with X, that is a strong
positive cause of Z and a strong negative cause of W (F prevents W). Finally, let F be strongly positively *correlated* with

179

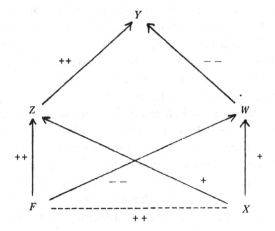

Figure 4.1

X, though neither is *causally* relevant to the other; F and X may be effects of a common cause.

In this example, X may have little or no causal impact on whether or not Y occurs. X is causally positive both for a positive and for a negative cause of Y, so X's two opposed causal tendencies for Y may exactly offset each other, and make X *causally neutral* for Y.[7] Because of the strong correlation between F and X, and because F strongly causes Z (the strong positive cause of Y), and strongly prevents W (the strong negative cause of Y), X may nevertheless sharply *increase the probability* of Y. However, this does not mean that X is *causally* positive for Y. This is because when X occurs, Y is typically caused by Z (or also by the absence of W), which typically occurs for reasons (the presence of F) other than X.

Does this mean that we should hold fixed the intermediate causal factor Z (and perhaps also W) in assessing X's causal

[7] We will consider other examples in which intermediaries have opposing tendencies for the effect factor in question. I argue, especially in Section 4.4, that in such cases, the causal significance of a causal factor for an effect factor should turn on the *average* probabilistic impact of the former on the latter, across intermediaries (but within background contexts). For now, this is simply assumed.

180

role for Y, or that we should hold it fixed on the particular, token, occasions on which it occurs independently of X? No. In this example, it obviously suffices to hold F fixed. If we make comparisons of the relevant conditional probabilities both in the presence of F and in the absence of F, then X's true causal role for Y should be revealed, which, as mentioned above as a possibility, may be causal neutrality. In the presence of F, the probabilities of Y conditional on X and conditional on $\sim X$ will both be high, but they can be close or equal; and in the absence of F, the probabilities of Y conditional on X and conditional on $\sim X$ will both be low, but they may be close or equal.

In this example, X is probabilistically positive for Y, overall. But because the intermediate factors Z and W can occur causally independently of X, we cannot conclude from the positive *probabilistic* significance of X for Y that X is *causally* positive for Y. However, by holding fixed the factor F, which is causally relevant to Z and W, but is not intermediate, we control for the possibility of independently occurring intermediate factors. In cases in which intermediate factors can occur for reasons other than the causal factor in question, we need not ever hold fixed the intermediate factors themselves; *rather we hold fixed the "other reasons,"* the causes of the intermediate factors. As Sober and I (1983) have claimed, by backing up like this, we pay intermediate factors their due.

In the example just considered, holding F fixed is, in fact, already *required* by the theory, since it is an independent positive cause of Y.[8] Of course, in other examples, the causes of the intermediate factors may not themselves be positive causes of the effect factor in question (as F was for Y in the previous example). They may be independent negative or mixed causes, or they may be causally independent factors with which the causal factor in question interacts with respect to the effect factor in question. In each case, the theory re-

[8]This example satisfies the sufficient conditions to be laid down in Section 4.3 for transitivity of causal chains, in this case from F to Y.

quires that they be held fixed. And if the causes of the intermediate factors are *not* independently positive, negative, mixed, or interactive, for the effect in question, they then should not be held fixed; but in that case I cannot see how the problem of "independently occurring" intermediate causal factors can arise in the first place. If the *causes* of the intermediate factors make no difference to the relevant conditional probabilities (of the later factor given the presence and absence of the earlier one), then it would seem that the intermediate factors themselves could not make a difference either.

Cartwright (1988a), however, has described an example in which she claims the account just described gives the wrong answer. She objects that the approach involves a bad kind of "averaging." The kind of averaging she has in mind is averaging, *within* background contexts, the causal significances that one factor has for another *through* the various intermediate causal factors. Evaluation of this kind of averaging, which I wish to defend, is crucial to understanding the role of intermediate causal factors in property-level probabilistic causation, and it is, of course, relevant to the issue of transitivity taken up in Section 4.3. So it will pay to examine Cartwright's example in some detail. In Section 4.4, we will look at what Hesslow (1976), Dupré (1984), and Otte (1985) have had to say about the issue of averaging across intermediaries.

Cartwright's alleged counterexample involves a detailed specification of probabilities for an example given earlier by Hesslow (1976). In the example,[9] X is the factor of taking birth control pills, Z is having a certain thrombosis-causing chemical in the blood, W is being capable of becoming pregnant, and Y is having thrombosis. X causes Z and $\sim W$; Z causes Y; and $\sim W$ causes $\sim Y$. Alternatively, we may say that X is positive for Z and negative for W, Z is positive for Y, and W is positive for Y. See Figure 4.2. Suppose the factors

[9]In my description of the example, I will use different notation from Cartwright's. In addition, I will introduce additional notation, to distinguish the presence of a condition at one time from the presence of the same condition at another.

182

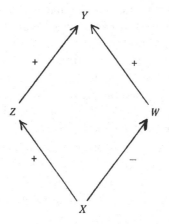

Figure 4.2

mentioned are the only ones relevant to assessing the causal impact of X on Y. In the example, Z and W can occur or fail to occur for reasons other than the presence or absence of X. So the question is whether, or just how, to hold fixed the intermediate factors Z and W.

Cartwright considers two cases. In case 1, Z and W are thought of as not necessarily intermediate: They are factors that may occur before or at the time of X. Here, it seems, we are to think of Z and W as "chemical present in blood *at the time of* X" and "woman capable *at the time of* X of becoming pregnant," respectively. Let us denote these factors by "Z'" and "W'." Since these factors are causally relevant to Y, and since they are not intermediate, they should, everyone agrees, be held fixed. The probability values assumed in Cartwright's example are

$Pr(Y/Z'\&W'\&X) = 0.666\ldots \quad < 0.8 = Pr(Y/Z'\&W'\&\sim X);$
$Pr(Y/Z'\&\sim W'\&X) = 0.4 \quad = 0.4 = Pr(Y/Z'\&\sim W'\&\sim X);$
$Pr(Y/\sim Z'\&W'\&X) = 0.5333\ldots \quad > 0.4 = Pr(Y/\sim Z'\&W'\&\sim X);$
$Pr(Y/\sim Z'\&\sim W'\&X) = 0.2666\ldots \quad > 0 = Pr(Y/\sim Z'\&\sim W'\&\sim X).$

183

The reason for the inequalities is that – except for the case in which the chemical is already present in the blood and the woman is already incapable of becoming pregnant – the introduction of X can still make a difference in whether or not the chemical will *later* appear in the blood and in whether or not a woman will *later* be capable of becoming pregnant. That is, even holding fixed Z' and W', the introduction of X can (except in the case of $Z'\&\sim W'$) make a difference in Z and in W, and Z and W are also causally relevant to Y.[10]

So in case 1, because of these inequalities, everyone agrees that X is causally *mixed* for Y. And this is despite the fact that overall, X may be *probabilistically* neutral for Y. In Cartwright's example, $Pr(Z'\&W') = 0.3$, $Pr(Z'\&\sim W') = 0.4$, $Pr(\sim Z'\&W') = 0.3$, and $Pr(\sim Z'\&\sim W') = 0$. Assuming these values, and assuming, plausibly, that these four combined factors are independent of X, it follows that

$$
\begin{aligned}
Pr(Y/X) &= Pr(Z'\&W'/X)Pr(Y/Z'\&W'\&X) \\
&\quad + Pr(Z'\&\sim W'/X)Pr(Y/Z'\&\sim W'\&X) \\
&\quad + Pr(\sim Z'\&W'/X)Pr(Y/\sim Z'\&W'\&X) \\
&\quad + Pr(\sim Z'\&\sim W'/X)Pr(Y/\sim Z'\&\sim W'\&X) \\
&= (0.3)(0.666\ldots) + (0.4)(0.4) \\
&\quad + (0.3)(0.5333\ldots) + (0)(0.2666\ldots) \\
&= 0.2 + 0.16 + 0.16 = 0.52,
\end{aligned}
$$

and,

$$
\begin{aligned}
Pr(Y/\sim X) &= Pr(Z'\&W'/\sim X)Pr(Y/Z'\&W'\&\sim X) \\
&\quad + Pr(Z'\&\sim W'/\sim X)Pr(Y/Z'\&\sim W'\&\sim X) \\
&\quad + Pr(\sim Z'\&W'/\sim X)Pr(Y/\sim Z'\&W'\&\sim X) \\
&\quad + Pr(\sim Z'\&\sim W'/\sim X)Pr(Y/\sim Z'\&\sim W'\&\sim X)
\end{aligned}
$$

[10] All this seems very natural, except that if having the chemical in the blood and being incapable of becoming pregnant may be temporary conditions, then even in the presence of $Z'\&\sim W'$, the pill may still have an impact on thrombosis, by having an impact on Z and W (the conditions at the later time). But how "natural" any of this seems is really beside the point, since of course the probabilities given are consistent and it is certainly imaginable that they describe the causal and probabilistic facts in some real case.

$$= (0.3)(0.8) + (0.4)(0.4) + (0.3)(0.4) + (0)(0)$$
$$= 0.24 + 0.16 + 0.12 = 0.52.$$

In case 1, averaging across the four background contexts, as in the calculations above, would give us the answer that X is neutral for Y. But averaging across background contexts is bad; and there is agreement that in case 1, X is *mixed* for Y, *not neutral*.

Cartwright's case 2 is supposed to be just like case 1 with respect to the probabilistic and causal relations between X and Y and with respect to the question of whether or not averaging is bad. But in case 2, the question is whether or not to average over *intermediate* "contexts," or combinations of the factors Z and W. In case 2, the factor of having that chemical in the blood at a given time and the factor of being capable at a given time of becoming pregnant have "a shorter causal life, so that they can act only if they occur after [X]" (1988a, p. 89). So we should not (or need not) hold fixed whether or not they occur at the time of X, since their occurrence then is causally irrelevant to Y later. And Cartwright says that "when they do [act after X], they act *in exactly the same way* as they do in Case 1" (p. 89). This means that the conditional probabilities of Y, conditional on each of the four combined factors determined by whether or not the chemical is in the blood *at a causally relevant time* and whether or not a woman is *at a causally relevant time* capable of becoming pregnant, are the same in the two cases. These four intermediate factors are $Z\&W$, $Z\&\sim W$, $\sim Z\&W$, and $\sim Z\&\sim W$, and Cartwright considers a background context B of women in which we hold fixed not these factors themselves, but just their causes, other than X.

Suppose further that in case 2, if it is $\sim X$ that occurs at the earlier time (that is, the time of X or $\sim X$), then the probabilities of having the chemical in the blood or of being capable of becoming pregnant are not affected between the earlier time

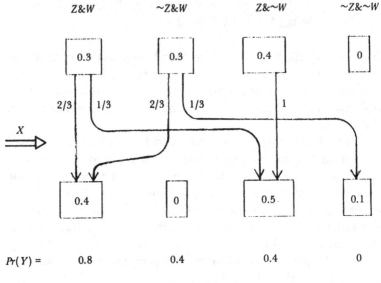

Figure 4.3

and the later time (the time of Y or $\sim Y$). This would seem to be Cartwright's rationale for the values:

$$
\begin{aligned}
Pr(Z\&W/\sim X) &= Pr(Z'\&W'/\sim X) &= 0.3; \\
Pr(Z\&\sim W/\sim X) &= Pr(Z'\&\sim W'\cdot\sim X) &= 0.4; \\
Pr(\sim Z\&W/\sim X) &= Pr(\sim Z'\&W'/\sim X) &= 0.3; \\
Pr(\sim Z\&\sim W/\sim X) &= Pr(\sim Z'\&\sim W'/\sim X) &= 0.
\end{aligned}
$$

The occurrence of X, however, can alter the probabilities of the four combined factors determined by whether or not Z and whether or not W. Figure 4.3, reproduced (with some changes) from Cartwright's 1988a essay, shows how X affects the distribution of these four factors. The top row of boxes gives the distribution of the four factors among those who do not have X. The bottom row of boxes shows the distribution for those who do have X. What about the arrows? The idea seems to be that, for example, among the $Z\&W\&\sim X$'s in

186

the first row, two-thirds of them would remain $Z\&W$'s were they to become X's, and one-third of them would be $Z\&\sim W$ were they to become X's. At any rate, the following certainly seems to be intended. Among the actual $Z\&W\&X$'s in the bottom row, for example, one-half of them are $Z\&W$ *token causally independently of X* (and they are two-thirds of the 30 percent of the X's that would have been $Z\&W$ if not X), and the other one-half of them are *otherwise* $\sim Z\&W$'s that are $Z\&W$'s *because of the action of X* (and they are two-thirds of the 30 percent of X's that would have been $\sim Z\&W$ if not X).[11]

The very bottom row in Figure 4.3 gives the probability of Y in each of the four combinations of whether or not Z and whether or not W, the same in each combination given X and given $\sim X$. These are the same conditional probabilities as those given for case 1, given $\sim X$. That these may also be the values for case 2, conditional on $\sim X$, is plausible given case 1 and the assumption that it is only the *presence* of X that could make the distribution of combinations of whether or not Z and whether or not W differ from the distribution of combinations of whether or not Z' and whether or not W'. And that in case 2 these probabilities should be the same both conditional on X and conditional on $\sim X$ is plausible given the *Markov condition*, which Cartwright assumes for the example.[12]

Figure 4.3 gives all the probabilities needed to calculate all the conditional probabilities relevant to case 2 of the example. It turns out that in our background context B, $Pr(Y/X) = Pr(Y/\sim X) = 0.52$, just as in case 1:

---

[11] Below, I will object that this interpretation of the arrows conflicts with the idea that we are operating within an appropriate causal background context B; but for now, on with the description of case 2.

[12] The "Markov condition" is, roughly, the stipulation that if X, Z, and Y are specifications of states of a system at consecutive times t_1, t_2, and t_3, respectively, then Z and $\sim Z$ each screen off Y from X: $Pr(Y/Z\&X) = Pr(Y/Z\&\sim X)$ and $Pr(Y/\sim Z\&X) = Pr(Y/\sim Z\&\sim X)$. The intuitive idea is that complete information about intermediate states renders information about earlier states irrelevant to the probability of still later states.

$$Pr(Y/X) = Pr(Z\&WX)Pr(Y/Z\&W\&X)$$
$$= Pr(Z\&\sim W/X)Pr(Y/Z\&\sim W\&X)$$
$$= Pr(\sim Z\&W/X)Pr(Y/\sim Z\&W\&X)$$
$$= Pr(\sim Z\&\sim W/X)Pr(Y/\sim Z\&\sim W\&X)$$
$$= (0.4)(0.8) + (0.5)(0.4) + (0)(0.4) + (0.1)(0)$$
$$= 0.32 + 0.20 = 0.52;$$

and,

$Pr(Y/\sim X) = 0.52$ (same calculation as in Case 1, except that we use "Z" and "W", instead of "Z'" and "W'", respectively.)

So, on the account advocated above, X turns out to be causally neutral for Y in case 2. But these two probabilities are averages across the four intermediate factors. And Cartwright asks why we should average across the four intermediate factors in case 2 but not across the four background contexts in case 1: "What difference should it make whether [Z] and [W] are already in place before the decision about birth control pills is taken, or whether they befall you afterwards, so long as their doing so is completely independent of the pills?" (1988a, p. 89).

Evidently, what Cartwright has in mind is the idea that, for example, among the pill takers who have $Z\&W$, one-half of them have it token causally independently of taking the pill (Figure 4.3): they are the two-thirds of otherwise $Z\&W$ pill takers who are "still" $Z\&W$ despite the action of the pill, where the remainder of $Z\&W$ pill takers are the two-thirds of the otherwise $\sim Z\&W$'s pill takers for whom the pill did make a difference, giving them the chemical in the blood. Similarly, it seems that four-fifths of the $Z\&\sim W$ pill takers are $Z\&\sim W$ token causally independently of taking the pill. Thus, one-half of the $Z\&W\&X$'s and four-fifths of the $Z\&\sim W\&X$'s are the way they are with respect to Z and W token causally independently of their being X's. Likewise, two-thirds of the $Z\&W\&\sim X$'s and all the $Z\&\sim W\&\sim X$'s are the way they are with respect to Z and W token causally independently of their being $\sim X$'s. All this is what seems to follow given the interpretation, explained above, of the arrows in Figure 4.3.

I think it is correct to average in case 2, even though it is incorrect in case 1. To see why, it is important to bear in mind that the probability values given in Figure 4.3 are supposed to pertain to a given causal background context B in which all appropriate factors are held fixed. In particular, all factors other than X, and other than property-level effects of X or of $\sim X$, that are causally relevant to Z or to W are held fixed in B. All the "other reasons" (than X or $\sim X$) for which Z or W may be present or absent at the intermediate time must already be held fixed in B; and these "other reasons" that are held fixed may even be factors that are present or absent after the time of X or $\sim X$ (as long as they are causally independent of X). This particular constellation, B, of background factors confers the probabilities in the top row of boxes in Figure 4.3 on $Z\&W$, $Z\&\sim W$, $\sim Z\&W$, and $\sim Z\&\sim W$ when X is absent; and B confers the probabilities in the bottom row of boxes in Figure 4.3 on these four factors when X is present. The top row of boxes gives the probabilities of the four combined intermediate factors conditional on $B\&\sim X$; and the next row of boxes gives the probabilities of the four combined intermediate factors conditional on $B\&X$.

Note also that it must obviously be an indeterministic matter in the example which (if any) individuals in B will get Z and which (if any) will have W.[13] For if this were a deterministic situation, then holding fixed, in B, all factors except for X, and except for effects of X and $\sim X$, that are causally relevant to Z or to W, and then adding X or $\sim X$, would make for all 0–1 probabilities in Figure 4.3. The context B, together with whether or not X, is all that is causally relevant to which

---

[13] By "indeterministic" here, I mean that the nonextreme probabilities given in the example must be taken seriously, that the *kind* we associate with the token population in question yields the nonextreme probabilities given. Of course, this is consistent with the possibility that relative to a different kind that we may associate with the token population, these probabilities may turn out to be all 0–1.

189

combination of "whether or not Z" and "whether or not W" a woman will get at the intermediate time.

Thus, given that all the chains are Markovian (as Cartwright assumes for her example), the context B, together with whether or not X, is all that is causally relevant to whether or not a woman will get Y – since B, and whether or not X, are all that is causally relevant to the intermediate factors. $B\&X$ and $B\&{\sim}X$ specify (or hold fixed in one way) everything that is causally relevant to whether or not Y will occur. Thus, since $B\&X$ and $B\&{\sim}X$ confer exactly the same probability on Y, it seems appropriate to conclude that X and ${\sim}X$ *make no causal difference,* in background context B, to whether or not a woman gets Y.

But what about the idea that there are certain individuals who take the pill and, for example, are $Z\&W$ *but not due to* X (the two-thirds of the otherwise $Z\&W$'s that are still $Z\&W$ despite taking the pill) while there are others who are $Z\&W$ *because of* X (the two-thirds of the otherwise ${\sim}Z\&W$'s that have $Z\&W$ presumably because of the pill)? Consider some individual of the first kind mentioned, presumably one that belongs in the lower left-hand box in Figure 4.3 and that perhaps we picture as "coming from" some hypothetical individual in a hypothetical upper left-hand box. So this individual is an X, a Z, and a W. What would this individual have been if not an X? Would she still have been a $Z\&W$?

Contrary to the way we may be tempted to interpret the arrows in Figure 4.3, we *cannot* answer of any such individual, "Yes, she would still have been a $Z\&W$ even if not X," and then conclude that she is a $Z\&W$ but not due to any token causal action of X. Nor can we say the following of any individual in the lower left box: *Either* she would have been a $Z\&W$ if not X *or* a ${\sim}Z\&W$ if not X, with the same 50 percent chance, since the arrows indicate that only X's that are otherwise $Z\&W$'s or otherwise ${\sim}Z\&W$'s, with the same chances, can be $Z\&W$'s. The arrows cannot be interpreted to mean this. *Since B contains all the information, other than whether or not*

190

X, that is causally relevant to the four combined factors, all we can say is that the *probability* would have been 0.3 (rather than 0.4) that she would still have been a *Z&W*, that the *probability* would have been 0.3 (rather than 0) that she would have been a ~*Z&W*, that the probability would have been 0.4 (rather than 0.5) that she would have been a *Z&~W*, and that the probability would have been 0 (rather than 0.1) that she would have been a ~*Z&~W*.

That is, if she had not been *X*, then the probabilities in the top row of boxes, rather than those in the bottom row, would give her chances of exemplifying the various combinations of the presence or absence of *Z* and the presence or absence of *W*. Because of *the indeterministic setting,* and because of *the assumption that context B holds fixed all factors that it must in order to be a causal background context,* we cannot interpret the arrows of Figure 4.3 as pointing to where various individuals or proportions of individuals "would go" or "where they came from" because of a counterfactual or actual action of *X*. In fact, because of these two features of the example, it is hard to see how the arrows in Figure 4.3 could be interpreted in *any* interesting and legitimate way. (Below, however, I will explain how dropping the assumptions allows some natural ways of interpreting the arrows, but gives quite different kinds of examples.)

Consider an individual that is, say, *B&Z&W&X*. There can be nothing about this individual in virtue of which it *would* have had any probabilities other than those in the top row of boxes in Figure 4.3 for the four combined factors, or any probability for *Y* other than 0.52, *had* the individual started out as a *B&~X* rather than as a *B&X*. Again, this is because *B* specifies *everything* (except for *X*) that is causally relevant to the four combined factors, or to *Y*. And *B* is assumed to confer the probabilities given in Figure 4.3 to the various factors. Of course, there may be things about the way an individual actually is at times between the occurrence or non-occurrence of X and the time of the occurrence or nonoccurrence of Y that affect the probability of Y in the presence and absence of X, in context B. But what *happens* to be true of individuals at intermediate times is irrelevant to the question of the property-level causal or probabilistic significance of the *earlier* factor X, *in context B,* for the later factor Y.[14]

Cartwright has the intuition that averaging and looking at just $Pr(Y/B\&X)$ and $Pr(Y/B\&\sim X)$ conceals part of the true causal role of the factor X for the factor Y in context B. My intuition is that X simply makes no difference for Y in B. Perhaps one source of Cartwright's intuition is that X *does* make a causal difference for the various "ways" in which Y may come about, in B. For example, X makes it more probable that you will get Y "via" (or perhaps I should just say "together with") $Z\&W$ than $\sim X$ does; and X makes it less probable that you will get Y "via" (or together with) $\sim Z\&W$ than $\sim X$ does. But this is quite different from X's making a causal difference in simply *whether or not one will get Y,* in B.[15]

Despite what I have just argued, let us now try out the idea that there *are* determinate ways individuals or proportions of individuals would be with respect to the four combined factors, were they different from the way they are with respect to X. This will help clarify just how this assumption is in conflict with the indeterministic setting or with the idea that B is a genuine causal background context. It seems that the most natural way of formulating the assumption, for this example, is to say that there are the following five kinds of individuals, whose exemplifications are causally independent of X, and that correspond to the five arrows in Figure 4.3:

---

[14] In Chapter 6, we will see that this is not true for token-level causation, where what happens to occur at intermediate times can be highly relevant to what token causal relation holds between two particular events.

[15] Of course, the theory advocated here is itself sensitive to the possibility that a factor X can make a causal difference for the various "ways" in which a factor Y may come about, in any context B, even if X is judged by the theory to be univocally causally positive, negative, neutral, or mixed, for Y. Just evaluate, using the theory, the causal significance of X for the various factors that may be, according to the theory, causally significant in the various ways the theory describes, to Y, in B.

I_1: If $\sim X$, then $Z\&W$; and if X, then $Z\&W$ (this has probability 0.2 and corresponds to the arrow in Figure 4.3 with the leftmost tail);

I_2: If $\sim X$, then $Z\&W$; and if X, then $Z\&\sim W$ (this has probability 0.1 and corresponds to the arrow in Figure 4.3 with the second to leftmost tail);

I_3: If $\sim X$, then $\sim Z\&W$; and if X, then $Z\&W$ (this has probability 0.2 and corresponds to the arrow in Figure 4.3 with the third to leftmost tail);

I_4: If $\sim X$, then $\sim Z\&W$; and if X, then $\sim Z\&\sim W$ (this has probability 0.1 and corresponds to the arrow in Figure 4.3 with the second to rightmost tail);

I_5: If $\sim X$, then $Z\&\sim W$; and if X, then $Z\&\sim W$ (this has probability 0.4 and corresponds to the arrow in Figure 4.3 with the rightmost tail).

If we assume that each individual is of exactly one of these five kinds, then we have a clear interpretation of the arrows in Figure 4.3: they indicate, for each of the five kinds of individuals, how they would be different with respect to the four combined factors, were they different with respect to X. For example, the arrows indicate that two-thirds of the $Z\&W\&\sim X$'s are of kind I_1 (and thus would still be $Z\&W$ if they had been X) and that one-third of the $Z\&W\&\sim X$'s are of kind I_2 (and thus would be $Z\&\sim W$ if they had been X).

Note first that the assumption that $I_1, \ldots, I_5$ partitions the population violates indeterminism, in a way: each of the I_i's together with whether or not X confers 0–1 probabilities on the four combined factors, though not on Y and $\sim Y$. Below I will consider an indeterministic version of the partition of I_i's. More important, however, is this: the assumption that the I_i's partition the population *is in conflict with the idea that B is a causal background context*. Within B, *X interacts with the partition* $\{I_1, \ldots, I_5\}$, *with respect to Y as the effect factor*, in the general sense of interaction explained in Chapter 3.[16] This,

[16] Cartwright (1988a) criticized the approach to intermediate causal factors advocated by Elliott Sober and me (1983) that suggested simply that we hold fixed *causes of*

together with the natural idea that the I_i's are causally independent of X, means that the I_i's must be held fixed in background contexts, so that exactly one of them will have probability 1 within any appropriate causal background context.

To see that there is interaction, just note the values:

$$Pr(Y/I_1\&X) = 0.8 = Pr(Y/I_1\&\sim X) = 0.8;$$
$$Pr(Y/I_2\&X) = 0.4 < Pr(Y/I_2\&\sim X) = 0.8;$$
$$Pr(Y/I_3\&X) = 0.8 > Pr(Y/I_3\&\sim X) = 0.4;$$
$$Pr(Y/I_4\&X) = 0.0 = Pr(Y/I_4\&\sim X) = 0.4;$$
$$Pr(Y/I_5\&X) = 0.4 = Pr(Y/I_5\&\sim X) = 0.4.$$

The five pairs of probability values are distinct. Thus, the assumption that the I_i's partition the population of B's, where the I_i's have the probabilities noted above and are causally independent of X, violates the assumption that B is a causal background context. Since X interacts with the I_i's, with respect to Y, the I_i's must be held fixed, giving us five background contexts $B\&I_i$.

In this case, I agree that X is causally mixed for Y, in the population of B's. However, two things should be noted. First, to get this correct answer, *we did not hold fixed any intermediate causal factors:* the I_i's are not intermediate, but rather causally independent of X. And second, the assumption that the I_i's partition the population is far from the spirit of the original example, in which X is a *probabilistic cause* of Z and $\sim W$: as noted above, within each I_i, X confers 0–1 probabilities on Z and on W. There is nothing wrong with considering this possibility as an example, of course, but even for this version of the example, where the I_i's partition the population and B is not a genuine causal background context, *it is not*

intermediate causal factors; in Eells and Sober (1983), we made no mention of holding fixed *interactive factors*. Now it appears that the idea of holding fixed interactive factors, discussed in Chapter 3 and in Eells (1986) and Eells (1988a) in relation to other issues, is pertinent to the problem of causal intermediaries as well. I note also that the approach to Cartwright's example being developed here replaces my diagnosis given in Eells (1988c). This also supplements my treatment of Hesslow's example, given in Eells (1987b) and (1988b), which is criticized in Dupré and Cartwright (1988) and in Cartwright (1989).

necessary to hold fixed intermediate causal factors to get the right answer.

I note that a less deterministic, and perhaps more realistic, version of the example can be described by considering the following six counterfactual factors:

C_1: If X, then $Pr(Z\&W) = 2/3$ and $Pr(Z\&\sim W) = 1/3$;
C_2: If X, then $Pr(Z\&W) = 2/3$ and $Pr(\sim Z\&\sim W) = 1/3$;
C_3: If X, then $Pr(Z\&\sim W) = 1$;
C_4: If $\sim X$, then $Pr(Z\&W) = 1/2$ and $Pr(\sim Z\&W) = 1/2$;
C_5: If $\sim X$, then $Pr(Z\&W) = 1/5$ and $Pr(Z\&\sim W) = 4/5$;
C_6: If $\sim X$, then $Pr(\sim Z\&W) = 1$.

From the two rows of boxes in Figure 4.3, it is natural to assign probabilities 0.3, 0.3, and 0.4 to C_1, C_2, and C_3, respectively, and 0.4, 0.5, and 0.1 to C_4, C_5, and C_6, respectively. Then consider the factors $J_{i,j} = C_i\&C_j$, for $i = 1, 2, 3$, and $j = 4, 5, 6$. If these factors form a partition, then again we have interaction. And again, given that the $J_{i,j}$'s are causally independent of X, they will have to be held fixed in genuine causal background contexts, so that again X will be causally mixed for Y. Here the interactive factors do not determine 0–1 probabilities for Z and W, conditional on X and $\sim X$. But as before, B does not qualify as a causal background context. The assumption that the $J_{i,j}$'s form a partition gives us nine genuine causal background contexts, $B\&J_{i,j}$, partitioning B (that is, nine assuming nothing else needs to be held fixed). Also note that, as before, the right answer of mixed causal relevance of X for Y in this version of the example was not obtained by holding fixed intermediate causal factors, for the $C_i\&C_j$'s are causally independent of X.

Thus, it appears that there is this dilemma for the evaluation of Cartwright's alleged counterexample to the suggestion that the role of independently occurring intermediate causal factors is properly accommodated by holding fixed just their causes, and not the intermediaries themselves. Either B is a genuine causal background context or it is not. If it is, then, given just the way the example has been described, there is no telling how an individual would have been with respect to Z and W were it not for the action of X or of $\sim X$: all we can do is say what the *probabilities* of the four combined factors would be, were the individual different from the way it is with respect to X (where these probabilities are given in the two rows of boxes in Figure 4.3). This is because, if B is a genuine causal background context, then it already holds fixed all factors, aside from X and effects of X and $\sim X$, that have a role in determining the chances of the four combined factors. This makes it hard to see even how to apply condition $\star$. Also, in this case, as we have seen, it seems that the correct verdict on X's causal role for Y is that X is causally neutral, not mixed, for Y, and that anyway we should not hold fixed, in any way, the intermediate causal factors Z and W.

On the other hand, we can drop the assumption that B is a genuine causal background context and consider factors such as the I_i's and $J_{i,j}$'s, described above. If these factors partition the population, then they must be held fixed (given the natural assumption that they are causally independent of X). X does turn out to be causally mixed for Y in this case, but we do not hold fixed intermediate factors to arrive at this correct answer. In either case, whether B is an appropriate context or we must use the $B\&I_i$'s or $B\&J_{i,j}$'s, Cartwright's example fails to show that we must sometimes hold fixed intermediate causal factors.

In this section, the focus has been on causal intermediaries and causal background contexts, on the condition that causal intermediaries should never be held fixed in causal background contexts. I have defended the condition, and I have argued that we should, in background contexts, average causal significance across intermediaries. It has been suggested that this kind of averaging can result in loss of sensitivity to some aspects of the causal significance of one factor for another. For example, if taking the pill is causally *positive* for thrombosis by causing that chemical to be in the blood, and also causally *negative* for thrombosis by preventing pregnancy, then averaging across the two intermediaries, it has been suggested, makes the theory insensitive to the "*dual nature*" of the cause. After clarifying the issue of transitivity of causal chains in Section 4.3, I will argue in Section 4.4 that this and other criticisms of the idea of averaging over intermediaries are mistaken. The discussion of transitivity, however, will require a resolution of a more general version of the issue of intermediate causal factors and causal background contexts. I turn to this in the next section.

4.2 CAUSAL SUBSEQUENTS AND CONTEXTS

We have seen that in assessing the causal role of a factor X for a factor Y, there are certain factors that we *must* hold fixed in background contexts. These are factors, other than X and Y, that are causally independent of X, and that are either (1) themselves causally positive, negative, or mixed for Y, or (2) such that X interacts with them, with respect to Y. We also have seen that there are certain factors that we *must not* hold fixed in background contexts. These are factors causally intermediate between X and Y. We know that factors of such and such kinds *must* be held fixed and that factors of another kind *must not* be held fixed. What about other factors? Does it matter whether or not we hold them fixed?

In stating and applying the theory of probabilistic causation so far, we have constructed causal background contexts just from factors that *must* be held fixed. Although that is satisfactory for the purpose of *explicating* probabilistic causal connection, it will at times be convenient for the purpose of *investigating certain formal properties* of probabilistic causal connection to allow more to be held fixed, as we shall see in exploring the question of transitivity in the next section.

197

Thus, it is important to ask whether there are factors, beyond intermediate causal factors, that *must not* be held fixed.

A careful look at the intuitive rationale for not holding fixed intermediate factors suggests factors of other kinds that must not be held fixed. The reason why we must not hold fixed intermediate causal factors is that doing so prevents the causal factor from displaying, in the form of the relevant conditional probabilities, its causal significance for the effect factor in question. If X causes Y, and if the only way it does so is by causing the intermediate causal factor Z, which in turn causes Y, then we expect that X will increase the probability of Y *by increasing the probability of Z*, which in turn raises the probability of Y, and we expect that this is the only way in which X can increase the probability of Y. But if we hold Z fixed in a context, then in that context Z already has probability 1 or 0, both in the presence and in the absence of X, so that X cannot raise the probability of Z in that context, and hence cannot raise the probability of Y.

It is easy to see how this problem can be generalized. Suppose there were a factor V that is not an intermediate causal factor but is such that, by holding it fixed, we increase or decrease the probability of the intermediate factor Z – perhaps to 1 or 0, or perhaps to some nonextreme value. For example, suppose V is an *effect* of Z that is not itself a cause of Y; see Figure 4.4 (a concrete, intuitive example of this will be given below). In this case, holding V fixed positively or negatively can be expected to increase or decrease the probability of Z, and this will distort the conditional probability relations from which we wish to read the causal significance of X for Y. The distortion may not be as severe as holding Z itself fixed (which would raise or lower the probability of Z all the way to 1 or to 0), but there is distortion nonetheless. Of course, what I have in mind here is a distortion of the "average degree of causal significance" of X for Y, defined in Section 2.2.

In the nonextreme case, X may still have *some* probabilistic impact on Z, and in the right direction, when we hold V

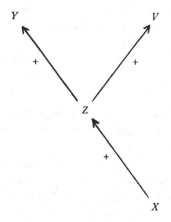

Figure 4.4

fixed. But the information that V is present, or that it is absent, will "dampen" X's probabilistic impact on Z, and thus also X's probabilistic impact on Y. Both X and V are positively probabilistically relevant to Z. But it is possible that V is much more strongly probabilistically relevant to Z than X is, and what is crucial here is that V may raise the probability of Z both in the presence and in the absence of X while $\sim V$ lowers the probability of Z both in the presence and in the absence of X. In that case, $Pr(Z/V\&X)$ and $Pr(Z/V\&\sim X)$ can both be very high and close in value; and $Pr(Z/\sim V\&X)$ and $Pr(Z/\sim V\&\sim X)$ can both be very low and close in value. And this is consistent with there being a much larger ·difference between less extreme values for $Pr(Z/X)$ and $Pr(Z/\sim X)$. So holding V fixed can rob X of much of its probabilistic impact on Z, and thus much of its probabilistic impact on Y. While holding the intermediate factor Z fixed renders X completely probabilistically irrelevant to Z, holding Z's effect, V, fixed may only render X *less* probabilistically relevant to Z.

Here is an intuitive illustration. It is a variation of the telephone example used earlier, in which X is the factor of your phoning me, Z is my telephone's ringing, and Y is my lifting the receiver on my phone. X causes Y just by causing Z. In the new example, I have a dog, and V is the factor of my dog's barking. This dog always (or almost always) barks when my telephone rings and never (or almost never) barks otherwise; the telephone's ringing strongly causes my dog to bark, and the telephone's ringing is just about the only cause of my dog's ever barking. The causal relations in this example are depicted in Figure 4.4.

Suppose first that Z is both a necessary and a sufficient cause of V, so that the probability of the biconditional between Z and V is 1. In this case, $Pr(Z/V) = Pr(\sim Z/\sim V) = 1$. Then holding V fixed has the same effect on the relevant conditional probabilities as holding Z fixed. In this case, it is obvious that V *must not* be held fixed in assessing X's causal role for Y; yet V is not itself a causally intermediate factor. And even if Z is not strictly necessary or sufficient for V, V may still be strongly positively probabilistically relevant to Z, both in the presence and in the absence of X. So it is still clear that holding V fixed would *distort* (in fact, "dampen") the magnitude of X's probabilistic significance for Y, so that it would not appropriately reflect the strength of causal connection between X and Y (the "average degree of causal significance" of X for Y, as defined in Section 2.2). In this case again, it is clear that V should not be held fixed.

In this example, the intermediate factor Z is a *positive causal factor for V*. I leave it to the reader to verify that the same kind of problem can arise when Z is causally *negative* for V. If the intermediate causal factor Z is causally *mixed* for a factor V, then holding V fixed can make the causal significance of X for Y look either mixed or negative, when the truth is positive causal relevance. I have in mind a situation like that depicted in Figure 4.4, except that Z is mixed (instead of positive) for V, and in which Z interacts with X with respect to V. One way in which we may have this kind of interaction is if

200

$$Pr(V/X\&Z) > Pr(V/X\&\sim Z)$$

and

$$Pr(V/\sim X\&Z) < Pr(V/\sim X\&\sim Z).$$

From this it follows that

$$Pr(Z/X\&V) > Pr(Z/X\&\sim V)$$

and

$$Pr(Z/\sim X\&V) < Pr(Z/\sim X\&\sim V).$$

By examining these last two inequalities, and contemplating the possible inequalities between the two left-hand sides and the two right-hand sides, it is easy to see that holding V fixed can make X *look* either mixed or negative for Z, when in fact X is causally positive for Z.[17] And for this reason, if Z is strongly enough positive for Y, then holding V fixed can make X *look* either mixed or negative for Y, when in fact X is positive for Y.

I take such examples to show that we *must not hold fixed factors to which intermediate causal factors are causally relevant* (where, as always, causal relevance includes positive, negative, and mixed causal relevance).

It is worth noting that in the telephone and dog barking example discussed above, X is a *cause* of V. Your calling me causes my dog to bark by causing my phone to ring. So there may be the temptation to try to solve the problem at hand by simply requiring that we not hold fixed any factor to which X is causally relevant, rather than by requiring separately that intermediate causal factors and effects of intermediate causal factors not be held fixed. However, this will not do, for it is possible for an intermediate causal factor Z to be causally relevant to a nonintermediate factor V that the earlier factor

---

[17] Below we will see that we *must not* hold fixed effects of the candidate effect factor in question.

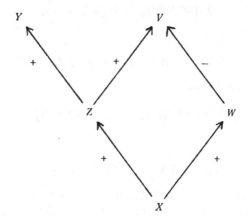

Figure 4.5

X is causally irrelevant to (causally neutral for) – where the same problem as in the examples above will arise if we allow V to be held fixed because X is causally irrelevant to it.

To see this, consider Figure 4.5. In this case, X is causally positive both for a positive cause of V and for a negative cause (a preventer) of V. As explained in the previous section and discussed more fully in the next, it is possible for opposing causal tendencies to exactly offset each other, resulting in causal neutrality of the earlier factor for the later one. So in this example, X may be causally neutral for V. Suppose it is. Then the intermediate factor Z is causally positive for V while X is causally neutral for V. It is clear that V must not be held fixed: If the strengths of the various causal connections are right, then both in the presence of X and in the absence of X, V could make Z almost certain and $\sim V$ could make $\sim Z$ almost certain, so that holding V fixed robs X of much of its probabilistic impact on Z, through which it affects Y.

In cases like those represented in Figures 4.4 and 4.5, the problem arises from V's probabilistic impact, in the presence and in the absence of X, on the intermediate factor Z. This arises from *Z's being causally relevant to V,* and such a *V* can have the relevant kind of probabilistic impact on *Z* regardless of what the causal or probabilistic relations are between *X* and *V*.

Another kind of nonintermediate causal factor that should not be held fixed is the obvious case of effects of the effect factor in question. In assessing *X*'s causal role for *Y*, we must not hold fixed effects of *Y*. For example, if *Y* is causally necessary and sufficient for an effect *V* of it, then *V* will screen off *Y* from *X*. Holding *V* fixed renders *X* probabilistically irrelevant to *Y*, even if *X* is, for example, causally positive for *Y*. Another variation of the telephone example further illustrates this point. My lifting the receiver is a positive causal factor for my having a telephone conversation with someone. But if we hold fixed whether or not I will speak with someone on the phone, your calling me does not affect the probability of my lifting the receiver. If I do have a telephone conversation, then I have lifted the receiver whether or not you have called me. And if I do not have one, then I have not lifted the receiver, whether or not you have called me. So, to avoid erroneously concluding that your calling me at one time is causally neutral for my lifting the receiver at a later time, we should not hold fixed effects of my lifting the receiver – for example, the factor of my speaking with someone on the phone at a still later time.

In this example, my lifting the receiver is causally *positive* for my having a telephone conversation. I leave it to the reader to verify that the lesson is the same whether the effect factor is positive, negative, or mixed for the still later factor. Examples like this show that we must not hold fixed factors to which the effect factor in question is causally relevant.

Thus, three kinds of factors that must not be held fixed are intermediate factors, effects of intermediate factors, and effects of the effect factor in question. Let us now consider effects of the causal factor *X* in question that are not also of any of the three kinds we have already considered. Figure 4.6 depicts a causal chain from *X* to *Y*, along with effects of

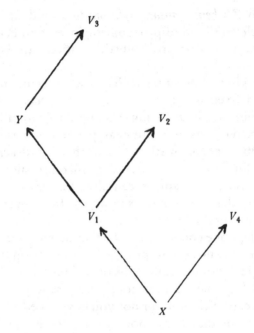

Figure 4.6

factors in the chain. V_1 is an intermediate causal factor, V_2 is a nonintermediate effect of an intermediate causal factor, V_3 is an effect of the effect factor Y in question, and V_4 is an effect of the causal factor X in question and is not of any of the other three kinds of factors we have considered. Must V_4 also be left unfixed?

For factors like V_1, V_2, and V_3 in Figure 4.6, there is a unified and straightforward explanation of the harm in holding them fixed, an explanation that does not apply to factors like V_4. If we hold fixed one of V_1, V_2, or V_3, we are holding fixed a factor that either is in the causal chain from X to Y or is probabilistically relevant, conditional on X and conditional on $\sim X$, to a factor in the causal chain from X to Y. Let us say that every factor is probabilistically, but not causally, relevant to itself. Then holding fixed one of these three factors is holding fixed a factor that is not causally relevant to, but is, conditional on X and conditional on $\sim X$, probabilistically relevant to some factor in the chain from X to Y.

This is like holding fixed, though in some cases in a partial or probabilistic sense, an intermediate or final factor in a causal chain. Holding fixed positively or negatively one of V_1, V_2, or V_3 has the effect, both in the presence and in the absence of X, of increasing or decreasing the probability of a factor in the causal chain from X to Y. Of course, only in the case of V_1 need this increase or decrease be extreme. This effect on the probabilities, conditional on X and conditional on $\sim X$, of factors in the causal chain from X to Y, blocks, in a full or partial way, the causal significance of X for Y from revealing itself in the probabilities of the final factor Y conditional on X and conditional on $\sim X$. Holding these factors fixed thus distorts the connection between the causal significance of X for Y and the probabilistic significance of X for Y.

This explanation of the harm of holding fixed any of V_1, V_2, and V_3 does not apply in quite the same way to the factor V_4 of Figure 4.6. The reason why one of V_1, V_2, and V_3 may be probabilistically relevant to an intermediate, or the final, factor in the causal chain is that they are either *effects of an intermediate or final factor* (in the cases of V_2 and V_3) or *an intermediate factor itself* (in the case of V_1). However, for a somewhat different reason, factors like V_4 should not be held fixed in assessing the causal role of X for Y. We must not forget about the possibility of interactive forks, explained in Section 2.4.

It is possible that X and V_4 may form, with an event in the causal chain from X to Y, an *interactive fork*. Thus, *even conditional on X or conditional on $\sim X$, V_4 may still be probabilistically relevant to events in the causal chain from X to Y. For example, holding V_4 fixed can act in the same way as holding V_2 fixed: if X, V_4, and V_1 form an interactive fork, then holding V_4 fixed is like holding the intermediate factor V_1 fixed, in a partial way, just as we saw that holding V_2 fixed is

205

like holding the intermediate V_1 fixed in a partial way. Other possibilities that show how holding V_4 fixed can distort the connection between the probabilistic and causal relationships between X and Y (possibilities left to the reader to explore), are that X, V_4, and V_2 form an interactive fork, or that X, V_4 and Y form one, or that X, V_4, and V_3 do so.

Let us say that, relative to assessing the causal significance of a factor X for a factor Y, a factor V is a *subsequent causal factor* if either (i) V is causally intermediate between X and Y, or (ii) V is an effect of an intermediate causal factor, or (iii) V is an effect of Y, or (iv) V is an effect of X that lies outside the causal chain from X to Y. Of course, all intermediate causal factors arc subsequent causal factors. Then to generalize the main idea discussed in the previous section, we have the rule: Subsequent causal factors *must not* be held fixed.

Our understanding of causal background contexts now is this: In assessing the causal role of a factor X for a factor Y, (i) we *must* hold fixed all *nonsubsequent* causal factors that are either causally relevant to Y or are elements of a partition with which X interacts with respect to Y, and (ii) we *must not* hold fixed *subsequent* causal factors. As to factors not covered in clauses (i) and (ii), the theory is silent on whether or not to hold them fixed: it does not matter.[18]

4.3 TRANSITIVITY, INTERMEDIARIES, AND SUBSEQUENTS

Under what conditions can we infer what the causal significance of a factor X is for a factor Y, when we know what the causal roles are of X for intermediate causal factors Z and what the roles are of the intermediate factors Z for Y? This is the "question of transitivity."

[18] It would be of interest if it could be shown that, as long as we hold fixed factors that must be held fixed and leave unfixed factors that must not be held fixed, then, conditional on X and conditional on $\sim X$, Y is probabilistically independent of any other (nonsubsequent) factor.

206

In this section, we will be concerned just with what happens at three times: t_1 (the time of X or $\sim X$), t_2 (the time of the presence or absence of intermediate factors Z), and t_3 (the time of Y or $\sim Y$).[19] All factors that will enter into the discussion, including those that must be held fixed in background contexts, will be factors that take place, or fail to take place, at one of the times t_1, t_2, or t_3. If factors earlier than t_1 have an impact on the causal relations between X, Y, and intermediate causal factors Z, then this impact will be accommodated in causal background contexts that specify possible states at time t_1. Also, of course, the discussion in this section can easily be adjusted to the case in which factors earlier than t_1 are explicitly considered: The discussion in this section should, in that case, simply be imported into background contexts that explicitly hold earlier factors fixed in the appropriate ways.

Although in this section there will always be exactly one time under discussion intermediate between the times of the earlier factor and the later factor, it will also be clear how to generalize the results in this section to the case in which more than one intermediate time is considered. The results of this section can simply be "iterated," in the obvious way, to deal with more than one intermediate time. Even though we will restrict our attention to just one intermediate time, the discussion will include the possibility of there being more than one causally intermediate factor at that time. As before, I will use X for the earlier (t_1) factor, Y for the later (t_3) factor, and Z, W, Z_1, Z_2, and so on, for intermediate (t_2) causal factors.

We will, of course, have to be careful to take into account the important role of causal background contexts in the discussion of transitivity. The role of background contexts can be somewhat intricate in investigating causal chains. For example, in a causal chain from X to Z to Y, it can turn out that what works just fine as a background context for assessing X's causal impact on Z will not work in assessing X's or Z's impact on Y.

[19] In Chapter 5, I will take up in detail the idea of indexing factors by times.

This is because we need not hold fixed independent (of X or of Z) causes of Y in assessing X's role for Z, but we must do so in evaluating X's or Z's role for Y. Also, it might be permissible to hold X fixed in evaluating Z's role for Y (for example if X in fact is a cause of Y), but of course we cannot hold X fixed in evaluating X's role for Z or for Y. In addition, there may be causes of Z that must be held fixed in assessing X's causal role for Z, where these causes of Z are not causes of Y and hence need not be held fixed in assessing X's or Z's causal role for Y. Matters get more complicated still when there is more than one intermediate factor in question.

Fortunately, however, a simplification is possible. As we saw in the previous section, the theory of probabilistic causation only contains requirements of the forms "factors of such and such a kind *must* be held fixed in background contexts," and "factors of such and such a kind *must not* be held fixed"; and the theory neither requires nor prohibits holding further factors fixed. As we shall now see, this will considerably simplify accommodating the role of causal background contexts in the discussion of the question of transitivity.

Let us turn first to the case in which there is just one factor Z at t_2 that is causally intermediate between X and Y (that is, only one factor Z at t_2 to which X is causally relevant and that is causally relevant to Y). Later we will consider the general case of any number of intermediate causal factors at t_2. Let $F_1, \ldots , F_n$ be all factors, *other than* X, that *must* be held fixed in assessing the causal significance *either* of X for Z, *or* of Z for Y, *or* of X for Y. That is, if a factor *must* be held fixed in evaluating any of these three relations (from X to Z, from Z to Y, or from X to Y), then it is an F_j – unless that factor is X, in which case it is not among the F_j's. Of course the F_j's may include factors that need not be held fixed in assessing some of the three relevant causal relations.[20]

[20] For example, they may include some causes of Y that must be held fixed in assessing the causal role of X or of Z for Y but need not be held fixed in assessing the causal role of X for Z; and they may include some causes of Z that must be held

It is obvious that the F_j's *include* all factors that must be held fixed in assessing X's role for Z or X's role for Y, since we do not hold fixed X in assessing its causal role for any other factor. And, with the possible exception of X, the F_j's include all factors that must be held fixed in assessing Z's role for Y. As to the possible exception, if X is a cause of Y (or if Z interacts with X with respect to Y), then the theory says we should hold X fixed in assessing Z's causal role for Y.

Also, the F_j's do *not* include any factor that must *not*, according to the theory, be held fixed in assessing the causal role of X for Z, Z for Y, or X for Y. That is, they include no subsequent causal factors, relative to any of the three relevant causal relations. This is because neither Z, nor effects either of Z or of Y, *need* to be held fixed in assessing any of the three relevant causal relations. Recall that the F_j's only include all factors, except for X, that *must* be held fixed in assessing any of the three relevant causal relations; and recall also that all the factors under discussion (including those that must or must not be held fixed) take place either at t_1, t_2, or t_3. I should also note also that there can be no causes of Y that are effects of X and that need to be held fixed in assessing Z's causal role for Y (which factors would be a kind of subsequent causal factor relative to the causal role of X for Y); this is because in the special case under consideration, there is *only one* causally intermediate factor between X and Y, namely Z. Similarly, Z will not interact with any effect of X with respect to Y; such a factor would be an additional intermediate factor between X and Y, where that factor would be either positive, negative, or mixed (interacting with Z) for Y.

Now let K_i's be the consistent maximal conjunctions of F_j's and $\sim F_j$'s. *The K_i's will serve just fine as causal background contexts for assessing X's causal role for Z and X's causal role for Y.*[21] For assessing Z's impact on Y, either the K_i's will do, or, if X

fixed in assessing the causal role of X for Z but need not be held fixed in assessing the causal role of X or of Z for Y.

[21] In comparing the $Pr(Z/K_i \& X)$'s with the $Pr(Z/K_i \& \sim X)$'s, and the $Pr(Y/K_i \& X)$'s with the $Pr(Y/K_i \& \sim X)$'s, we simply ignore pairs in which there is an undefined conditional probability (where one of $Pr(K_i \& X)$ and $Pr(K_i \& \sim X)$ is undefined).

must be held fixed, the factors $K_i\&X$ and $K_i\&{\sim}X$ will do.[22] And *whether or not X must* be held fixed in this case, *the $K_i\&X$'s and $K_i\&{\sim}X$'s will serve just fine as causal background contexts for assessing Z's causal role for Y*, since in this case X is *not* a factor that *must not* be held fixed (it is not a subsequent causal factor, as defined in the previous section, relative to Z's causal role for Y). This is the simplification of the role of causal background contexts for the discussion of transitivity in the case in which there is just one factor at t_2 that is causally intermediate between X and Y.

Although I will focus on the case in which X is causally *positive for Z* and Z is causally *positive* for Y, there are actually nine components of a complete description of the property of transitivity of probabilistic causation. Everything I will say about the first component in what follows carries over straightforwardly to the others. In the following characterization of transitivity, arrows labeled with a "+", "−", and "0" mean positive, negative, and neutral causal significance, respectively.

Probabilistic causation is transitive if and only if, for any three factors X, Z, and Y,

(i)   If $X \xrightarrow{\ +\ } Z \xrightarrow{\ +\ } Y$, then $X \xrightarrow{\ +\ } Y$;

(ii)   If $X \xrightarrow{\ -\ } Z \xrightarrow{\ -\ } Y$, then $X \xrightarrow{\ +\ } Y$;

(iii)   If $X \xrightarrow{\ +\ } Z \xrightarrow{\ -\ } Y$, then $X \xrightarrow{\ -\ } Y$;

(iv)   If $X \xrightarrow{\ -\ } Z \xrightarrow{\ +\ } Y$, then $X \xrightarrow{\ -\ } Y$;

(v)   If $X \xrightarrow{\ 0\ } Z \xrightarrow{\ 0\ } Y$, then $X \xrightarrow{\ 0\ } Y$;

(vi)   If $X \xrightarrow{\ +\ } Z \xrightarrow{\ 0\ } Y$, then $X \xrightarrow{\ 0\ } Y$;

(vii)   If $X \xrightarrow{\ 0\ } Z \xrightarrow{\ +\ } Y$, then $X \xrightarrow{\ 0\ } Y$;

(viii)   If $X \xrightarrow{\ -\ } Z \xrightarrow{\ 0\ } Y$, then $X \xrightarrow{\ 0\ } Y$;

(ix)   If $X \xrightarrow{\ 0\ } Z \xrightarrow{\ -\ } Y$, then $X \xrightarrow{\ 0\ } Y$.

Note, incidentally, that the question of transitivity of *probabilistic causation in general* is distinct from the question of the transitivity of *particular causal chains of factors*. Probabilistic

[22] Again, in comparing the relevant conditional probabilities, we ignore pairs containing an undefined conditional probability.

causal connection itself is transitive if the forgoing (which is a universally quantified nine-part conjunction over three variables X, Y, and Z) is true of the relation. Let us say that any particular probabilistic causal chain of three factors, from X to Z to Y, is transitive if X, Y, and Z satisfy both the antecedent and the consequent of one the (unquantified) conditionals (i)–(ix) above. Of course, if *all* causal chains are transitive in this sense, then so is probabilistic causation in general.

We have already seen examples that show that probabilistic causation is not transitive in general. In case 2 of Cartwright's detailed version of Hesslow's example, discussed in Section 4.1, taking the pill, X, was causally positive for having the thrombosis-causing chemical in the blood, Z, and having the chemical in the blood was positive for thrombosis, Y; but X was neutral for Y. This violates part (i) of transitivity. The diagnosis, of course, is that X was *also* causally *negative* for another positive cause (ability to become pregnant, W) of Y. It is easy to imagine violations of each of the other clauses characterizing transitivity, by including other causal chains. For example, (ii) would be violated if, in addition to X's being negative for the negative cause Z of Y, X was also strongly causally positive for a strong negative cause W of Y. And (vi) would be violated if, in addition to X's being positive for Z, which is neutral for Y, X was also positive for a positive (or a negative) cause of Y.

These examples all involve more than one causal factor intermediate between X and Y. We will consider causal chains with multiple intermediaries, and see a sufficient condition for their transitivity, later. Here is a simple numerical example in which transitivity fails and in which only one causally intermediate factor is (explicitly) mentioned. Suppose the only relevant factors are X, Y, and Z. For example, X can be being a smoker at time t_1, Z having a heart attack at time t_2, and Y having heart pains at time t_3.[23] Suppose the population

[23] As in an example given in Eells and Sober (1983). Of course, it is not medically plausible that X, Y, and Z include all factors causally relevant to any of them, but more contrived examples will have this feature.

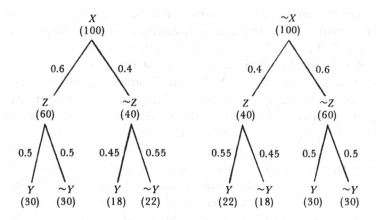

Figure 4.7

in which we wish to assess causal relations consists of 100 smokers (X's) at t_1 and 100 nonsmokers ($\sim X$'s) at t_1, and suppose the actual frequencies of all factors mirror exactly the corresponding probabilities. Figure 4.7 shows how the smokers at t_1 and nonsmokers at t_1 distribute themselves for heart attacks (Z) at t_2 and heart pains (Y) at t_3.

Smoking causes heart attacks in the example, according to the theory of probabilistic causation: $Pr(Z/X) = 0.6 > Pr(Z/\sim X) = 0.4$. Also, heart attacks at t_2 clearly cause heart pains at t_3: Whether X is held fixed or not, Z increases the probability of Y, as the following calculations show.

$Pr(Y/X\&Z)$ $= 30/60 = 0.5 > Pr(Y/X\&\sim Z) = 18/40 = 0.45$;
$Pr(Y/\sim X\&Z) = 22/40 = 0.55 > Pr(Y/\sim X\&\sim Z) = 30/60 = 0.5$;

and,

$Pr(Y/Z)$ $= 52/100 = 0.52 > Pr(Y/\sim Z) = 48/100 = 0.48$.

So X causes Z and Z causes Y. However, X *lowers* the probability of Y:

$Pr(Y/X) = (30+18)/100 = 0.48$
$< Pr(Y/\sim X) = (22+30)/100 = 0.52$.

212

So if Z is the only causally intermediate factor at t_2, then probabilistic causation is not in general transitive for the case of just one causally intermediate factor.

However, there is a peculiarity in the example. If Z at t_2 is the *only* causally intermediate factor, and if X at t_1, Z at t_2, and Y at t_3, are the *only* relevant factors in the example (as assumed), then saying whether Z is present or absent at t_2 should capture everything about what happens at t_2 that is relevant to whether or not Y will occur at t_3. And if saying whether or not Z happens at t_2 *does* capture everything at t_2 that is causally relevant to what happens at t_3, then what happens at t_1 should be irrelevant to what happens at t_3, once the presence or absence of Z is fixed at t_2: Y should be probabilistically independent of X both conditional on Z and conditional on $\sim Z$. We would expect this probabilistic independence if we believed in "*no direct causation at a temporal distance,*" or "*temporal locality of causation*" – that is, if we believed that, if two events X and Y are temporally separated, then the earlier one, X, can have an effect the later one, Y, *only* by affecting the way things Z are at intermediate times.[24]

However, in the example, we have

$$Pr(Y/Z\&X) \quad = 0.5 < Pr(Y/Z\&\sim X) = 0.55; \text{ and}$$
$$Pr(Y/\sim Z\&X) = 0.45 < Pr(Y/\sim Z\&\sim X) = 0.5.$$

Thus, it would be natural to suspect that, besides producing heart attacks at t_2, smoking at t_1 also produces some other medical condition at t_2, say W, that is also causally relevant to heart pains at t_3. And if Z and W are the *only* the factors at t_2 that X at t_1 is causally relevant to, and that are causally relevant to Y at t_3, then we should expect each of the four possible *combined* intermediate factors, specified by saying both whether or not Z and whether or not W, to screen off Y from X. Conditional on each of the combined intermediate factors, Y should be probabilistically independent of X. Figure 4.8 shows one way

[24] This is the Markov condition for singly connected chains of Eells and Sober (1983).

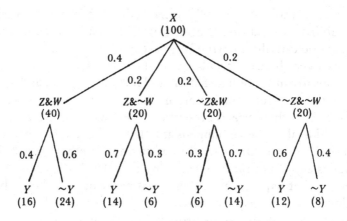

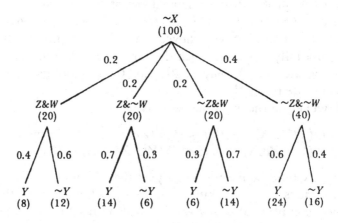

Figure 4.8

in which this can happen for our example (that is, consistent with the statistics in Figure 4.7). The figure gives possible probabilistic relations among X, Z, W, and Y such that each of the four possible combinations of presence and absence of Z and W screen off Y from X. And Figure 4.9 shows the probabilistic relations just among X, W, and Y.

214

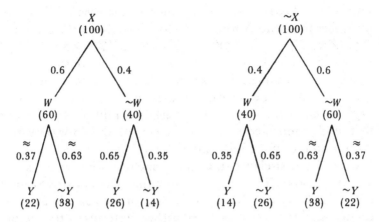

Figure 4.9

Note that in Figure 4.9, X is causally positive for W ($Pr(W/X) = 0.6 > Pr(W/\sim X) = 0.4$), but W is causally *negative* for Y:

$$Pr(Y/X\&W) = 22/60 \approx 0.37$$
$$< Pr(Y/X\&\sim W) = 26/40 = 0.65;$$
$$Pr(Y/\sim X\&W) = 14/40 = 0.35$$
$$< Pr(Y/\sim X\&\sim W) = 38/60 \approx 0.63;$$

and

$$Pr(Y/W) = 36/100 = 0.36 < Pr(Y/\sim W) = 64/100 = 0.64.$$

Thus, the intransitivity of the causal chain from X to Z to Y, in which X is positive for Z and Z is positive for Y yet X is negative for Y, is, as in previous examples, plausibly explainable by X's *being positive for a second intermediate factor W that is negative for Y*. Indeed, although X is equally positive for Z and for W, W is more strongly negative for Y than Z is positive for Y. That is why X turns out to be negative for Y overall.

So if we believe in "no action at a temporal distance," or "temporal locality of causation," and if we are interested in causal chains involving three times in which there *really is* exactly one causally intermediate factor, then it is natural to

215

impose the *Markov condition,* which in the case of chains of three factors, from X to Z to Y, comes to this: $Pr(Y/Z\&X) = Pr(Y/Z\&\sim X)$ and $Pr(Y/\sim Z\&X) = Pr(Y/\sim Z\&\sim X)$. Of course, when background contexts have to be considered, the Markov condition holds when it holds within each background context. Let us call a causal chain from X to Z to Y *singly connected* if it satisfies the Markov condition. It turns out that single connectedness is a sufficient condition for transitivity of causal chains (Eells and Sober 1983).

Consider first clause (i) of the characterization of transitivity given above. Thus, suppose first that X is a positive causal factor for Z. Given the simplification of the consideration of background contexts described earlier, in terms of the factors F_j, this means

(I) $Pr(Z/K_i\&X) > Pr(Z/K_i\&\sim X)$, for each i.

Suppose also that Z is a positive causal factor for Y. This means

(II) $Pr(Y/K_i\&X\&Z) > Pr(Y/K_i\&X\&\sim Z)$, and
$Pr(Y/K_i\&\sim X\&Z) > Pr(Y/K_i\&\sim X\&\sim Z)$, for each i.

Single connectedness gives

(III) $Pr(Y/K_i\&Z\&X) = Pr(Y/K_i\&Z\&\sim X)$, and
$Pr(Y/K_i\&\sim Z\&X) = Pr(Y/K_i\&\sim Z\&\sim X)$, for each i.

To show that single connectedness is a sufficient condition for transitivity of causal chains (for clause (i) of the characterization of transitivity above), we need to show that it follows from (I), (II), and (III) that X causes Y, that is, that

(IV) $Pr(Y/K_i\&X) > Pr(Y/K_i\&\sim X)$, for each i.

To establish (IV), let K_i be arbitrary, and for readability, let PR be the probability function Pr conditional on K_i, that is, $Pr(-) = Pr(-/K_i)$.

Now,

216

$$PR(Y/X) = PR(Z/X)PR(Y/X\&Z)$$
$$+ PR(\sim Z/X)PR(Y/X\&\sim Z),$$

and

$$PR(Y/\sim X) = PR(Z/\sim X)PR(Y/\sim X\&Z)$$
$$+ PR(\sim Z/\sim X)PR(Y/\sim X\&\sim Z).$$

To simplify these expressions, we may, by *assumption (III)*, set

$$u = PR(Y/X\&Z) = PR(Y/\sim X\&Z)$$

and

$$v = PR(Y/X\&\sim Z) = PR(Y/\sim X\&\sim Z).$$

Let us also set

$$a = PR(Z/X)$$

and

$$b = PR(Z/\sim X).$$

Then $PR(Y/X) > PR(Y/\sim X)$ if and only if

$$au + (1 - a)v > bu + (1 - b)v.$$

This is equivalent to

$$a(u - v) > b(u - v).$$

It is *assumption (II)* that allows us to divide both sides by $u - v$ without changing the direction of the inequality, yielding $a > b$. And *assumption (I)* guarantees that $a > b$. Since an arbitrary K_i was chosen in the first place, this establishes that single connectedness is a sufficient condition for transitivity of causal chains – in sense (i) of transitivity explained earlier.

Single connectedness, (III) above, also guarantees the other clauses in the characterization of transitivity of probabilistic causation given above. To verify this, simply vary the in-

217

equalities between a and b above and between u and v above. For X's being positive, negative, or neutral for Z is the same as $a > b$, $a < b$, and $a = b$, respectively (for all contexts, of course). And in view of (III), Z's being positive, negative, or neutral for Y is the same as $u > v$, $u < v$, and $u = v$, respectively (in all contexts, of course). To verify clause (ii), for example, just note that $a < b$ and $u < v$ implies that $au + (1 - a)v > bu + (1 - b)v$. For (iii), $a > b$ and $u > v$ implies that $au + (1 - a)v < bu + (1 - b)v$. For (iv), $a < b$ and $u > v$ also implies $au + (1 - a)v < bu + (1 - b)v$. And for (v)–(ix), it is easy to see that if either $a = b$ or $u = v$, we have $au + (1 - a)v = bu + (1 - b)v$.

It is worth noting that while single connectedness is sufficient for transitivity of causal chains involving exactly one causally intermediate factor, it is not, strictly speaking, necessary. For the case of clause (i) of the characterization of transitivity above, consider Figure 4.9 again and consider the relations among X, W, and $\sim Y$. If, in this example, W in fact really is the only causally intermediate factor between X and $\sim Y$, then the chain fails to satisfy the Markov condition and hence is not, according to the strict definition, "singly connected." Nevertheless, the chain itself is transitive: X is positive for W, W is positive for $\sim Y$, and X is positive for $\sim Y$. Also, it is not difficult to invent chains in which the other clauses of the characterization of transitivity are satisfied, yet in which single connectedness fails. Although violations of the Markov condition may be surprising, or implausible, or hard to imagine in real life, it is not part of the definitions of the various kinds of probabilistic causal connection that causal chains be Markovian. Rather, it is a separate plausible constraint.

Let us now turn to the more interesting case of more than one causally intermediate factor in a chain from X to Y. Suppose there are exactly n factors, $Z_1, Z_2, \ldots, Z_n$, at t_2, that are causally intermediate between X at t_1 and Y at t_3. Let us say that such a chain is *multiply connected* if

218

(1) X is causally positive for each of the Z_i's and each of the Z_i's is causally positive for Y,

(2) the chain is Markovian (within each background context, Y is independent of X conditional on each way of holding fixed, positively or negatively, each of the Z_i's), and

(3) the Z_i's are mutually independent conditional on X and conditional on $\sim X$.

It has been shown (Eells and Sober 1983) that multiple connectedness is a sufficient condition for transitivity – for it to be true that X causes Y. I will not repeat the long and intricate demonstration of this here.

It is perhaps more interesting that multiple connectedness is *not necessary* for transitivity, given our definitions of positive, negative, and neutral causal significance. More exactly, what is of interest is that even a suitable generalization of condition (1) (intended to appropriately accommodate the rationale behind all the clauses in the characterization of transitivity above) is not necessary for transitivity. It has been argued by a number of philosophers, in a number of ways, that this feature of the theory is objectionable. After clarifying the issue in the remainder of this section, I will in the next section explain and address the objections.

We have already seen that the Markov condition is not necessary for transitivity of causal chains with just one causally intermediate factor, so it is not surprising that multiple connectedness is not necessary for transitivity of causal chains with multiple intermediaries either. However, it would not be very interesting if failure of the Markov condition, clause (2) of multiple connectedness, were the only reason why a chain could be transitive without being multiply connected: the Markov condition is too plausible. Also, the conditional mutual independence condition is obviously not necessary: Slight deviations from strict conditional mutual independence will, not surprisingly, preserve transitivity of a chain.

219

The conditional mutual independence condition, clause (3) of multiple connectedness, is merely a simplifying assumption.

Furthermore, it would not be surprising if clause (1), as stated, is violated in cases of transitive chains. Clause (1) would be violated, and we would still have transitivity, if (2) and (3) were satisfied and X were causally *negative* for all the Z_i's and all the Z_i's were all causally *negative* for Y. So it is of interest that condition (1) can be weakened to accommodate the more general idea of transitivity embodied in the characterization given above involving just one intermediary.

Condition (1) of multiple connectedness has been called the condition of *unanimity* (Eells and Sober 1983), or *unanimity of intermediaries* (Dupré 1984). But condition (1) only mentions the case in which X is *positive* for intermediate factors and the intermediate factors are *positive* for Z. Corresponding to the relations shown between X and Z and between Z and Y in the nine clauses of the characterization of transitivity given above, there are nine ways in which a causal chain with multiple intermediaries can be called unanimous. For example, in the case of clause (ii) on p. 210, we have a kind of unanimity if X is causally negative for all the Z_i's and all the Z_i's are causally negative for Y; and in the case of clause (vi), we have another kind of unanimity of X is causally positive for all the Z_i's and all the Z_i's are causally neutral for Y. Let us call these kinds of unanimity, *unanimity*$_{(i)}$ through *unanimity*$_{(ix)}$.

Further, we can say that a chain from X through Z_i's to Y is *positive-unanimous* if, for any given Z_i, the relation between X and Z_i and that between Z_i and Y are *either* like those shown between X and Z and between Z and Y in clause (i) *or* like those shown between X and Z and between Z and Y in clause (ii) of the above characterization of transitivity. *Negative-unanimity* can be similarly defined disjunctively in terms of clauses (iii) and (iv) of the above characterization of transitivity. And *neutral-unanimity* can be similarly defined disjunctively in terms of clauses (v)–(ix). Note that a chain can be positive-, negative-, or neutral-unanimous without exhibiting any of unanimity$_{(i)}$ through unanimity$_{(ix)}$; but a chain that exhibits one of unanimity$_{(i)}$ through unanimity$_{(ix)}$ must be either positive-unanimous (in the case of unanimity$_{(i)}$ and unanimity$_{(ii)}$), negative-unanimous (in the case of unanimity$_{(iii)}$ and unanimity$_{(iv)}$), or neutral-unanimous (in the case of unanimity$_{(v)}$ through unanimity$_{(ix)}$).

Let us say that a causal chain with multiple intermediaries is simply *unanimous* (in terms of intermediaries) if it is unanimous in any of the ways just described; and let us *now* say that a chain is *multiply connected* when clause (1) is replaced by this more general condition of unanimity. The proof of the sufficiency of multiple connectedness for transitivity given in Eells and Sober (1983) can easily be adjusted to show that this more general condition of multiple connectedness is sufficient for transitivity.

Let us now focus on the condition of unanimity (of intermediaries). This condition is clearly not necessary for transitivity of probabilistic causal chains, given the explication of positive, negative, and neutral causal relevance presented earlier in this book. Consider again the example of Figure 4.8, and let us think about the causal relations between X, Z, and $\sim Y$ on the one hand, and between X, W, and $\sim Y$ on the other, as summarized in Figures 4.7 and 4.9. From the calculations given above, it is clear that

> X is positive for Z and Z is negative for $\sim Y$;
> X is positive for W and W is positive for $\sim Y$; *and*
> X is positive for $\sim Y$.

That is, X is *positive* for $\sim Y$ even though, in addition to being causally *positive* for a *positive* causal factor (W) for $\sim Y$, X is also causally *positive* for a *negative* causal factor (Z) for $\sim Y$. On the explication of probabilistic causation given in the previous chapters, X causes $\sim Y$ even though it is not unanimous for $\sim Y$.

So unanimity of intermediaries is not necessary for transitivity of causal chains, on the theory of probabilistic causation endorsed here. This is perhaps a puzzling feature of the theory. Indeed, a number of philosophers have, more or less explicitly, objected to this feature of this kind of theory, arguing, for example, that it conflicts with other basic requirements of the theory or that it makes the theory insensitive to certain kinds of causal facts. In the next section, I will take up these criticisms and argue that in the test cases the theory delivers the natural and correct verdicts about what causes what in which populations.

4.4 ON UNANIMITY OF INTERMEDIARIES

According to the theory of probabilistic causation endorsed here, a factor X can be, *univocally,* a positive cause of a factor Y, even if X causes factors that prevent Y, or prevents factors that cause Y (all this relative to a single given population, of course). Negative causal factors for a factor Y can be positive for some positive causes of Y, as well as negative for some negative causes of Y. And factors causally neutral for a factor Y can be positive for factors negative for Y, positive for factors positive for Y, and so on. The same goes for mixed causal factors.

In general, suppose that all we know about factors X, Z, and Y are: (i) that Z is causally intermediate between the earlier X and the later Y, and (ii) the way in which X is causally relevant to Y (positive, negative, neutral, or mixed). Then we cannot, on the basis of this and the theory, eliminate *any* of the (16) possible combinations for the causal significance of X for Z and the causal significance of Z for Y. All of this is because of the theory's *rejection* of the unanimity of intermediaries condition as a *necessary* condition for the various kinds of probabilistic causal connection. Some philosophers have found this feature of theories like the one endorsed here to be perplexing or objectionable. It is worthwhile addressing several discussions pertinent to this feature of the theory, both for the purpose of exploring the implications and versatility of the theory, and for the purpose of answering criticisms.

222

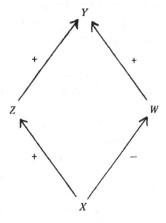

Figure 4.10

Most discussions of this issue have used Hesslow's (1976) pill/thrombosis case as a working example, an example I discussed in some detail in Section 4.1 in connection with Cartwright's argument about holding fixed intermediate causal factors. I will now return to this example to explain and assess criticisms regarding our theory's rejection of the unanimity of intermediaries condition as a necessary condition of probabilistic causal connection.

As before, X will be the factor of taking birth control pills at an earlier time t_1, Z will be the factor of having a thrombosis-causing chemical in the blood at t_2, W is being capable at time t_2 of becoming pregnant, and Y is having thrombosis at t_3.[25] Recall that X is causally positive for Z and negative for W; and both Z and W are causally positive for Y. The causal structure of the example is reproduced in Figure 4.10. Note, incidentally, that if we consider the causal relations between X, Z,

[25] Hesslow does not explicitly discuss the factor Z of having a thrombosis-causing chemical in the blood, but it is clear that, if we reject direct action at a temporal distance, then some such factor must be relevant – if X is to have, intuitively, one causal role for Y through W and, independently of this, a different causal role for Y.

223

$\sim W$, and Y, then we have a situation, like the one discussed near the end of the previous section, in which the earlier factor is causally positive for two intermediaries, where one of the intermediaries is positive and the other is negative for the later factor. So Hesslow's example is, in essence, the same in structure: a causal chain, with two intermediaries, in which unanimity fails.

There are, as far as I know, basically four kinds of objection to the kind of probabilistic theory of causality elaborated above that have made use of this example or examples like it. First, we have already considered Cartwright's use of the example in arguing that intermediate causal factors should not always be left unfixed; I argued in Section 4.1 that her argument fails and that, in her detailed version of the example, Z and W should be left unfixed in background contexts. That issue was not directly connected with the issue of unanimity, and it is easy to see how her argument could be carried out for an example in which the causal chain *is* unanimous.[26] In any case, it is not necessary to reconsider Cartwright's objection here. The other three objections attempt to show that the kind of probabilistic theory elaborated above fails to accommodate certain natural intuitions about Hesslow's example, and examples like it. I will first state the three objections and then show how the probabilistic theory in fact does accommodate the intuitions behind them. The way in which the probabilistic theory accommodates the relevant intuitions is very similar for each of the three objections.

The first of the three objections was raised by Hesslow himself (1976). In a brief paragraph, he uses the example to argue that, in cases in which an earlier factor has opposing

[26] For example, consider a population in which, as it happens, tokens of the intermediate factors always occurred token causally independently of the earlier factor, yet in which the chain is, at the property level, unanimous. An example of this is the telephone example I described in Section 4.1, in which there is a 10 percent chance that any call from you is interrupted and my phone rings independently of what you do. In fact, this example involves a *singly* connected chain, which is automatically unanimous.

224

causal tendencies for a later one, *a cause can lower the probability of an effect of it,* which is contrary to the basic idea of the theory of property-level probabilistic causation. If, in the example, the efficacy of the pill in *preventing* thrombosis (by preventing pregnancy), is *stronger* than the efficacy of the pill in *producing* thrombosis (by producing that chemical in the blood), then we can have $Pr(Y/X) < Pr(Y/\sim X)$. Hesslow observes that, "in a population which lacked other contraceptives this would appear a likely situation" (p. 291). However, according to Hesslow, we cannot for this reason deny that contraceptive pills can cause thrombosis: It is, after all, part of the example that the pill causes that chemical to be in the blood, which in turn causes thrombosis, and that the pill in fact does have the *two, opposing* causal tendencies.

Second, John Dupré (1984) has argued that "there is a considerable tension, if not an outright incompatibility between" requiring context unanimity and not requiring unanimity of intermediaries (p. 169). He argues that when unanimity of intermediaries fails, then context unanimity will also fail, or at least will be extremely likely to fail. If an earlier factor raises the probability of some intermediaries that are positive for a later factor, and also raises the probability of other intermediaries that are negative for the later factor, then, according to Dupré, there will be (or at least there will very likely be) some *background contexts* in which causal chains of the first kind, but not those of the second, are fully or partially blocked, and other background contexts in which causal chains of the second kind, but not those of the first, are fully or partially blocked. In that case, the causal significance of the earlier factor for the later one is reversed across the two kinds of contexts, and context unanimity fails.

Dupré illustrates this with an example of a hypothetical gene that causes both exercising and smoking, where exercising is negative for a heart attack and smoking is positive for a heart attack. Exercising and smoking are causally intermediate between the gene and a heart attack, and unanimity of intermediaries fails. Of course, in the general population there will be an average probabilistic effect, across intermediaries, of the gene on heart attacks, where through one intermediary the gene is positive for heart attacks and through the other intermediary the gene is negative for heart attacks. Dupré then finds one background context in which the causal impact of the gene on heart attacks through one intermediary predominates, and another background context in which the causal impact of the gene on heart attacks through the other intermediary predominates. In a context in which tobacco is readily available, the gene's average impact on the probability of a heart attack (across the two intermediaries) may be positive, while in a context in which tobacco is not available, the gene will be negative for a heart attack, since its impact on heart attacks through the intermediate factor of exercising must there predominate over its impact on heart attacks through the intermediate factor of smoking. So context unanimity fails, in the example.[27]

Of course, the same point can be made in terms of the pill/thrombosis example. In a background context in which some factor that prevents pregnancy is already present, the pill may not affect the probability of pregnancy, so that the only way in which the pill can affect the probability of thrombosis is by way of its impact on that chemical in the blood: in such a context, the pill will be positive for thrombosis. On the other hand, in a context in which some factor is present that prevents the formation of the thrombosis-causing chemical in the blood, the effect of the pill on thrombosis through the intermediate factor of pregnancy will predominate: In such a context, the pill will be negative for thrombosis.

[27] Actually, there is even more symmetry to be found in the example. Both in contexts in which nobody smokes and in contexts in which everybody smokes, the effect of the gene on heart attacks through the intermediary of exercise would predominate: in these two kinds of contexts, whether or not one smokes is, in effect, held fixed. In addition, if the factor of exercise were, in effect, held fixed, then the effect of the gene on heart attacks through the intermediary of smoking would predominate.

If unanimity of intermediaries fails, and if, for each opposing causal tendency, a background context can be found in which it predominates over the others, then context unanimity will fail as well. Thus, according to Dupré, if we endorse context unanimity, we should also endorse unanimity of intermediaries, as a necessary condition for (unmixed) probabilistic causal significance.

The third and final objection was raised by Richard Otte (1985). He claims that by adopting the kind of probabilistic theory of causation elaborated above, "we would end up denying the dual nature of many causes" (p. 123). According to this objection, which is similar to Hesslow's, one factor X can be both a positive cause and a negative cause of another factor Y. For example, in the pill/thrombosis case, taking oral contraceptives both causes and prevents thrombosis, according to Otte. And Otte elaborates his point as follows:

If we require that a cause be either a positive cause or a negative cause, but not both, our analysis of the thrombosis case will be unintuitive. In this case we want to say both that the taking of oral contraceptives contributes to the possibility of thrombosis, and that it helps prevent thrombosis. Some women may take oral contraceptives in order to prevent thrombosis. Their reason for doing so would be that oral contraceptives is a negative cause of thrombosis. Other women may not take oral contraceptives because they are concerned that they may cause thrombosis. I think that both of these arguments are reasonable, because we recognize that oral contraceptives can be both a positive and a negative cause of thrombosis. (pp. 122–3).

However, according to the basic idea of the approach to probabilistic causality elaborated above (Otte has Cartwright's 1979 theory specifically in mind), positive causes raise, and negative causes lower, the probability of the effect; so on any such approach, it seems that a cause cannot be both positive and negative for the same effect.

These three objections are closely related. If the impact of the pills on thrombosis through their effect on pregnancy predominates in general over their impact on thrombosis through the chemical in the blood, as Hesslow envisions, then the probability of thrombosis given the pill is less than the probability of thrombosis in the absence of the pill. Hesslow says that taking oral contraceptives is nevertheless a positive cause of thrombosis in this case. However, he would, presumably, agree that taking oral contraceptives is in this case also a negative cause of thrombosis, because of its negative connection with pregnancy, a cause of thrombosis. His reasons for saying that the pill is positive for thrombosis are matched by perfectly symmetric reasons for saying that the pill is also negative for thrombosis. So his position would seem to be close to Otte's. And of course Dupré agrees that in examples in which unanimity of intermediaries fails, there will be (or at least there will very probably be) some contexts in which factor X is causally positive for factor Y as well as other contexts in which X is causally negative for Y.

Notice that the three objections each appeal, in one way or another, and whether explicitly or not, to the idea that there are different kinds of individuals, or different *subpopulations,* among or within which the impact of the causal factor X for the effect factor Y may be different. Hesslow reinforces the idea that the pill can lower the probability of thrombosis by envisioning a population in which the pill is the only kind of contraceptive available; and it seems that he must also have in mind the possibility that in other populations, the pill may raise the probability of thrombosis. Dupré explicitly considers different *contexts* across which the probabilistic (and, he would no doubt agree, *causal*) impact of one factor on another differ. These contexts correspond, of course, to different kinds of individuals, or different subpopulations of the overall populations he considers.

Otte likewise envisions two kinds of women, those who take the pill to prevent thrombosis and those who stay away from the pill to prevent thrombosis. Of course, what is relevant to the theory of causation is not the women's rationales for what they do, but rather what causes what in which women. The women who take the pill to avoid thrombosis would have an effective strategy if, for them, the impact of the pill on thrombosis through pregnancy predominates over the impact of the pill on thrombosis through the chemical in the blood; and those who stay away from the pill to prevent thrombosis would have an effective strategy if, for them, the impact of the pill on thrombosis through the chemical in the blood predominates over the impact of the pill on thrombosis through pregnancy.

As in the case of several other criticisms of the probabilistic theory of causation that we have already encountered, the three objections just described can be handled by carefully attending to the importance of *populations* in the theory: Population-level probabilistic causation is a relation between, in addition to a causal factor X and an effect factor Y, also a *population P* (considered to be of a certain kind) within which X is causally positive, negative, neutral, or mixed for Y. We must also bear in mind the distinction between property-level causation and token-level causation, described earlier and elaborated more fully in Chapter 6. With all this in mind, let us now consider the three objections in turn.

In a population of the kind Hesslow considers – for example, one in which the pill is the only kind of contraceptive available – the theory says that taking oral contraceptives is a negative causal factor for thrombosis. This is a population in which the pill's impact on thrombosis through pregnancy predominates over its impact on thrombosis through the chemical in the blood. But if we take a *different* population, one in which the pill's impact on thrombosis through the chemical in the blood predominates over its impact on thrombosis through pregnancy, then in *that* population, the pill will be a positive causal factor for thrombosis.[28] And if

[28] This is somewhat similar to the idea of looking separately at a subpopulation of women who do become pregnant and a subpopulation of women who do not become pregnant, but it is not quite the same idea. The later idea will be discussed below.

229

we combine the two populations, then the pill will be causally mixed for thrombosis in the combined population. This is one way in which the theory is sensitive to the fact that the pill can be causally positive for thrombosis even though it lowers the probability of thrombosis in some population: the pill may lower the probability of thrombosis *in one population*, yet raise it *in other populations*.

Hesslow claims that, even in a population of the kind he considers, "there are cases in which [contraceptive pills] caused [thrombosis]" (1976, p. 291). This may be true enough, but notice the reference to *cases*. In the relevant population there are *cases* in which the pill caused thrombosis. But, of course, these "cases" are singular, token-causal situations. And it is entirely consistent with the theory of property-level causation that a population P contains individuals for whom having a property X was a *token cause* of their having a property Y, even if the property X is a *negative causal factor* for the property Y in population P. Again, as more fully explained in Chapter 6, X's being causally negative for Y at the population level (relative to some population P) does not imply that there are no cases in which the possession of X token causes the possession of Y. In Hesslow's example, the pill is a negative causal factor for thrombosis, even though there may be cases in which the pill token caused thrombosis.[29]

Dupré's objection can be handled in a similar way. This will be even easier, though; for whether or not Hesslow's objection was based in part upon a conflation of property-level and token-level causation, this seems not to be a factor in Dupré's objection. Consider first Dupré's claim that when unanimity of intermediaries fails, context unanimity will also fail (or at

[29] In Chapter 6, we will see in more detail how an actual instantiation of a factor X at a time can *cause* the instantiation of a factor Y at a later time, even if the factor X *lowers the probability of* the factor Y in the relevant background context. It is perhaps also worth mentioning here that, in Hesslow's example, I am assuming, unless otherwise indicated, causally homogenous populations, as may be determined by causal background contexts as characterized in Chapter 3. Again, this is because my argument is just that we should average across intermediaries *within given background contexts*.

least will be very likely to fail). I am inclined to agree that context unanimity is outright incompatible with failure of unanimity of intermediaries, if all this means is that, if there is some population in which X is not unanimous for Y in terms of intermediaries, then there will be some population in which X is neither context unanimous, nor unanimous in terms of intermediaries, for Y. This, of course, is consistent with satisfaction of context unanimity together with failure of unanimity of intermediaries in some populations.

Consider again Dupré's example involving the gene, exercise, smoking, and heart attacks. In Dupré's *combined* population, including individuals who live where tobacco is available as well as individuals who live where it is not available, the gene is a *mixed* causal factor for heart attacks. But in the subpopulation of individuals who live where tobacco is available (and the impact of the gene on heart attacks through its effect on smoking predominates), the gene is a *positive* causal factor for heart attacks – and, at least relative to factors under discussion, the gene is context unanimously positive for heart attacks. And in the subpopulation of individuals who live where tobacco is unavailable (and the impact of the gene on heart attacks through its effect on exercising must predominate), the gene is a *negative* causal factor for heart attacks – and, at least relative to the factors under discussion, the gene is context unanimously negative for heart attacks.

Also, in the first subpopulation, where tobacco and exercise are *both* available, it seems that the gene may be contextually unanimous for heart attacks but *not* unanimous in terms of intermediaries. In the second subpopulation, where exercise is available but tobacco is not, it seems that the gene is both context unanimous *and,* since the causal chain is in effect singly connected, unanimous in terms of intermediaries as well. In any case, it is clear that context unanimity is fully compatible with failure of unanimity of intermediaries – indeed, Dupré only claims that failure of context unanimity is "likely", if unanimity of intermediaries fails. (Of course, whether this is

231

"likely" or not would depend on *what kinds of populations* one may have in mind, for the satisfaction of the two kinds of unanimity conditions depends not only on the causal factors and effect factors in question, but also on the population.) The application of Dupré's idea to the pill/thrombosis case can, of course, be handled in a parallel way.

I turn now to Otte's claim that the probabilistic theory is insensitive to the "dual nature" that many causes have. Otte himself points to the way in which the theory accommodates this, by considering two arguments a woman may make with regard to the efficacy of taking the pill for thrombosis. One argument is to the effect that one should *take* the pill to prevent thrombosis, and the other is to the effect that one should *stay away from* the pill to prevent thrombosis. These two arguments point to the two subpopulations already mentioned above in connection with Otte's objection. Roughly, the first subpopulation consists of those for whom the first argument is correct, and the second subpopulation consists of those for whom the second argument is correct.

The first subpopulation consists of all women who are such that the pill's impact on thrombosis through its effect on pregnancy dominates its impact on thrombosis through its effect on whether or not that chemical will be present. And the second subpopulation consists of all women who are such that the pill's impact on thrombosis through its effect on whether or not that chemical will be present dominates its impact on thrombosis through its effect on pregnancy.

The probabilistic theory will give the right ("dual") answers that the pill is a negative causal factor for thrombosis in the *first* subpopulation and a positive causal factor for thrombosis in the *second*. In the *combined* population, the pill is a mixed causal factor for thrombosis. And should there be women for whom the impacts of the pill on thrombosis through its effects on the two intermediaries exactly balance each other off, then *in the subpopulation of these women*, the pill will be causally neutral for thrombosis, according to the theory. Not only is the probabilistic theory sensitive to the *dual* nature of many causes, it properly accommodates the fact that, relative to different populations and subpopulations, a causal factor can have *four* kinds of causal significance for a second factor: positive, negative, mixed, and neutral.

So far, we have supposed that there is both a subpopulation in which the impact of birth control pills through pregnancy dominates its impact through the chemical, as well as a subpopulation in which the pills' impact through the chemical dominates. In this case, the dual nature of the cause is revealed, by the probabilistic theory of causation, by different statistics in these two subpopulations. It may be objected, however, that it is a possibility that for *all* women, the pill's impact on thrombosis through pregnancy dominates its impact through the chemical. Suppose this is true. (Of course, the same objection could be framed in terms of the supposition that it is the other causal chain, through the chemical in the blood, that dominates, for all women.) Thus, in the overall population of all women, the pill prevents thrombosis by preventing pregnancy, where the pill only weakly causes the presence of the chemical or the chemical is only a weak cause of thrombosis. How, in this case, can the dual nature of the cause be revealed, using the probabilistic theory of causation?

Another way in which the dual nature of the cause reveals itself in Hesslow's example is by its roles in these different populations: (1) the overall population, (2) the subpopulation of women who become pregnant (some despite taking the pill), (3) the subpopulation of women who do not become pregnant (whether or not they take the pill), (4) the subpopulation of those who have the chemical in the blood (for whatever reason), and (5) the subpopulation of those who do not have the chemical in the blood (for whatever reason).[30] We have supposed that in the overall population, population (1),

[30] Compare Otte (1981, pp. 182–3), Salmon (1984, p. 194) and Eells (1988b, pp. 114–15).

the pill is causally *negative* for thrombosis. In populations (2) and (3), we have, as we may put it, "held fixed" the potential impact of the pill on thrombosis through the factor of pregnancy, so that in these two subpopulations, only the pill's impact through the chemical will be revealed. In these two populations, the pill will be a *positive* causal factor for thrombosis. In populations (4) and (5), the factor of the chemical's being in the blood is fixed, so that only the pill's impact through pregnancy will be revealed. And in these two populations, the pill will be a *negative* causal factor for thrombosis.

Thus again, by relativizing probabilistic causal significance to populations, the probabilistic theory of causation is sensitive to the various kinds of causal impact one factor can have on another through different intermediaries. Note that this approach is not in conflict with the idea that we should not hold fixed intermediate causal factors. We are investigating causal relations *in subpopulations* determined by factors that are, *in a larger population,* causally intermediate (in the population-level sense, of course); and this is quite different from holding fixed intermediate causal factors in investigating causal relations in the larger population.

Dupré and Cartwright (1988) have objected to this strategy. They consider the population of women who do not become pregnant, population (3) above, and they say,

We have held pregnancy fixed, and hence do not expect to see changes depending on it. But the conclusion [that the pill causes thrombosis in this subpopulation] is, nonetheless, false. Suppose that pregnancy is highly conducive to thrombosis whereas the chemical effect of the pills is only slightly so. Then, most women will in fact have been saved from thrombosis by the pills; without the pills they would probably have become pregnant and run a high risk of thrombosis. What is true in this population is that contraceptives have prevented thrombosis for most women, but caused it for a few. (p. 529)

I agree that it is natural to suppose that the causal facts are as described here, but the causal facts described are all about

234

what *token* causes what, in the actual subpopulation of women who do not become pregnant. For most of these women, the pill is a token cause of not becoming pregnant, which in turn is a token cause of not suffering from thrombosis. This, however, is consistent with the fact that, in this population in which these token-causal relations obtain (and in which the incidence of thrombosis is relatively low), the pill is a positive causal factor for thrombosis. The fact that the pill's property-level causal impact through pregnancy dominates its property-level impact through the chemical, in the Dupré–Cartwright example, is revealed by probability comparisons in populations in which neither the factor of pregnancy, nor the factor of the chemical, is held fixed.

We have seen that the fact that unanimity of intermediaries is not necessary for transitivity of causal chains, and the fact that the probabilistic theory averages across intermediaries (within background contexts, of course), do not have the kinds of unintuitive consequences that have been alleged. First, it is consistent with the pill's being a negative causal factor for thrombosis that in some cases the pill token causes thrombosis; second, failure of unanimity of intermediaries does not imply failure of context unanimity; and third, the probabilistic theory of causation is sensitive to the "multiple natures" of many causes. At this point, however, it may be wondered whether anything *positive* can be said for the idea of averaging across multiple intermediaries.

In assessing the causal impact of a factor X at t_1 on a factor Y at t_3, we are interested in what difference the occurrence or nonoccurrence of X at t_1 makes for Y at the later time t_3. Holding everything else fixed, what difference does X or $\sim X$ make for Y across the time between t_1 and t_3? Of course, whether it is X or $\sim X$ that occurs at t_1 will make a difference in what will likely occur at an intermediate time t_2. X and $\sim X$ confer different probability distributions on intermediate causal factors. And, if we are careful to hold all the appropriate factors fixed in a background context, then it will be exactly the influence of X or of $\sim X$ that is responsible for a distribution, just after t_1, of probabilities of t_2-causally intermediate factors.

By averaging over intermediate causal factors, we are, in effect, treating the problem as if there were only one intermediate causal factor, namely, the post-t_1 probability distribution, produced just by X or by $\sim X$ at t_1, for the intermediate factors that will occur or not at t_2. If the theory rules that X is, say, causally positive for Y, then, even if unanimity of intermediaries fails, in terms of the various intermediate factors Z, it will nevertheless be true that X is positive for a *single* intermediate factor of a particular *probability distribution* for intermediate factors Z, which distribution in turn is positive for Y.

Here is a way of making this more precise. Suppose we are interested in the causal impact of X on Y in background context K_i, and suppose that the theory rules that X is causally positive for Y, though unanimity of intermediaries may fail, in terms of intermediate causal factors Z. Now let R_i be the factor, "the post-t_1 probability distribution for intermediate causal factors Z is given by $Pr(-/K_i \& X)$." Then, in context K_i, $\sim R_i$ is probabilistically equivalent to, "the post-t_1 probability distribution for intermediate causal factors Z is given by $Pr(-/K_i \& \sim X)$." Focusing on the single factor R_i, in K_i, instead of on the intermediate causal factors Z themselves, we in effect have a singly connected causal chain in K_i, a causal chain in which unanimity holds: X is positive for R_i which in turn is positive for Y.

In any case, a causal factor X at t_1 will, almost invariably, have some effects at t_2 that are positive for Y at t_3 and others at t_2 that are negative for Y at t_3, where it is clear that some of the possible intermediate factors will have to be *discounted to some degree or other* in assessing X's causal role for Y. For example, if we would naturally judge X at t_1 to be a *positive causal factor* for Y at t_3, then it will almost invariably be the case that X is *positive for some positive causal factors for Y* as well as *positive for some negative causal factors for Y.*

An example that shows this clearly is the question of the causal role of jumping out of a tenth story window at t_1 (X) for dying on the sidewalk at t_3 (Y). The most likely consequence of X at t_1, for what will be happening at all times after that but before t_3, is constant acceleration towards the sidewalk. However, another series of possible consequences for the intermediate times is this: grabbing a flagpole on the eighth story, climbing in an eighth story window, and then slowly and safely descending the stairs. Factor X clearly raises the probability of, and is causally positive for, each of these latter, "deviant" event types; and each of these factors in turn clearly lowers the probability of, and is causally negative for, dying in the sidewalk at t_3 (in the relevant context, or population). So unless we are to give up the probability-increase idea altogether, or else say that jumping out a tenth story window is not simply causally positive for dying on the sidewalk, then we must somehow *discount* the sequence of possible consequences just described. *Averaging causal impact over the possible intermediaries, weighting the average by the probabilities of the intermediaries, given X and given ~X, is, of course, the natural way to do this.*

In this chapter, we have seen that the natural Markov condition is a sufficient condition for transitivity of causal chains with just one intermediate causal factor. For chains with multiple intermediaries, unanimity of intermediaries, together with the Markov condition and conditional mutual independence (among the intermediaries conditional on the presence and on the absence of the causal factor), is a sufficient condition for transitivity. Unanimity of intermediaries is not necessary, however, since the theory of property-level probabilistic causation "averages" across intermediaries, within background contexts. While some philosophers have thought that this feature of the theory has undesirable or unintuitive consequences, I have argued that the alleged difficulties stem *either* from failing to appreciate and exploit the fact that a population is one of the items related in the relation of probabilistic causation *or* from conflating the property and token levels of causation. I also have tried to provide an intuitive rationale for the correctness of averaging probabilistic impact *across intermediaries, within background contexts.*

5

Temporal priority, asymmetry, and some comparisons

We turn finally to the role of time in the theory of property-level probabilistic causation. This will be dealt with in Sections 5.1 and 5.2, and this will complete the theory of property level probabilistic causation offered in this book. Section 5.3 offers some comparisons and contrasts between this theory and several others.

First, recall the problem that, at the beginning of Chapter 2, I called "the problem of temporal priority of the cause to the effect." This problem arises from the fact that probabilistic correlation is symmetric. Leaving aside qualifications having to do with causal background contexts, if a factor X is a genuine probabilistic cause of a factor Y in a population, then X raises the probability of Y in that population. This implies that Y raises the probability of X in the same population. But we cannot infer that Y is a cause of X in the population, for while correlation is symmetric, causation is not.

Second, if we agree that property-level probabilistic causation is *asymmetric,* then we may want to capture this by saying that one factor can only be a cause of "later" factors, and that it can only be caused by "earlier" factors. But what is it for one factor itself – that is, one event type or one property – to be earlier or later than another? How can we make sense of the idea that such abstract things as factors (or types, or properties) enter into temporal relations among themselves?

In this chapter, I will deal with these problems. I will assume, without much argument, that causes always precede their effects, and I will argue that the theory of property-level probabilistic causation needs a separate, explicit requirement

239

to this effect. We shall see that it is necessary to understand property-level probabilistic causation as a relation among not only a causal factor, an effect factor, and a population (and a kind associated with the population), but also temporal parameters. I will clarify the idea of one factor's preceding another in time, and this will allow a simple formulation of the requirement that property-level probabilistic causes must precede their effects.

I note here at the outset that I will not in this chapter further discuss Einstein–Podolsky–Rosen (EPR) phenomena or the requirements of special relativity, which were touched upon at the end of Chapter 2. Below, I will show how to associate not only times, but also places, with factors. Thus, we can say that the discussion in this chapter is restricted to cases in which the space-times associated with factors are *timelike separated*.

The temporal precedence requirement will rule out both *symmetry* of causation and *simultaneous* causation. Although examples have been discussed in which it seems that causation exhibits symmetry, or in which causes seem to be simultaneous with their effects, it seems to me that these are mistaken ways of viewing the cases.

Several authors have suggested that causation among factors can be symmetric. Von Wright (1974, pp. 11–12) suggests the possibility that rain can cause flooding and flooding can cause rain (if the evaporation from the flood causes the formation of rain clouds). And Wayne Davis gives this example: "Suppose Jack and Jill regularly give each other colds. Then Jack's getting a cold causes Jill to get one, and Jill's getting a cold causes Jack to get one" (1988, p. 146). It seems to me mistaken, however, to view these as genuine cases of symmetry of property-level causation. Consider the example about Jack and Jill (the rain and flooding example is completely parallel). In evaluating the causal significance of Jack's having a cold for Jill's having a cold, we have to hold fixed all causally independent (of Jack's having a cold) causes of Jill's

240

having a cold, and this will include whether or not Jill already has a cold *before* Jack has one. But when we hold fixed (positively or negatively) the factor of Jill's having an earlier cold, then Jack's having a cold's increasing the probability of Jill's having a cold is equivalent to Jack's having a cold increasing the probability of Jill's having a *later* cold. Thus, Jack's having a cold being a positive causal factor for Jill's having one is just Jack's having one being causally positive for Jill's having a later one. And there is no symmetry here, since Jill's having a later cold is not causally positive for Jack's having an earlier one. In any case, it is intuitively obvious that one person's having a cold can only be a positive causal factor for another's having a *later* cold, and this causal relation obviously does not exhibit symmetry.

As to the possibility of simultaneous causation, it has been suggested that in some cases of "pushing and pulling" the cause is *simultaneous* with the effect: "the application of a force to an object causes a simultaneous acceleration of that object" (Horwich 1987, p. 136). Familiar examples that have been discussed (see especially Richard Taylor 1966, pp. 35–6) include a locomotive pulling a caboose on a tightly connected train (in which the motion of the locomotive is supposed to cause the simultaneous motion of the caboose), a hand moving a pencil (where the motion of the hand is supposed to cause the simultaneous motion of the pencil), and the plate on a record player moving a record on its surface (in which the rotation of the plate is supposed to cause the simultaneous rotation of the record). Also there is Kant's famous example of a ball resting on a cushion, where the presence of the ball seems to cause the simultaneous depression in the cushion. However, it seems more reasonable to consider, as urged by A. David Kline (1980), that in the cases involving one motion's causing another, energy and momentum are transferred from one object (the locomotive, the hand, the plate) to another (the caboose, the pencil, the record). This, we know (from special relativity theory), takes (a very small amount of) time, so that the present motion of the second object (caboose, pencil, record) is caused by a slightly preceding motion of the first (locomotive, hand, plate). The example about the ball on the cushion can be understood in a similar way. If the ball were removed (or vaporized), it would take (a very little) time for the elastic energy stored in the cushion to result in the beginning of the disappearance of the depression.[1]

It is perhaps worth noting that in *all* of the examples discussed earlier in this book, there have been assumptions at work, implicit if not explicit, about temporal relations among the relevant factors. Even though it has not been explained just how to frame these assumptions, or exactly what they mean, it is easy to see that some such assumptions have been operating. In Chapter 4, for example, when discussing transitivity, causal intermediaries, and causal subsequents, I explicitly restricted attention to what happens at three *times, t_1, t_2,* and *t_3* – without being at all explicit about what it *means* for factors to be "at times."

Consider also the early example about approaching cold fronts, falling barometers, and rain, in which no temporal assumptions were explicit. It was assumed that the approach of a cold front is a positive causal factor for rain. Clearly, the intent is that the approach of a cold front (given reasonable ways of precisely defining this factor) is a positive causal factor for rain *soon,* within some interval of time in the *near future,* where the amount of time involved is just (typically) *several dozen hours* following the time of the approach of the front. The factor of the approach of a cold front (reasonably understood) obviously is *not* a positive causal factor for rain within the *next few milliseconds,* or for rain sometime within the *next several decades.* Although it may be difficult to specify the relevant temporal magnitudes and relations in a precise way, it is obvious that the causal claim in question can only be

[1]See Kline (1980) for more detail on some of these examples and others like them, and for further references on the issue of the possibility of simultaneous causation.

true given some kind of implicit or explicit understanding of the temporal features of, and the temporal relations among, the relevant factors.

In another example discussed previously, ingesting an acid poison, without also ingesting an alkali poison, was a positive causal factor for death. But, of course, this can be true only if understood as meaning, roughly, that ingesting acid poison, without alkali poison, is causally positive for *imminent* death. Ingesting acid poison is not, for example, causally positive for death within the next ten thousand years, or for a person's being mortal. Another example of this kind is this. Certain medical treatments for certain diseases may be causally negative for good health in the very short run, causally positive for good health in a longer run (or for longevity), and causally neutral for good health in the very, very long run.

All this suggests that property-level probabilistic causal claims must include a temporal component. Perhaps the form of these claims should be understood to be something like this: "X, at time t, is causally positive (negative, neutral, or mixed) for Y, at time t' (in some population P of some kind Q)." In the probabilistic theory of causality of Suppes (1970, 1984), for example, the notation for the event types that enter into property-level probabilistic causal relations includes temporal subscripts. In this chapter, I will offer a detailed proposal as to how to incorporate the temporal component into property-level probabilistic causal claims; this will include, along with "time point" causal factors, "time interval" causal factors.

In Section 5.1, I argue that a separate temporal precedence requirement must be explicitly incorporated into the analysis of property level probabilistic causation. Taking the temporal component of property-level probabilistic causal claims into account is crucial to the resolution, and indeed to the very formulation, of what I have labeled "the problem of temporal priority of the cause to the effect," for the property-level theory. In Section 5.2, I offer a natural and precise way of incorporating the temporal component. This gives an obvious and straightforward formulation of the temporal precedence requirement. The inclusion of this requirement completes the theory of property-level probabilistic causation offered here. Finally, in Section 5.3, I briefly describe several other approaches to property-level probabilistic causation that have been offered in that last 30 years or so, and I compare and contrast these with the theory offered here.

5.1 THE PROBLEM OF TEMPORAL PRIORITY

Setting aside for now details involving causal background contexts, if all it meant for X to be a probabilistic cause of Y were that X raises the probability of Y, then property-level probabilistic causation would have to be *symmetric:* If X causes Y, then Y causes X. This is because the relation of "raising the probability of" is symmetric. But, of course, the causal relation is not symmetric. It is a natural guess, however, that this symmetry in the "basic probability-increase idea" for property-level probabilistic causation will disappear when the details involving causal background contexts are taken into account. The guess is based on the idea that, given the analysis of the previous chapters, the factors that have to be held fixed in evaluating the causal role of an "earlier" factor X for a "later" factor Y will in general be different from the factors that have to be held fixed in evaluating the causal role of Y for X. Let's examine this idea.

Suppose that X is a positive causal factor for Y (in some population of some kind), so that Y is (I assume) causally neutral for X. If K_i's are the appropriate causal background contexts for evaluating the causal role of (the "earlier") X for (the "later") Y, and if K'_j's are the appropriate causal background contexts for evaluating the causal role of Y for X, then the guess is that it will always turn out that $Pr(Y/K_i\&X) > Pr(Y/K_i\&\sim X)$ for all i, while $Pr(X/K'_j\&Y) = Pr(X/K'_j\&\sim Y)$ for all j. Thus, the suggestion is that if the right

244

factors are held fixed, then the analysis of the previous chapters would be adequate to the problem of temporal priority of causes to their effects. This guess, however, is off the mark.

Consider this example, described by Wayne Davis (1988). Becoming pregnant, X, is a positive causal factor for having a baby, Y. To simplify things, let us imagine a narrow population, P, in which all factors – other than X, Y, and effects of X, $\sim X$, Y, and $\sim Y$ – that have a causal bearing either on becoming pregnant or on having a baby are fixed in one way. For example, the population may be the class of healthy women who are now trying to become pregnant. Let us assume that the constellation of factors fixed in population P does not necessitate either that a woman will become pregnant or that a woman will not become pregnant. And let us assume that, in this population, becoming pregnant does not necessitate having a baby – nor, of course, does pregnancy necessitate not having a baby. Then we will find, in population P, both women who do become pregnant and women who do not become pregnant. Also, among those who do become pregnant, we will find both some who have a baby and some who do not have a baby. However, I suppose that becoming pregnant is a *causally necessary condition* for having a baby (in P), so that among those who do not become pregnant, there are none who have a baby.

Of course, X (pregnancy) raises the probability of Y (having a baby), in P. And, in our special population P, this is just what it is for X to be a positive causal factor for Y, according to the theory of property-level probabilistic causation as developed so far, since all the other relevant factors are already fixed in one way in the special population P. By symmetry of correlation, Y raises the probability of X, in P. And, in the special population P, there are no factors that need to be held fixed in assessing the causal role of Y for X: Again, all the relevant factors are already fixed in one way in the special population P. This means that, on the theory of property-level probabilistic causation as developed so far, having a

245

baby is a positive causal factor for becoming pregnant, in population P. The theory as developed so far gives the wrong answer in this case, and this example illustrates the problem of temporal priority of the cause to the effect.

And in this example, it will not do to say that, if we hold fixed all causes of pregnancy, then whether or not one has a baby will not affect the probability of pregnancy. This will not do, because it is plausible to assume that, in at least some of the resulting causal background contexts, the probability of pregnancy will not equal 1. And this is the kind of arrangement of background factors that I have assumed to be already in place throughout the entire (possibly very narrow) population P of the example. But if the probability of pregnancy is not 1, then having a baby *must* increase the probability of pregnancy. This is because pregnancy is a *causally necessary condition* for having a baby (as assumed in the example). So the probability of pregnancy conditional on having a baby is 1, and hence strictly greater than the unconditional probability of pregnancy, and thus also greater than the probability of pregnancy conditional on not having a baby – all in the population P, of the example.

The problem of temporal priority of causes will arise whenever (i) an "earlier" factor X is causally necessary for a "later" factor Y, and (ii) no arrangement of causal background factors is causally sufficient for X (in the sense of making the probability of X equal to 1). Then, in all causal background contexts, the later factor Y raises the probability of the earlier factor X (to 1). So the theory as developed so far says that Y is a positive causal factor for X, while in fact it cannot be, since X precedes Y. And if (ii) above is changed to "*some* arrangement . . . is *not* causally sufficient for X . . ." (in the sense stated), then the changed (ii), together with (i), and the theory as developed so far, will incorrectly rule that Y is a *Pareto-positive* causal factor for X, in the relevant population (where the "Pareto" modifier here is the one explained in Chapter 2). Here are some other examples of the same gen-

246

eral pattern: enrolling in Philosophy 101 (X) is a positive causal factor for passing the course (Y), but not vice versa; the decay of a radioactive substance (X) is a positive causal factor for a decay's being correctly registered on a geiger counter (Y), but not vice versa; exposure to radiation (X) causes radiation sickness (Y), but not vice versa; and low air pressure (X) causes accurate barometers to fall (Y), but not vice versa. Of course, not all cases in which an earlier factor is correlated with a later factor will exactly fit the pattern described above. For one thing, X need not be quite absolutely causally *necessary* for Y. Also, X's being genuinely causally negative or mixed for Y may lead to the wrong answer that Y is causally negative or mixed for X.

I believe that the only way to solve this problem for the probability-increase theory of property-level probabilistic causation is to require, explicitly, in a separate condition, that *the effect factor be later than the causal factor.* A factor X can be causally positive, negative, or mixed for a factor Y only if Y is later than X; otherwise, X is causally neutral for Y. In addition to the requirements involving causal background contexts, this temporal precedence requirement also must be included in the theory of property-level probabilistic causation; furthermore, of course, the requirements involving causal background contexts, which themselves involve the idea of (independent) property-level probabilistic causal factors, must be adjusted to take account of the ideas behind the temporal precedence requirement. Thus, in the first place, the very idea of temporal precedence of one factor to another must be explicated, in order to enable us to include in the explication of a property-level probabilistic causal claim the very idea that the causal factor X precedes the effect factor Y. In addition, we have seen that such a component of a property-level probabilistic causal claim should indicate *by how much* the causal factor precedes the effect factor, and we should allow for the association of time *intervals,* and not just time *points,* with the factors involved.

As noted above, however, causes and effects in the relation of property-level probabilistic causation are *abstract entities:* properties, or factors, or types. And it is not immediately obvious how to understand such items as falling in a natural temporal order in a population (independently of causal ideas, at least). So we need somehow to adjust our understanding of factors in a population in order to accommodate a temporal requirement. In the next section, I offer a suggestion.

5.2 THE TEMPORAL REQUIREMENT

The key to understanding the idea of a temporal order among causal factors in a population, and to formulating the temporal precedence requirement for property-level probabilistic causation, is, I think, to appreciate the following two points. First, individuals in a population (a population appropriate as a relatum in the relation of property-level probabilistic causation) are *temporally extended entities* (and hence so is the population itself). And second, an individual's exemplifying a *factor F* at a given *time t* is itself the exemplification, by the relevant temporally extended individual, of a certain property – though this is not exactly the same as the individual's exemplifying the property *F*, as explained more fully below.

As mentioned in Chapter 1, the idea of an *individual* in a population is not always completely straightforward. Sometimes, the relevant individuals may be understood in a fairly simple and natural way, for example as human beings, laboratory mice, rolls of a die, days, and so on. In other cases, however, individuals need to be understood as *complexes* of what we would usually call individuals themselves, or as *situations*. In this section, I will elaborate the two ideas mentioned above for various kinds of "individuals" that can make up a population – including cases in which, relative to an appropriate *kind* that may be associated with a population, individuals, or their spatial or temporal parts, may be specified by *time and space points or intervals*. (The latter idea will be used in the development of a theory of token probabilistic causation in the next chapter.)

I turn first to the case in which the individuals of a population may be understood, intuitively and for the purpose of explicating property-level probabilistic causal facts about the population, as simple, "undivided," entities (though the individuals may, for other purposes, be better understood as complex). Being temporally extended, these individuals, a, are composed of their temporal stages, which we may write as a_t, for the various times t of an individual a's duration. To make the temporal extension of an individual a explicit in notation, we may write: $a = <a_t>_t$, or $a = \{ a_t \; : \; t$ is a time in the duration of $a \}$. Then a's exemplifying a factor F at a time t can be written: $F(a_t)$. Strictly speaking, this is not the same as a's exemplifying F; rather, $F(a_t)$ means that the t-temporal stage of a exemplifies F. Nevertheless, $F(a_t)$ *does* attribute a property to *the individual a*, though not exactly the property F. The property attributed can be expressed as "having F at time t." To incorporate the temporal parameter into the notation for the property, we may define $F_t(a)$ to be equivalent to $F(a_t)$. Then F_t is the property of *having F at time t*, which, of course, is different from the property F, strictly speaking.

Then, fully unpacked, property-level probabilistic causal claims can be understood as asserting that a certain relation obtains among (i) a property X, (ii) a time t for X, (iii) a property Y, (iv) a time t' for Y, as well as (v) a population P and (vi) a kind Q associated with P. And the explication of positive, negative, mixed, and neutral causal factorhood given in the previous chapters remains the same, except that: first, we must replace factors X and Y with factors X_t and $Y_{t'}$, for some times t and t' (thus making explicit the temporal parameters); second, we must include the further qualification that in order for X_t to be causally positive, negative, or mixed for $Y_{t'}$, the time t for X must be earlier than the time t' for Y, where otherwise (that is, if t is the same as or later than

249

t'), X_t is causally neutral for $Y_{t'}$; and third, we should also be more explicit about the times of the factors held fixed in causal background contexts. As to the causal background contexts K_i used to evaluate the causal role of a factor X_t for a factor $Y_{t'}$, they should hold fixed all factors $Z_{t''}$ that are causally independent of X_t and are either separate causes of $Y_{t'}$ or interactive factors for $Y_{t'}$, with respect to X_t – where t'' is any time earlier than t', and where causally independent separate causes and interactive factors are as explained in Chapters 2 and 3.

Note that the sketch of an interpretation of probability given in Chapter 1, in which times were not explicitly associated with factors, carries over straightforwardly to time-indexed factors, since, as noted above, factors like X_t and $Y_{t'}$, in which the temporal parameters are explicit, are indeed properties of temporally extended individuals in a population.

I note also that it may be objected that property-level probabilistic causal claims may not always be naturally construed as referring to *particular* times. For example, the claim "ingesting an acid poison is a positive causal factor for death (in a population P of kind Q)" may naturally be understood as meaning that ingesting an acid poison, *no matter when it occurs,* is a positive causal factor for (imminent) death. I agree. If a generalization over times is indeed what is intended in making a given property-level probabilistic causal claim, then the explication of the claim can remain the same as above, except that "t" and "t'" above may be understood as *variables, ranging over times* (rather than as *constants, denoting particular times*). And then the form of the causal claim can be understood as *universally quantifying* over the result – that is, "for all t and t' (where t and t' stand in such and such a temporal-order relation to each other), ..." This way of explicating the intent of property-level probabilistic causal claims, where no *particular* times are intended, may be borne in mind when temporal *intervals* are discussed below, and when, below, not only times, but also *places,* are incorporated into the factors, in

250

which case we may wish to quantify over places as well as times.

It is worth stressing that the causally independent (of X_t) separate causes and interactive factors that must be held fixed may pertain to any times t'' earlier than the time t' *of the effect factor* $Y_{t'}$, including times intermediate between the times t and t' of the cause and effect factors X_t and $Y_{t'}$. It will not always do to hold fixed only such factors that pertain to times earlier than (or earlier than or simultaneous with) the time t of the causal factor X_t.[2] That holding fixed such *temporally* intermediate factors can make a difference can be seen from the following simple example.

I have an urn containing an equal number of green, white, and red balls; I will draw one ball at random; and then my winnings will be determined by the roll of a die as follows. If a *green* ball is drawn, then I win \$1 if one of 2–6 comes up on the die and nothing otherwise (probability of winning = 5/6); if a *white* ball is drawn, then I win \$1 if an even number comes up on the die and nothing otherwise (probability of winning = 1/2); and if a *red* ball is drawn, then I win \$1 if an ace comes up on the die and nothing otherwise (probability of winning = 1/6). Before drawing, I have a choice: either to paint all the balls white, and thus be sure that I win the \$1 if and only if the roll of the die comes up even, or to leave the balls alone. X_t is painting all the balls white; $Y_{t'}$ is winning the \$1; and $G_{t'}$, $W_{t'}$, and $R_{t'}$ are, respectively, the temporally intermediate factors of drawing a green, white, and red ball. See Figure 5.1a, which indicates the relevant conditional probabilities.

What is the causal significance of painting all the balls white for winning the \$1? (Of course, the relevant *population P* in this example is a class of plays of the game, and the relevant *kind Q* of *P* is given by the description here of the game.)

---

[2]In this way, the theory offered here conflicts with suggestions (or at least the spirit of suggestions) offered in Reichenbach (1956, p. 204), Good (1961–2), Suppes (1970), and Eells and Sober (1983). In Section 5.3, I will discuss the theories of Good, Reichenbach (briefly in notes), and Suppes (among others).

Simple calculation reveals that $Pr(Y_{t'}/X_t) = Pr(Y_{t'}/\sim X_t') = 1/2$. $((1)(1/2) = (1/3)(5/6) + (1/3)(1/2) + (1/3)(1/6) = 1/2$.) This suggests that X_t is *causally neutral* for $Y_{t'}$. However, there are factors at work in the example that have not yet been incorporated into the description of the case. Let $G'_{t'}$, $W_{t'}$, and $R'_{t'}$ be the temporally intermediate factors (between X_t and $Y_{t'}$, and simultaneous with $G_{t'}$, $w_{t'}$, and $R_{t'}$) of drawing a ball that was *originally* (before time t), green, white, and red, respectively. The way in which the throw of the die determines whether or not I win the \$1 is still determined by the overt, surface color of the drawn ball. But there are nevertheless these factors pertaining to the original color that we may take account of. Figure 5.1b depicts the example when these factors are taken into account. As in the depiction of the example in Figure 5.1a, my winnings are determined by the roll of the die together with the *overt, visible* color of the ball, and not by the color under the paint if the ball is painted.

It is clear that X_t *interacts* with the partition $\{G'_{t'}, W'_{t'}, R'_{t'}\}$, with respect to $Y_{t'}$: Given $G'_{t'}$, X_t is negative for $Y_{t'}$ (lowering the probability of $Y_{t'}$ from 5/6 to 1/2); given $W'_{t'}$, X_t is neutral for $Y_{t'}$ (leaving the probability of $Y_{t'}$ at 1/2); and given $R'_{t'}$, X_t is positive for $Y_{t'}$ (increasing the probability of $Y_{t'}$ from 1/6 to 1/2). Also, it is clear that the factors $G'_{t'}$, $W'_{t'}$, and $R'_{t'}$ are *causally independent of X_t*: I assume that painting the balls does not affect the way a ball is selected. Thus, the theory requires that we hold fixed these factors in assessing X_t's causal role for $Y_{t'}$, giving us the answer that X_t is *causally mixed* for $Y_{t'}$. This seems to be the intuitively correct answer; and this example shows that holding fixed the appropriate *temporally intermediate* factors can make a difference in the answer the theory delivers.[3] In particular, it shows that *it does not suffice, for constructing appropriate causal background contexts, to hold fixed all and only factors that are earlier than or simultaneous with the causal factor in question,* and, thus, it shows that this natural sugges-

---

[3]As always, we never hold fixed *causally* intermediate factors; and, as noted, the factors $G'_{t'}$ and $R'_{t'}$ $W'_{t'}$ are causally independent of X_t.

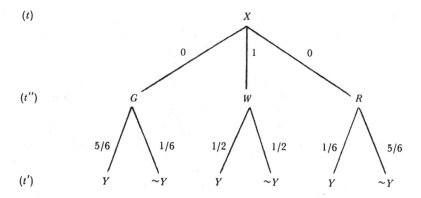

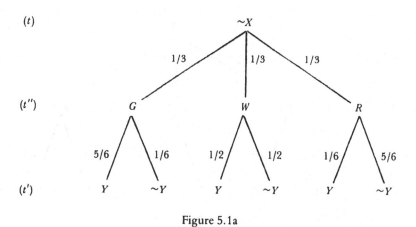

Figure 5.1a

tion cannot succeed in rendering the theory of property-level probabilistic causation *noncircular.*

It is worth noting that the temporal precedence requirement as formulated here – in terms of particular times t, t', and so on – gives a theory that is more versatile than one that simply required that the cause precede the effect. The fact

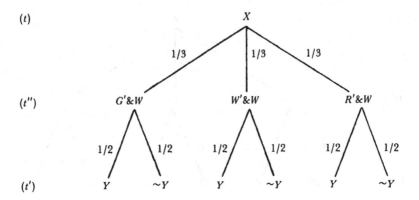

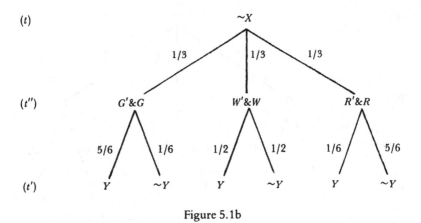

Figure 5.1b

that the theory incorporates temporal parameters *referring to particular times* allows the theory to be sensitive to the fact that, in a given population, a causal factor X_t can be, for example, causally positive for a factor Y at one later (than t) time t' ($Y_{t'}$), neutral for the same factor Y at a still later time t'' ($Y_{t''}$), positive for Y at an even later time t''' ($Y_{t'''}$), neutral for Y

254

at other times, and so on. Also, of course, when we vary t in X_t, then the X_t's can have different causal significances for $Y_{t'}$, for a given later (than the t's) time t'. (Examples illustrating these kinds of possibilities were described near the beginning of this chapter).

In addition, it is conceivable that it is not only the temporal "distance" between times t and t' that is relevant to the causal significance of a factor X_t for a factor $Y_{t'}$ in a population – even when t is earlier than t'. The fact that the (positive) "distance" $t' - t$ is equal to the (positive) "distance" $t''' - t''$ does not obviously imply that X_t has the same causal significance for $Y_{t'}$ as $X_{t''}$ has the $Y_{t'''}$. Populations may change over time with respect to the causal significance of a factor X at one time for a factor Y at a later time that is at a fixed temporal "distance" from the time of the earlier X. This can happen if the probabilities of background contexts K_i change over time. For example, the discovery of insulin in the 1920s dramatically changed the average temporal distance between onset of juvenile diabetes and its various complications. In this case, this is because the probability of the background context in which insulin treatment is available changed, over time, from 0 to positive values. (And the description of the relevant kind Q of the relevant token population P may include information about when various medical advances are made.) Again, the theory accommodates this kind of possibility by incorporating *particular times* into the analysis, rather than just the idea of temporal precedence, or just the idea of temporal distance.

In formulating the temporal precedence requirement, I have used the idea of *"particular times,"* t, t', and so on. This should not be understood as excluding the idea of particular *temporal intervals*. More generally, we may interpret t, t', and so on, as temporal intervals (where points can be understood as special kinds of intervals, $[t,t]$, for point times t). Then, we may interpret "X_t," "$Y_{t'}$," and so on, as "having X throughout the interval t," "having Y throughout the interval t'," and

255

so on. And we may then require that, in order for X_t to be causally relevant to $Y_{t'}$, t *must wholly precede* t'. Alternatively, and perhaps more plausibly, we may require that the beginning and end points of t each precede, respectively, the beginning and end points of t'.[4] The characterization of causal background contexts, taking into account time *intervals* of the independent separate causes and interactive factors that have to be held fixed, can then be understood as requiring that the time intervals, t'', of these factors, $Z_{t''}$, must wholely precede the time interval t' of the effect factor $Y_{t'}$ in question. Again, we may alternatively, and perhaps more plausibly, require that the beginning and end points of such a t'' each precede, respectively, the beginning and end points of t'.

This completes the incorporation of the temporal requirement into the theory of property-level probabilistic causation. Before concluding this section, however, I turn to the possibility, noted above and in Chapter 1, that the individuals in a population may not be appropriately understandable as "simple, undivided" entities. Consider the property-level probabilistic causal claim, "Careless smoking (in a forest) is a positive causal factor for a forest fire." The claim involves (at least) both people and forests. It asserts that a *person's* being a careless smoker (in a forest) is a positive causal factor for a *forest's* catching on fire. Unlike the causal claims discussed earlier, this claim does not assert that a single, "simple," individual's exemplification of one factor is a causal factor for the *same* individual's exemplification of another factor.[5]

---

[4] That is, if $t = [t_b, t_e]$ (or $t = (t_b, t_e)$, or $t = [t_b, t_e)$, or $t = (t_b, t_e]$), and $t' = [t'_b, t'_e]$ (or $t' = (t'_b, t'_e)$, and so on), where t_b, t_e, t'_b, and t'_e are the beginning and ending point times that define the intervals t and t', we may require that both t_b is earlier than t'_b and t_e is earlier than t'_e.

[5] One possibility for analyzing the claim would be to say that it asserts that the factor of a person's being a careless smoker (in a forest) at a time (X_t) is a positive causal factor for the person's *token* causing a forest fire at a later time ($Y_{t'}$). Understood in this way, the claim does assert that the presence of one factor in an individual (a person) is a positive causal factor for the presence of another factor in the same individual (the same person). However, it is also possible to understand the claim entirely at the *property* level, as I shall show below.

In this case, we may understand the individuals of the relevant population as *situations* of a kind. Where p is any person and f is any forest, we may understand the population to consist, to put it roughly, of pairs $<p,f>$, for times in which person p is in the proximity of forest f. More exactly, the individuals can be understood as: $a = <p,f> = <p_t,f_t>_t = \{<p_t,f_t> :$ the person p is in the proximity of the forest f at time $t\}$, where $<p_t,f_t>$ is the t-temporal stage of the "individual" $<p,f>$, and, as before, p_t and f_t are the t-temporal stages of the "less divisible" individuals p and f. (Here I am using notation like "a" and "$<p,f>$" as short for the more detailed expressions. This is only one way of understanding the relevant individuals, and I do not deny that there are other ways of carving up temporal stages of persons and forests, in spatial proximity to each other, into individuals of a population that would be appropriate for explicating the causal claim in question.)

Now it is a property of such an "individual", $<p,f>$, that its ith component (the ith member of the ordered pair, for $i = 1$, 2) has such and such a property at such and such a time. For $i = 1,2$, let $I_i(<p_t,f_t>)$ be the ith component of $<p_t,f_t>$ (that is, $I_1(<p_t,f_t>) = p_t$ and $I_2(<p_t,f_t>) = f_t$). Let X be the property of being a careless smoker and let Y be the property of being on fire. Then $XI_1(<p_t,f_t>)$ means that the person-component of the pair $<p,f>$ is a careless smoker at time t (that is, $X(p_t)$, or $X_t(p)$)), and $YI_2(<p_t,f_t>)$ means that the forest-component of the pair $<p,f>$ is on fire at time t (that is, $Y(f_t)$, or $Y_t(f)$)). These both assert properties of the pair $<p,f>$, and we can shift the temporal and pair-component notation into the notation for the property as follows: Define $X_{t,1}(a)$ (or $X_{t,1}(<p,f>)$, and so on, for the other ways of denoting the "individual" a), to mean the same as $XI_1(<p_t,f_t>)$ (that is, $X(p_t)$, or $X_t(p)$); and define $Y_{t,2}(a)$ (or $Y_{t,2}(<p,f>)$, and so on, for the other ways of denoting the "individual" a), to mean the same as $YI_2(<p_t,f_t>)$ (that is, $Y(f_t)$, or $Y_t(f)$). For times t at which one of p and f does not exist, we may say either that $X_{t,1}$ and $Y_{t,2}$ are unde-

257

fined, or simply that a does not exemplify these factors. Of course, X, X_t, and $X_{t,1}$ are all quite different kinds of factors, and the same goes for Y, Y_t, and $Y_{t,2}$.

Then the property-level probabilistic causal claim in question can be understood as follows (when we include temporal parameters, t and t', for times of careless smoking and forest fires, respectively): $X_{t,1}$ is a positive causal factor for $Y_{t',2}$, in the population of pairs $<p,f>$ described above (where, of course, in order for the claim to be true, t must precede t'). Of course, this kind of analysis can be generalized to the case of more complex kinds of individuals, $<a_1, \ldots, a_n>$, as necessary. And, as before, it is possible to interpret the times specified – t and t' for $<p,f>$, or $t_1, \ldots, t_n$ for the more general case of $<a_1, \ldots, a_n>$ – as time *intervals,* rather than as point times.

There is yet another way of understanding and representing individuals in a population, a way that will be used extensively in the next chapter, on token-level probabilistic causation.[6] If the description of the relevant kind, Q, of the population in question includes information about *where certain things are when,* then an individual can be described by saying *how things are at various times and places.* For example, the description of the kind, Q, of the population, P, of person-in-proximity-to-forest pairs of the example above may include specifications of where these people and forests are at various times. In this case, a proposition like "X is exemplified at (the time-place) $<t,s>$" may be equivalent (given this kind of description of the kind Q of P) to a proposition like $X(p_t)$ (or $X_t(p)$, or $X_{t,1}(<p,f>)$). Detectives know this. If person p is what is present at place s at time t, then "Careless smoking is taking place at $<t,s>$" (that is, "X is exemplified at $<t,s>$") is the same as "p is carelessly smoking at t" (that is, $X(p_t)$, or $X_t(p)$, or $X_{t,1}(<p,f>)$). Also, if we know that forest f is what is

---

[6]Some of what follows is somewhat intricate (as well as somewhat arbitrary) in formal detail. The reader may not wish to get tangled up in it, and may, without much loss, skip the rest of this section.

present at places s' at time t', then "The factor of being on fire (Y) is exemplified at $<t',s'>$" is the same as "f is on fire at t'" (that is, $Y(f_{t'})$, or $Y_{t'}(f)$, or $Y_{t',2}(<p,f>)$).

Where a is a pair $<p,f>$ from the population of the example described above, we can track the history of the person p as $<t_p,s_p>_{t,s} = \{ <t_p,s_p> \quad : \quad$ the person p is at place s_p at time $t_p \}$, and the history of the forest f as $<t_f,s_f>_{t,s} = \{ <t_f,s_f> \quad : \quad$ the forest f is at place s_f at time $t_f \}$.[7] Then individuals a of population P can be represented as, $a = <p,f> = <<t_p,s_p>,<t_f,s_f>>_{t,s} = \{ <<t_p,s_p>,<t_f,s_f>> \quad : \quad$ the time-place $<t_p,s_p>$ of person p is in proximity to the time-place $<t_f,s_f>$ of forest $f \}$. (Again, this is only one way of understanding the relevant individuals, and I do not deny that there are other ways of carving up temporal and spatial stages of persons and forests in proximity to each other to get individuals of a population that would be appropriate for explicating the causal claim in question.) Now, where $<t,s>$ is a particular time-place, we can write $X_{t,s,1}(a)$ (or $X_{t,s,1}(<p,f>)$), and so on, for the different notations for the "individual" a), to mean that things are X at the time-place $<t,s>$ of p (where, of course, if $<t,s>$ is not encountered by p, then a does not exemplify the factor $X_{t,s,1}$). Here, we have shifted not only the temporal and pair-component notation into the notation for the property, but the spatial component as well. Similarly, $Y_{t,s,2}(a)$ (or $Y_{t,s,2}(<p,f>)$), and so on, for the different notations for the "individual" a), means that things are Y at the time-place $<t,s>$ of f (where, again, if f does not encounter $<t,s>$, then a does not exemplify $Y_{t,s,2}$).

If we include both temporal and spatial parameters in the

---

[7]Note that, because of the kind of description assumed here of the kind Q of our population P, it is not unnatural to shift the emphasis from the "natural" constituents of the "individuals" of P (the constituents p and f of the "individuals" a of the example) to their *times* (t) and their *places* (s). This shift of emphasis is reflected in notation here by exchanging the relevant "main line" and subscript notation – specifically, by changing "p" and "f" from main line to subscripts, and by changing "s", "t" from subscripts to main line.

259

causal claim in question, we have, "Careless smoking at $<t,s>$ is a positive causal factor for a forest fire at $<t',s'>$, in population P of kind Q" (where it is understood, as explained above, that the description of the kind Q includes the time-places of the individuals a, or $<p,q>$, of P). And the claim can now be expressed as follows: $X_{t,s,1}$ is a positive causal factor for $Y_{t',s',2}$, in population P of kind Q. Again, in order for the claim to be true, t must precede t'. And again, the relevant times, t and t', and places, s and s', may be understood either as points or as intervals.

It is perhaps worth noting that, since *particular times and places (or time and place intervals)* are included in the factors involved in population-level probabilistic causal claims, as construed in the last several paragraphs, the relevant populations P may then be naturally construed as *singleton* (actual) populations, restricted to single time-places of individuals a, though they need not be so construed. (Related possibilities apply to the earlier kinds of representations of individuals, where only *times* were included in the factors: Here, a population may consist of *time stages* of the "natural" individuals a, or of their "natural" components, p, f, and so on.) Construing the relevant (actual) populations as such singletons is consistent with nontrivial probabilistic appraisals of property-level causal relevance, because of the hypothetical relative frequency interpretation of probability sketched in Chapter 1 (used as a "working model" here). And the relevant populations need not be singleton, even if each factor involved in a population-level probabilistic causal claim includes a particular time and place, since, of course, such a factor can be exemplified in a numerous population.

If x and y are actual particular events (however we wish to understand actual particular events), and if $<t_x,s_x>$ and $<t_y,s_y>$ are their time-places, then we can consider the singleton population $P = \{<<t_x,s_x>, <t_y,s_y>>\}$ (relative to some kind Q that P exemplifies). And we thus now have an understanding of the idea of "its being X at $<t_x,s_x>$ is some kind of a property-level probabilistic causal factor for its being Y at $<t_y,s_y>$ (in the actual population P of kind Q)." This is just (using ideas explained above): $X_{t_x,s_x,1}$ is a causal factor for $Y_{t_y,s_y,2}$ (in population P of kind Q). This includes, of course, an understanding of probabilities for, and probabilistic relations among, the factors of "things' being X at $<t_x,s_x>$" and "things' being Y at $<t_y,s_y>$." As indicated above, this idea – of things' being X at $<t_x,s_x>$ being a property-level probabilistic causal factor for things' being Y at $<t_y,s_y>$ – will be used extensively in the next chapter. Although this is perhaps an intuitive idea anyway, we now have one way of making it more precise.

These refinements complete the theory of property-level probabilistic causation offered here. The synopsis given at the end of Chapter 3 is modified here only by being explicit about *times of the relevant factors,* by the explicit incorporation of *the temporal precedence requirement,* and by being explicit about the idea that *causal background contexts hold fixed all the relevant factors at times earlier than the effect factor in question.* Actually, these ideas do not involve any new conditions, or any revisions, in the theory of property-level probabilistic causation as described in previous chapters. They just make explicit what was implicit all along in the theory. Again, the three ideas are these: (i) an understanding of the *notation* "X_t," $Y_{t'}$," "$Z_{t''}$," "$X_{t,1}$," "$X_{t,s,1}$," and so on, (ii) making explicit the requirement that a factor X_t can be a positive, negative, or mixed causal factor for a factor $Y_{t'}$ *only if t precedes t'* (where otherwise X_t is causally neutral for $Y_{t'}$), and (iii) making more explicit the requirement that causal background contexts K_i hold fixed all factors $Z_{t''}$ that are causally independent of X_t, and are either separate causes of $Y_{t'}$ or interactive for $Y_{t'}$ with respect to X_t (in the senses explained in Chapters 2 and 3), *and where the time t'' of $Z_{t''}$ is earlier than the time t' of $Y_{t'}$.* And, of course, the important unanimity requirements for positive, negative, and neutral causal relevance remain intact as stated in Chapter 2.

261

5.3 COMPARISONS

In this section, I very briefly sketch some of the main alternative approaches to property-level probabilistic causation that have been advanced in the last 30 years or so and describe some of the main differences between them and the approach advanced in the first five chapters of this book. (In describing contrasts, I will sometimes leave the temporal components, described above for the theory offered here, of factors X, Y, and so on, implicit.) I will begin with Cartwright, since the main lines of her approach are the closest to the ideas offered here, and I will then turn to the theories of Good, Suppes (and Reichenbach briefly in notes), and Giere. (In Section 6.6 of the next chapter I will return to Good, briefly discussing his theory of token-level probabilistic causation, and also briefly discuss the ideas of Reichenbach and Salmon in this connection.)

Cartwright

I have already in previous chapters extensively discussed several aspects of the theory of Cartwright's (1979). Some of her 1979 ideas were revised in her 1988a and 1989 writings, and some of this also was discussed above. Here I outline the basic approach, which is structurally quite similar to the theory of Chapter 2, and point out some contrasts.

The core of the theory of Cartwright's 1979 paper is her condition CC (p. 423), which I paraphrase as follows: X causes Y if and only if $Pr(Y/K_i\&X) > Pr(Y/K_i\&{\sim}X)$ for every state description K_i that specifies, positively or negatively, every factor, except for X and positive or negative effects of X, that are themselves positive or negative causes of Y. Thus, as discussed in Chapter 4, the original condition CC does not hold fixed factors causally intermediate between X and Y. Cartwright (1979) expressed a reservation about this feature of CC, her "condition (iv)," which she replaced in her (1988a) and

(1989) with her condition ★ (resulting in changing CC to what she now calls condition $CC^\star$): Roughly, if a causally intermediate factor is present in some individuals *token*-causally independently of X, then the intermediate factor should be held fixed "for these individuals" (where, otherwise, causally intermediate factors should not be held fixed). Thus, according to Cartwright, property-level probabilistic causal relations cannot be understood independently of token-level causal relations. This is a fundamental difference between Cartwright's theory and the one offered here, in which the main definition is closer to Cartwright's original CC. (All this was discussed in detail in Chapter 4.)

Other differences are these. First, there is no relativization of property-level probabilistic causal claims to *populations,* in the theory of Cartwright (1979). Indeed, Dupré and Cartwright (1988, p. 535) explicitly state that "relativization to a population . . . was not intended by Cartwright (1979)."[8] In her 1989 book, however, Cartwright allows that, for some purposes at least, certain kinds of relativizations to populations can be useful (p. 144). The theory offered here, on the other hand, *always* relativizes property-level probabilistic causal claims *both* to a *population P and* to a *kind Q* that the population exemplifies.

Second, Cartwright's conditions hold fixed, in what I have called causal background contexts, only factors that are *positive or negative causes* of the effect factor in question. Perhaps she would allow that, in addition to these positive and negative causes, *mixed causes* of the effect factor in question must be held fixed as well (when all these are causally independent, in the appropriate sense, of the causal factor in question). In any case, in the theory offered here, in addition to positive, negative, and mixed causes of the effect factor in question (which are causally indepenent of the causal factor in question), I also require (in Chapter 3) holding fixed factors with

---

[8]In fact, I am guilty of having in the past misinterpreted the theory on this point, as pointed out in Dupré and Cartwright (1988, pp. 527, 535).

which the causal factor *interacts,* with respect to the effect factor in question (when all these factors are causally independent of the causal factor in question). (Chapter 3 details Cartwright's and my different approaches to interaction.)

And finally, the theory offered here explicitly introduces a temporal component into property-level probabilistic causal claims, a component not explicit in Cartwright's conditions.

Good

I. J. Good (1961–2, 1983, 1985, for example) has offered a theory of property-level probabilistic causation that explicitly invokes the *time* of the *cause* factor (though not in notation) and that involves just a *single* causal background context for any given property-level probabilistic causal claim. The single causal background context in Good's theory is the conjunction of H, "all true laws of nature, whether known or unknown" (1983, p. 199), and U, "the 'essential physical circumstances' just before $[X]$ started" (1983, p. 199). Then X's being a positive causal factor for Y is, *qualitatively,* just X's preceding Y and X's being positively probabilistically relevant to Y conditional on $H\&U$ ($Pr(Y/H\&U\&X)$ > $Pr(Y/H\&U\&\sim X)$), temporal parameters for the relevant factors being not explicit in notation).

The core of Good's theory of property-level probabilistic causation, however, is his *quantitative* definition of the "causal support for $[Y]$ provided by $[X]$, or the tendency of $[X]$ to cause $[Y]$" (1983, p. 198). This value is (in notation that differs slightly from Good's):

$$Q(Y{:}X/U\&H) = \log[Pr(\sim Y/H\&U\&\sim X)/Pr(\sim Y/H\&U\&X)],$$

also called "the weight of evidence against $[X]$ if $[Y]$ does not happen" (1983, p. 207). Note that Q is positive if and only if Y is positively probabilistically correlated with X (conditional on $H\&U$), and Q is negative or 0 if and only if Y is negatively correlated with X or probabilistically independent

264

of X (conditional on $H\&U$), respectively. Thus, Q is a measure of statistical relevance (conditional on $H\&U$).

There are several differences and similarities between Good's theory and the one offered here that are worth noting. The most striking structural difference is Good's using just *one* causal background context, $H\&U$, while in Chapters 2 and 3, we incorporated *multiple* causal background contexts, K_i, into the theory. This difference has several consequences. First, of course, Good has no use for a unanimity requirement. Second, Good's theory does not allow for the category of *mixed* causal relevance of a factor X for a factor Y, since there are no multiple causal background contexts across which the probabilistic significance of X for Y can vary. And third, Good's definition of Q *differs from the definition, given in Chapter 2, of ADCS,* the "average degree of causal significance" of a factor X for a factor Y. This is because *ADCS* involves *multiple* causal background contexts K_i, where the degrees of correlation in the different contexts are *averaged*, weighted by the probabilities of the contexts, whereas Good's definition of Q involves just the *one* causal background context $H\&U$ – and also because I have used probability *differences* in the definition of *ADCS* as a measure of statistical correlation, whereas Good used the *logarithm* of a probability *quotient*. Good's choice of the logarithm of a probability quotient results in mathematically and intuitively attractive properties for the relations among the causal significances of logically complex causal factors and the causal significances of the components of the complex causal factors.

It is also worth noting that Good holds fixed just "the 'essential physical circumstances' *just before* [X] *started*," just the state of the world, or preexisting circumstances, *earlier than the cause factor X,* whereas in the previous section it was stipulated that causal background contexts should hold fixed all causally independent (of X) causes and interactive factors *earlier than the time t' of the effect factor $Y_{t'}$ in question.* And, as

265

we saw in the previous section, this difference in theory can make a difference in the answers the theories give about the probabilistic causal significance of a factor X for a factor Y: in some cases, holding fixed all appropriate factors earlier than Y_t will give the answer of mixed causal relevance, while holding fixed only factors earlier than X_t will give the answer of causal neutrality (and, in other cases, positive or negative causal relevance). And other kinds of different answers are possible, given this difference in theory. (Of course, as noted above, Good's theory cannot *ever* give the answer of mixed causal relevance.)

As to similarities, it should be noted that, at least for cases in which "confounding factors" are earlier than the cause factor in question, the two theories can end up holding fixed the same factors – though Good holds them fixed in just one way, positively or negatively, while the theory offered here holds them fixed, in separate causal background contexts, both positively and negatively. Second, the two approaches agree on the necessity of explicitly incorporating a temporal component in the analysis of property-level probabilistic causal claims. In fact, Good points out (1983, p. 197) that he was dissatisfied with an earlier attempt of his to give an interpretation of causation without making reference to time. Third, the two theories agree on the basic probability-increase idea that an essential component of positive, negative, and neutral causal relevance is positive correlation, negative correlation, and probabilistic independence.

Finally, Good was, as far as I know, the first to describe the distinction between *property-level* (or type- or population-level) probabilistic causation and *token-level* (or singular) probabilistic causation, and the first to develop different theories for the different levels.[9] He gives examples that vividly illustrate the difference between causal relations on the two

[9]At least it seems to be a reasonable *interpretation* of Good's theory to say that the distinction he draws is the same as, or close to, the one I describe – and, in fact, in Chapter 6, I use one of *Good's examples* to describe the distinction I have in mind.

levels (including the "Holmes–Moriarty–Watson" example, which will be described and analyzed in detail in the next chapter). Good calls causal relations at the two levels "the tendency of [X] to cause [Y]" (for the property level) and "the degree to which [X] actually caused [Y]" (for the token level). While Q, described above, is the measure for the degree of property-level probabilistic causal significance, Good develops a separate measure, *chi*, for the degree of token-level probabilistic causal significance. Of course, I concur with Good about the importance of this distinction. After developing my own theory of token-level probabilistic causation in the next chapter, I will briefly describe Good's approach and contrast it with the theory of the next chapter.

Suppes

As mentioned above, the probabilistic theory of causality of Patrick Suppes (1970, 1984) explicitly includes temporal parameters for event types, in the form of temporal subscripts in the notation for event types, X_t, $Y_{t'}$, and so on. And Suppes' theory explicitly requires that, in order for an event type X_t to be a cause of an event type $Y_{t'}$, t must be earlier than t'. And this theory also shares, with the theory offered here, the basic probability-increase idea that causes raise the probabilities of their effects, where, as in the theory offered here, this basic idea must be subject to further ideas involving causal background contexts.

The temporal requirement, the probability increase idea, and the qualification about causal background contexts, take shape in Suppes's theory in the form of two main definitions (here, my notation, and my formulation of the definitions, are somewhat different from Suppes's):

Definition 1. An event X_t is a *prima facie cause* of an event $Y_{t'}$ if and only if
  (i) t is earlier than t'

267

and

(ii) $Pr(Y_{t'}/X_t) > Pr(Y_{t'}/\sim X_t)$;

and

Definition 2. An event X_t is a *spurious cause* of an event $Y_{t'}$ if and only if X_t is a *prima facie* cause of $Y_{t'}$ and there is a partition, $\{Z_{1t''}, \ldots, Z_{nt''}\}$, such that

(i) t'' is earlier than t

and

(ii) $Pr(Y_{t'}/Z_{it''}\&X_t) = Pr(Y_{t'}/Z_{it''}\&\sim X_t)$, for all $i = 1, \ldots, n$.

(There is, presumably, an implicit *population* underlying the probability function *Pr*.) Expressed more compactly, we can say that one event is a *prima facie* cause of a second if the first precedes the second and the two events are correlated. And one event is a *spurious cause* of a second if it is a prima facie cause of the second but there is a partition of events earlier than the first such that each member of the partition screens the two events off from each other. Then *genuine causes* are defined as prima facie causes that are not spurious.[10]

There are several important structural differences between this proposal and the one offered here, as well as an important difference concerning the interpretation of the definitions. First, and most importantly, Suppes's definitions make it *a necessary condition for one factor's being a cause of another that the two be positively correlated.* However, we have seen examples (in Chapter 2) in which a property-level cause *lowers* the probability of its effect (overall, or on average), as well as examples in which the effect is *probabilistically independent* of the cause (overall, or on average). We have seen that *each* of

[10] I note that Reichenbach (1956, p. 204) offers a related proposal, according to which X_t is causally relevant to $Y_{t'}$ if the two events are correlated and there is no *set* of events earlier than or simultaneous with X_t conditional on every member of which $Y_{t'}$ is probabilistically independent of X_t. Given that "genuine" is the same as "not spurious" this definition is similar to Suppes's (1970) "Definition 2" of "spurious cause in sense one"; our "Definition 2," above, is a formulation of Suppes's (1970) "Definition 3" of "spurious cause in sense two," which is Suppes's official version, and the same as his (1984) "Definition 2."

positive, negative, neutral, and mixed causal relevance can combine with *each* of (overall) probability increase, probability decrease, and probabilistic independence. As Skyrms (1988, pp. 60–1) points out, on the Suppes (and Reichenbach) style definitions, positive causal factorhood cannot combine with (overall) "spurious independence" or (overall) negative correlation: "There is no *prima facie* cause."

A second structural difference between the two theories is that while Suppes *quantifies* over partitions, the proposal offered here seeks to identify a *correct* partition (where some partitions finer than the coarsest correct partition can also be correct). Third, the events $Z_{it''}$ of partitions used to test for spuriousness in Suppes's theory pertain to times earlier than the time t *of the cause factor* X_t in question.[11] But again, according to the theory offered here we should hold fixed all the appropriate causally independent (of X_t) factors pertaining to times earlier than the time t' *of the effect factor* $Y_{t'}$ in question, and we have seen how this difference in requirements can make a difference in what answer the theories give about the kind of causal relevance one factor has for another.

Finally, there is this difference in connection with the interpretation of the "events," X_t, $Y_{t'}$, $Z_{t''}$, and so on, in the two theories. In the first five chapters of this book, on property-level probabilistic causation, we have understood these events as abstract factors (or properties, or event types), and not as individual (or particular, singular, or token) events. Suppes carefully draws attention to this distinction (between what he calls "individual events" and "kinds of events"). After giving his definitions, and explaining and illustrating them, he says:

A deliberate equivocation in reference between events and kinds of events runs through the earlier systematic sections of this mono-

---

[11] This is similar to Good's conditionalizing only on conditions preexisting prior to the time of the earlier, causal event in question. And Reichenbach's temporal requirement, noted above, also tests for causal relevance by seeing whether events earlier than or simultaneous with the earlier, causal event in question screen off the later event from the earlier one.

graph. It is intended that the formalism can be used under either interpretation. (1970, p. 70)

Suppes gives examples of probabilistic causal claims that seem to force, or strongly suggest, one interpretation over the other, and he briefly discusses the logical and inferential relations between the two levels (1970, pp. 70–1; 1984, pp. 59–63). But Suppes seems to remain committed to the idea that the *analysis* of probabilistic causal claims is the same on the two levels. This is in sharp contrast to the ideas offered here. In the introduction, and in the next chapter, I argue that causal facts on the two levels are quite different from and independent of each other, both in terms of their analysis and in terms of what we can infer about facts on one level from facts on the other. And the theory of token-level probabilistic causation developed in the following chapter will be structurally very dissimilar from the theory delineated in the first five chapters, though there will be some interesting parallels.

Giere

Fine points aside, Ronald Giere's (1979, 1980, 1984a) theory of property-level probabilistic causation, explicitly intended to apply solely to the *population,* or property, level, is this:

X is a positive (negative, neutral) causal factor for Y in a population P if and only if there would be a higher (lower, the same) frequency of Y in P if 100 percent of the members of P were to have X than there would be if 0 percent of the members of P *were* to have X.

And Giere defines the *effectiveness* X for Y in population P as follows:

$Ef_P(X, Y) = Pr_X(Y) - Pr{\sim}_X(Y)$, where $Pr_X(Y)$ is the frequency that Y would have in P if 100 percent of the members of P were to have X, and $Pr{\sim}_X(Y)$ is the frequency that Y would have in P were 0 percent of the members of P to have X.

I note that Giere says his definitions are not intended to provide reductive analyses of the meanings of causal statements, but rather they are intended to help in understanding causal hypotheses in terms of the ways in which they are tested and applied. (I also note that the notation in the definitions above differs somewhat from Giere's formulations of the definitions.)

The obvious first things to notice, in the way of comparison and contrast to the theory advanced in the first five chapters of this book, are these. First, both theories explicitly make property-level probabilistic causal claims relative to a *population,* though Giere's theory does not explicitly introduce a *kind* for the population. Second, there are no explicit *temporal* parameters in Giere's theory, which is a difference between the theories. Third, the two theories offer different kinds of definitions of *degree* of population level probabilistic causal significance.

Fourth, there is the *counterfactual* element of Giere's theory (indicated by the words "would be" and "were" above). This is a similarity between the two theories given the sketch of an interpretation of objective probability suggested in Chapter 1, which involves sequences of *hypothetical* populations: Both theories involve, in one way or another, a modal element. A difference is in the way the modal element is included. The present theory (in the sketch of an interpretation of objective probability) uses the idea of hypothetical populations to characterize probability, while Giere characterizes causal relevance in terms of "actual" frequencies in two hypothetical populations (one a hypothetical version of a population P in which 100 percent of members have a factor X and another in which 0 percent have X).

Fifth, it is clear that the Giere style definition does not allow for the category of *mixed* causal relevance in a single population. The only kinds of causal relevance, on Giere's theory, are positive, negative, and neutral. On the other hand, mixed causal relevance in a population P may be characterized in terms of subpopulations of P across which causal significance differs; but the problem is to provide principled

271

ways of dividing up a general population P. And finally, the structures of the two theories are very different on the surface, Giere's involving just *one* frequency comparison *across two hypothetical populations,* and the present theory involving *multiple* probability comparisons *in multiple causal background contexts.*

Besides these obvious comparisons, however, Elliott Sober (1982, 1984a, 1985a) has pointed out, in a number of vivid examples, how the answers delivered by Giere's theory, about the property-level probabilistic causal significance of one factor for another in a population, can differ from the answers delivered by the style of theory offered here. He has given examples from evolutionary biology, as well as from some more common place arenas, that illustrate this difference between the two kinds of theories. Consider, for example (Sober 1985a), spectators at a baseball game, where the park stands are nearly full. Clearly, standing up is a positive causal factor for getting a good view (anybody's view would be better standing than sitting). But consider the question, from Sober (1985a): "What would happen if everyone did it?" If everyone (100 percent of the population P of fans at the ball park in question) stood up, then the average quality of view would be the same as if nobody (0 percent of the population P) stood up. Giere's theory apparently rules that standing up is not a positive causal factor for a good (or improved) view in this population P.

Another example (Sober 1982, discussed further in Mayo 1986 and in Eells and Sober 1987)[12] involves natural selection with frequency-dependent fitness values. There is a population P of butterflies that taste good to bluejays. Some members of P mimic the appearance of Monarch butterflies, which jays avoid because they taste bad. If jays choose their prey based on visual appearance, and if mimicry of Monarchs is rare in P, then mimicry is advantageous. In these circum-

[12] The rest of this paragraph is taken almost verbatim from Eells and Sober (1987, p. 292).

272

stances, mimicry is a positive causal factor for survival and the production of offspring in P. It is possible that, because of this advantage, mimicry will increase in frequency in P. But then, jays might increasingly take the shared visual appearance of the mimics in P and the Monarchs as an indication of a good-tasting butterfly (especially if there then are more mimics than Monarchs in the environment of P). As a result, the frequency of mimics in P can increase to the point where mimicry completely loses its advantage. Thus, average survival and reproductive success in P can be the same were there 0 percent mimicry in P as it would be with 100 percent mimicry in P: Giere's theory would then say that mimicry is causally *neutral* for survival and reproduction in P, regardless of the actual frequency of mimicry. (In this example, the effect factor, survival and production of offspring, is a quantitative property, plausibly measured in terms of number of offspring produced. To apply Giere's theory, we may look at the mean number of offspring in the two relevant hypothetical populations.) But, as noted above, when mimicry is rare, it actually is advantageous, a *positive* causal factor for survival and reproduction in P.[13]

I concur with Sober that these examples are genuine counterexamples to Giere's theory. Intuitively, standing up at a ball game would seem to be a positive causal factor for getting an improved view of the game, and in the population of butterflies in the second example, mimicry of Monarchs would seem to be a positive causal factor for survival and the production of offspring (when mimicry is rare). And Giere's theory seems clearly to deliver the answer of causal neutrality in both examples. Moreover, the kind of theory offered in the first five chapters of this book delivers the intuitively correct answers in examples such as these, as I will now very briefly and roughly explain.

[13] Sober (1982) has also described examples in which Giere's theory implies *negative* causal relevance, while intuitively the truth seems to be positive causal relevance.

According to the theory of property-level probabilistic causation advocated in this book, we must, when making the relevant probability comparisons, hold fixed factors that are causally independent of the candidate cause and that are themselves either causally relevant to the candidate effect factor in question or interactive for the candidate effect factor with respect to the candidate cause. This means that in the example about standing up at a baseball game, *the frequency of standers (in one's environment) must be held fixed.* This is because (roughly) a low frequency of standers is causally positive for a good view, while a high frequency of standers is causally negative for a good view. And no matter what the frequency of standers, those who are standing have a better view than those who are not standing. In the example about the population of butterflies, *the frequency of mimics of Monarchs (in the environment of a Monarch) must be held fixed.* This is because mimicry interacts with the frequency of mimics, with respect to survival and reproductive success: when the frequency of mimics is low, mimicry increases the probability of survival and reproductive success, and (roughly speaking) when this frequency is very high, mimicry is probabilistically irrelevant (or nearly so) to survival and reproductive success.

Giere's two hypothetical populations are supposed to be exactly the same as the actual population with respect to *all factors causally relevant to the candidate effect factor Y in question:* Only *the frequency of the candidate causal factor X in question* varies from 0 percent to 100 percent across the two hypothetical populations. But, of course, you cannot always keep the two hypothetical populations the same and different in these respects *when the frequency of the causal factor X in the population P is itself a causal factor for Y.*

Of course, there are fine points that I have not addressed concerning the application of the theory to these examples. For example, one's standing in a ball park would seem to be a positive causal factor for higher frequencies of standers; and since this frequency seems not to be causally independent of

274

standing, it may be argued that the theory advocated here cannot require that this frequency be held fixed. And when the frequency of mimics in the butterfly example is 0 percent or 100 percent in a causal background context, then there are either no mimics in the context or no nonmimics in the context, so that some of the relevant conditional probabilities may be undefined. In addition, it may be urged that mimicry is a positive causal factor for higher frequencies of mimics, so that again this frequency cannot be held fixed according to the theory (since it is not causally independent of the causal factor in question).

For these two problems, we should say, roughly, that what is held fixed is the frequency of standers or mimics *in the environment of* a given individual, in whom we want to assess the causal significance of standing or mimicking. This is somewhat vague perhaps, but I hope, nevertheless, that the leading idea of the application of this theory to Sober's counterexamples to Giere's theory are clear enough. For some details, see Eells and Sober (1987). See also Fetzer (1986), who argues, consistent with the approach sketched above to Sober's examples, that " 'frequency-dependent' causation . . . is nothing more than another manifestation of *context-dependent causation*" (p. 119). See also Eells (1986) for the application of the theory to examples like these.

Another kind of difficulty for Giere's theory has been raised by Gene Miller (1985). Miller argues that there are kinds of correlations that must be "separated" in order for the true causal relations to be laid bare, but which Giere's counterfactual theory is unable to "separate." Suppose that a factor Z is lawfully both necessary and sufficient both for a factor X and for a factor Y in a population P. This is, of course, consistent with supposing that X does *not* cause Y in P: X and Y can be *joint effects* of the *common cause, Z*, in P. But X is lawfully both necessary and sufficient for Y: an individual will have X if and only if it has Z, and this if and only if it has Y. It follows that there is a higher frequency of Y (namely, 1)

275

in the hypothetical version of P in which 100% have X than there is (namely, 0) in the hypothetical version of P in which 0 percent have X. Giere's theory identifies X as a positive causal factor for Y, even though the truth may be causal neutrality – Y is just perfectly, though spuriously, correlated with X.

To make the example closer to Miller's, suppose that Z is a gene and X and Y are phenotypic traits produced by Z. Suppose further that X is causally neutral for reproductive success but that Y is advantageous. Then Giere's theory will tell us that X is positive for reproductive success, since if all members of a population had X, they would all have the advantageous Y as well, and if they all lacked X, they would all lack Y as well. Miller also gives a nondeterministic example. In this example, there are two genes that are closely linked (located in close proximity on the same chromosome), so that they lawfully co-occur (this is Z) with a frequency of 90 percent. The first gene produces phenotypic trait X, which again is causally neutral for reproductive success, and the second gene produces the advantageous trait Y. Again, if all the individuals of P had X, then there would be a higher frequency of Y, and of reproductive success, than there would be if none had X. So again, Giere's theory gives the answer that X is a positive causal factor for Y, and for reproductive success.

However, in the way of comparison, the theory advanced in this book also is unable to cope with the *deterministic* example described above; the scope of this theory is limited to cases in which *a factor X does not itself necessitate either Y or $\sim Y$* – though, as explained in the introduction, this theory of population-level probabilistic causation is intended to be compatible with determinism in the sense that X, *combined with certain complex combinations of other factors,* necessitates one of Y or $\sim Y$, in an actual population P (of a very specific kind, Q, of it). For the *nondeterministic* example described by Miller, however, the theory advanced here will hold fixed the presence or absence of the gene that causes Y, when evaluating the causal role of X for Y, and for reproductive success (of course, the probabilities of contexts in which the two genes do not co-occur will be low). This will give the correct answer that X is neutral for Y, and for reproductive success. In order for Giere's theory to deliver the correct answer in this example, an appropriate interpretation of the relevant counterfactuals must be described, in which the lawful, 90 percent co-occurrence of the two genes is broken.

I must say, finally, that Giere and others have defended Giere's theory from what initially seem to be the natural interpretations of Sober's examples. The defenses involve subtle and complicated understandings of the relevant properties or populations, intricacies that I will not examine here. For more details on Giere theory, on the bearing of Sober's examples on the theory, and on defenses and suggested revisions of Giere's theory, see, in addition to the references cited above, Collier (1983), Giere (1984b), and Mayo (1986).

In this chapter, I have incorporated a temporal requirement into the theory of property-level probabilistic causation. After explaining the rationale for the requirement and and explicating the concepts involved in it, I then formulated the requirement and showed how these temporal ideas figure in the final formulation of the theory of property-level probabilistic causation that I offer here. I hope that the comparisons drawn in this final section of this chapter will help put this theory in perspective with other recent philosophical literature in the area, most of which differ not in spirit but mainly in some philosophically subtle but significant details from the ideas advanced here.

In the next chapter, on token-level probabilistic causation, the structure of the theory will be quite different, but we shall make extensive use of some of the concepts explicated above.

6

Token-level probabilistic causation

As explained in the introduction, the relation I am calling "token causation" is a relation between two actually occurring, concrete, token events, while type-level causation relates abstract entities called "properties," "types," or "factors." In the preceding chapters, I used upper case italicized letters to represent factors. Now we need to refer to token events, and I will use lower case italicized letters, x, y, z, and so on, for this purpose. As explained more fully below, the relation I wish to analyze in this chapter, in terms of probability relations, is roughly this (where x is of type X and y is of type Y): x's being of type X caused (atemporally) y's being of type Y. Another way of putting it is as follows. Where x takes place at time and place $<t_x,s_x>$ and y takes place at time and place $<t_y,s_y>$, the relation I wish to analyze is this: things' being X at $<t_x,s_x>$ caused things to be Y at $<t_y,s_y>$.[1]

The basic idea in the probabilistic theory of type level causation was that causes raise the probability of their effects. We saw that this idea needed several qualifications. The possibilities of spurious correlation and causal interaction had to be accommodated, and it was necessary to build into the theory the requirement that causes precede their effects in time. There is, I think, a basic probability-increase idea that is appropriate for token-level probabilistic causation, but we will see that for token causation, probability change must be un-

---

[1] As in Chapter 5, I will not in this chapter further discuss EPR phenomena or the requirements of special relativity, touched on in Chapter 2; and the discussion in this chapter can be thought of as restricted to the case in which $<t_x,s_x>$ and $<t_y,s_y>$ are timelike separated.

derstood differently from the way it is understood in the explication of type-level causation. To put it roughly, for the token-level theory, we must look at how the probability of the later event actually evolves around the time of the earlier event and between the times of the earlier and later events; for the token-level theory, it does not suffice to look just at the "static" conditional probability relationships among the relevant abstract factors. And as in the case of the type-level theory, several qualifications on the basic probablity-increase idea must be made.

Some of the qualifications of the basic probability-increase idea for the theory of token-level probabilistic causation are analogous to qualifications made in the type-level theory. And the idea of a causal background context is crucial in the theory of token level causation, as it is in the type-level theory. The roles of independent causes and interactive events are important in the characterization of a background context for the token-level theory, as they were in the type-level theory. However, there are also significant disanalogies between the theories.

As briefly discussed in the introduction, it is possible for the "relevant" type of a token cause to lower the probability of the "relevant" type of the token effect.[2] It is possible for a *positive token cause* to be of a type that is a *type-level, probability-decreasing, negative causal factor* for a type exemplified by the token effect. However, the most fundamental difference between the theories of type and token causation involves the way *probability change* should be understood in the two theories. And it is just because of the kind of possibility just noted that probability change must be understood differently in the theory of probabilistic token causation from the way it is understood in the probabilistic theory of type-level causation. A different understanding of probability change is required in order to make the kind of possibil-

[2] The issue of what are the "relevant" types will be discussed below, especially Section 6.1.

279

ity just noted consistent with the "basic idea," for the theory of token probabilistic causation, that token causation does, after all, coincide with probability increase, properly understood for the token level.

Another disanalogy I should note is that while the theory of type-level causation had a point both if determinism is true and if determinism is false, this is not the case for the theory of probabilistic token causation that I shall develop in this chapter. I am not sure how or whether the theory of probability may play a role in explicating token causation if determinism is true. In any case, the theory of token causation developed in this chapter assumes that the causal facts to be explicated involve only indeterministic causal relations.

I first examine in some detail several examples that show that the type or types of a token cause can lower the probability of, and be type-level negative causal factors for, the type or types of the token effect. This involves some subtle issues having to do with the "identity" of token events, the nature of probability, and the details of the examples themselves. I next explain a way of understanding the idea of probability change for token-level causation that makes the examples discussed previously consistent with the idea that token causation coincides with probability increase. We will then have a rough outline of a theory of probabilistic token causation – we will have the "basic probability-increase idea" for token level causation.

I then consider several genuine counterexamples to this rough probability-increase idea, and I show how they can be accommodated in a revised theory that makes the necessary qualifications analogous to those made in the theory of type-level causation. These are qualifications involving independent causes and interaction. Finally, the idea of *degrees* of token causal significance of one event for another is developed. This is of interest in itself, and also is necessary to accommodate some versions of the counterexamples to the "basic idea" dealt with earlier.

280

An important possibility that a probabilistic theory of token causation must accommodate is that types exemplified by a token cause can lower the probability of (and thus be type-level negative causal factors for) types exemplified by a token effect of the token cause. It is this kind of possibility that I want to clarify and establish in detail in this section. This means that the theory of token-level probabilistic causation should not maintain that a token cause must be of a type that is a positive causal factor for a type exemplified by the token effect.

An example of "probability-decrease token causation," based on an example of Dretske and Snyder's (1972), was briefly discussed in the introduction. A mean but healthy cat is in the pound, and all the preparations have been made to put this cat to sleep. At this point there is, say, an 80 percent chance that the cat will be dead in an hour, its only chance for survival being the improbable events of an escape or rescue. Let us say that the 80 percent chance of death within the hour (most likely in the pound) leaves a probability of 0.191 of the cat's escaping or being reclaimed by its owner, and a probability of 0.009 of its surviving the following. Improbably enough (1 percent chance), somebody breaks into the pound, "rescues" the cat, and then sets up the following "experiment." There is a box containing a randomizing device that, when activated, gives the box a physically irreducible 10 percent chance of immediately being in state S (and then remaining in this state), and a 90 percent chance of immediately being in state not-S (and then remaining there). Attached to the box is a loaded revolver that fires exactly when the box is in state S. The cat's "rescuer" points the revolver at the cat and activates the randomizing device. If the revolver fires immediately (the box is in state S), the cat will be dead; otherwise (box in state not-S), the cat will be kept alive for observation.[3]

[3] It may be helpful to describe the probabilistic structure of the example in more detail as follows (though this is not crucial to the intuitive point of the example).

What the rescuer has done (rescue an otherwise probably doomed cat and then start the experiment described) is a negative causal factor for death within the hour for cats in the same situation as this cat. If I were in the position of this cat, I would welcome the opportunity to exemplify the factor of having done to one what this person did to the cat. The probability of the cat's being dead within the hour decreases from 80 percent to 10 percent. It is 80 percent before the rescuer does anything, and it is 10 percent at the instant the experiment begins, when the rescuer finishes doing what he does to the cat. In this case, however, the revolver fires and the cat is dead within the hour after all. What the rescuer did *lowered the probability* of death within one hour, yet it is clearly what the rescuer did that *caused* the cat to be dead within the hour.

A different example of probability decrease causation has been described by I. J. Good (1961–2, 1983). Moriarty, Watson, and a loose boulder are at the top of a cliff. Right below is Holmes, unaware of what is happening above him. Watson has to leave right away, and he knows that if he does nothing before he leaves, Moriarty will carefully push the boulder off the cliff with the intention of crushing Holmes. The only thing that Watson can do in the way of trying to save Holmes is to try to get to the boulder and push it off himself before he leaves, trying to push it hard enough to overshoot Holmes. Watson succeeds in getting to the boulder, and he pushes it in his attempt to save Holmes. What Watson has done decreases the probability of Holmes's death by crushing, say from 80 percent to 10 percent. But, improbably enough, the attempt fails, and Holmes is crushed anyway. Again, it seems, it is the

The 80 percent chance of death in an hour breaks down as follows: 0.799 chance of death from poison in the pound and 0.001 chance of being "rescued" by our experimenter and shot to death by the revolver. The 20 percent chance of surviving the hour breaks down as follows: 0.191 chance of escape or reclamation by its owner and 0.009 chance of being rescued by the experimenter and surviving the experiment. This makes the chance of the cat's being rescued by our experimenter equal to 0.01 (= 0.001 + 0.009).

exemplification of a *negative causal factor* (a *probability decreaser*) that is the *token cause;* as Good puts it, the push "had a tendency to prevent" Holmes's death, yet "caused" it (1983, pp. 216–17).

Here is another example, a version of one discussed by Deborah Rosen (1978), Eells and Sober (1983), and Sober (1985b), all to make roughly the same point I am making here. A golfer swings at the golf ball and the ball begins rolling straight toward the cup for a birdie. But, improbably enough, a squirrel comes along and kicks the ball away. Kicks of exactly this kind have a tendency to prevent balls rolling exactly as this one was rolling from going into the cup. It is, let us assume, an irreducibly statistical fact about exactly this kind of setup – with the ball rolling exactly as it was and the squirrel kicking just as it did – that the probability of a birdie is lowered from 80 percent to 10 percent at the time of the kick. If this exact kind of setup were replicated 1,000 times, then the ball would fall into the cup in about 100 of the replications. We may assume that the golf course is sloped in such a way that, given the way the squirrel kicked the ball, there is only a 10 percent chance that the ball would take an alternative course that would land the ball in the cup. Of course, the example assumes physical indeterminism about what happens around the squirrel kick.[4]

In this particular case, the golfer was lucky enough that the ball came off the squirrel's foot on a new trajectory into the cup. After the squirrel's kick, the ball rolled, without any further interference, into the cup. In this example, the squirrel's kick lowered the probability of (and is a type-level negative causal factor for) a birdie, yet the token event of this squirrel's kick is what caused the token birdie.

Several reasons naturally come to mind for doubting that

[4]The example could be made more definite and realistic by including some truly indeterministic mechanism that the rolling ball encounters, say a mechanism involving events at the microlevel and governed by quantum mechanical laws, and that determines the angle and speed at which the ball leaves its encounter with the mechanism.

these really are cases of token causation with probability decrease. First, it has been suggested in connection with examples like this that, if we describe the token cause in enough detail, then we may find probability increase. Rosen makes this point in connection with her original golf example (which involved a ball colliding with a tree limb). She suggests that the idea of a background context is relevant here, and that in this case an appropriate background context would include information about "the angle and the force of the approach shot together with the deflection [in her example, by a tree limb]" (1978, p. 608). Including this information in the background context results in (what she calls) a "broader," "revised" causal picture of the situation, and "the results [the birdie in this case] are unlikely only from a narrow standpoint" (p. 608).

Rosen's suggestion then is roughly this: If we characterize a token cause in enough detail, then events of just the same kind as the token cause will increase the probability of an effect of the kind that actually occurred. Salmon (1980, 1984) has expressed skepticism about this approach, which he calls "the method of more precise specification of events," pointing out that "whether the ball will drop into the hole is extremely sensitive to minute changes in the conditions of the collision" and that "an unrealistically detailed description of the surface texture of the branch would be required to yield even a reasonable probability for the hole-in-one [or birdie]" (1984, pp. 194–5). I am inclined to agree with Paul Humphreys (1980), on the other hand, that this reason for rejecting Rosen's approach is "more appropriate to the pragmatics of explanation than [to] finding the causes of the birdie" (p. 311), but I also agree with both Salmon and Humphreys that the approach is nevertheless not generally applicable.

In the golf example involving the squirrel, it was assumed that the event of the squirrel's kick conferred a *physically irreducible low probability* on a birdie, and a lower probability for a birdie than there was before the squirrel kick. Because of the possibility of *indeterminism,* assumed in all the examples in this chapter, the approach suggested by Rosen cannot always succeed.

There is another reason for doubting that the examples given are really cases of causation with probability decrease, a reason that is related to Rosen's suggestion. In the examples, it may be urged, we have to take into account *the causes of the token cause,* as the time of the occurrence of the token cause approaches. In the golf example, when we assess the probability of a birdie at a time just before the squirrel kick, we have to take into account the fact that the squirrel is approaching the ball, that the squirrel cannot slow down to avoid a collision, and so on. In that case, the probability of a birdie is already very low before the actual kick, so that the kick itself does not, at the time of the kick, lower the probability of a birdie any further.

Similarly, in the example about the cat, if we take account of events preceding the rescue – the intentions of the rescuer, his ability to carry out his plan, and so on – then, it may be argued, the probability of the cat's being dead within the hour is already only 10 percent. And in the case of Moriarty, Watson, and Holmes, it may be urged that, in evaluating the probability, just before Watson pushes the boulder, of Holmes's being crushed, we have to take into account events that have already occurred, such as Watson's having formed the intention to push the boulder, his having succeeded in reaching the boulder, and so on. And if we do this, then the probability of Holmes's being crushed has already dropped to 10 percent, and Watson's actual push cannot further decrease the probability of Holmes's being crushed.

This point contains a valid idea. It is true that a negative causal factor will not lower the probability of the later factor (at least not by very much) if we hold fixed, positively, the *causes* of the negative causal factor, as long as these causes of the negative causal factor are *strong* enough, so that the probability of the negative causal factor is already quite high. But this is exactly why I include in the three examples above the stipulation that it is *improbable* that the token cause would occur (or, more precisely, improbable that the negative causal factor would be exemplified). In the example about the cat, it should be improbable, until the very moment that the rescuer does what he does, that he should succeed in doing that. In the example about Moriarty, Watson, and Holmes, it should be improbable, until the moment of Watson's push, that Watson should succeed in doing what he did. And in the golf example, it should be improbable, until the moment of the squirrel's kick, that the squirrel should kick the ball at all. These stipulations are just parts of the examples, to illustrate the *possibility* of token causation with probability decrease. But even if the types of the token causes in the examples were not improbable, still, assuming indeterminism, they would not have probability 1 – so that still the *actual occurrence* of the relevant type of the token cause has room to decrease the probability of an exemplification of the relevant effect type.

A different, somewhat more interesting, objection to the lesson I am drawing from these examples concentrates on how we should understand the token effects, rather than on how we should understand the token causes. It might be urged that types of token causes always raise the probability of types of token effects, if only we always choose the "right" type for the effect. In the example about the cat's death, if we understand the token effect to be of the type "death by gunshot wound" (or "death together with a gunshot wound"), then what the rescuer did *increases* the probability of the effect. Since had the rescuer not done what he did, the cat would probably have been put to sleep in the pound with a poison. What the rescuer did raised the probability of death by gunshot wound from some very small value (in fact, 0.001 in the example), all the way to 10 percent.

Similarly, if we understand the effect event in the example about Holmes's death to be of the type "crushed by boulder that has Watson's fingerprints on it," then what Watson did

286

increases the probability of the effect. And in the squirrel's kick example, if we understand the effect to be of the type "birdie with an approach to the cup from a direction different from the one given to the ball by the golfer," or "birdie with a squirrel's footprint on the ball," then, again, the squirrel's kick increases the probability of the effect.

For each of the examples, there is a type Y exemplified by the effect event such that the cause type *lowers* the probability of Y, *and* there are types Y', also exemplified by the effect event, such that the cause type *raises* the probability of Y'. But for each of the examples, it seems clear that the earlier event caused the later event to be of *each* of the types mentioned, whether the earlier event raised or lowered the probability of the type of the later event. For example, it seems clear that, in the case of the cat, what the experimenter did caused *both* death of the cat within the hour *and* death of the cat by gunshot wound. I do not deny that in cases of token causation we can always (or at least often) find types exemplified by the effect event such that the cause increases the probability of, and causes the exemplification of, these types. But this does not make the exemplification of the types whose probabilities are lowered any less an effect of the cause, in examples like the ones described above.

On the interpretation of objective probability sketched in Chapter 1, probability attaches to *types* – that is, to event types, factors, or properties – and not directly to *token events*. So, to evaluate token causal relations in terms of objective probability, we have to deal, of course, with types.[5] But that does not mean that we have to identify any "right" or "appropriate" types of the token events concerned. Rather, a question about the causal role of one token event for another should be understood as *coming with the "relevant" types*. To-

[5]We also have to deal with a *token population* and a *kind* that the token population exemplifies, as explained in Chapter 1. How this works in the theory of probabilistic token causation will be explained in the subsequent sections of this chapter. Roughly, the population and kind will be determined by the way things are in the particular case in question.

287

ken causal claims are about the causal role of the actual *exemplification of one type* for the actual *exemplification of another type*.

Although I shall not try to offer a metaphysics of events or a semantics of event names, I shall understand an event to be fully specified by its time and place of occurrence and the properties that are exemplified at that place and time. Thus, events x and y can be understood to be of the following form: $x = <<t_x,s_x>, X, X', X'', \ldots>$ and $y = <<t_y,s_y>, Y, Y', Y'', \ldots>$, where t and s give the time and place and the upper case letters give the factors exemplified then and there. Of course t and s need not be point times and places; they may be intervals. Also, it may be natural to restrict the factors involved to "locally" exemplified properties, so that they are not relational and they are exemplified just in virtue of the way things are at the relevant time and place (this issue is discussed further in Section 6.6). Then I shall understand questions about the causal role of one event for another to be of the form: "What is the token causal significance of x's being X for y's being Y?" Whenever an event x exemplifies a type X and an event y exemplifies a type Y, I assume it is "appropriate" to ask about the causal significance of x's being X for y's being Y.

As we shall see in the next section, the possible answers to such a question will be, roughly: "y's being Y was *because of* (or *token caused by*) x's being X," "y was Y *independently of* x's being X," and "y was Y *despite* x's being X." Of course, we should expect the answer to vary when we vary, in the question, the various types that the tokens x and y exemplify – that is, when we vary the types, all actually exemplified by x and y, that we ask about. For example, when x is a token cause of y, x will have *some* features X that are token causally relevant to y's having certain features Y, and x may have *other* features X' that are token causally *irrelevant* to y's having certain features Y' that it exemplifies. Thus, strictly speaking, "x is a token cause of y" must be understood as elliptical for

"x's exemplifying X is a token cause of y's exemplifying Y," for some factors X and Y.

In this section, we have seen that a token event x can cause a token event y, where, given different ways of understanding the events x and y (varying which exemplified types we associate with these tokens), x may either *increase* or *decrease* the probability of y. Depending on the types chosen, the probabilistic significance of x for y may be *either* positive *or* negative, all consistent with x (understood as being of a given type) being a cause of y (understood as being of a given type). This is perhaps reason for pessimism about the possibility of characterizing token causation in a probabilistic theory. In the next section, after mentioning what is perhaps a separate reason for pessimism about this, I will develop a different way of understanding probability change for the question of token causal significance. This understanding will provide the basic framework for a theory of probabilistic token causation. After two natural qualifications of the new basic probability increase idea are made in Sections 6.3 and 6.4, and after the theory is refined in Section 6.5, we will see that probabilistic token causal significance will, subject to these natural qualifications, coincide (basically) with direction of probability change.

6.2 TOKEN CAUSATION AND PROBABILITY TRAJECTORIES

Token causes sometimes raise and sometimes lower the probability of their token effects. This, roughly, is the lesson of the previous section. This suggests the possibility that, leaving aside complications involving separate causes and interaction (which will be dealt with in Sections 6.3 and 6.4), a token cause may be characterized as an event that *somehow affects* the probability of its token effect. More precisely, the suggestion is this: x's being X is a token cause of y's being Y if X is either positively or negatively probabilistically relevant to Y (and, of course, both x and y occur). However, this will not do, as an example will show.

To make a somewhat similar point – but concerning the issue of the relation between probability change and the *explanation*, rather than the *causation*, of token events – Cartwright (1979) describes a case that seems *completely parallel as far as the probabilities go* to the examples discussed in the previous section, but *different* in that we would *not* say that the earlier event *caused* the later one. Comparison of this example with those in the previous section is quite revealing about the relation between token causation and probability change, and would seem to provide further reason for pessimism about the possibility of characterizing probabilistic token causation in terms of probability change.

Consider a normal, healthy plant. The probability of its being alive and in good health in a year is 0.8. However, improbably enough, it is sprayed with defoliant. The label on can of spray said it was 90 percent effective in killing this kind of plant. So the probability of the plant's surviving for a year drops from 0.8 to 0.1. Nevertheless, the plant survives, and it is again healthy in a year. Although the probabilities given in this case are the same as those given in the examples of the previous section, this case differs from those in that in this case we would not say that the spraying caused the eventual survival and health of the plant, but in the other examples, the earlier event is clearly a token cause of the later event. (Note that it is again part of the example that the probability of the earlier event is low, until it happens.)

The examples of the previous section showed that token causation does not coincide with probability *increase* (relative to the relevant factors), and this case shows that token causation does not even coincide with probability *change* (relative to the relevant factors). This collection of examples further reinforces the idea, urged in the introduction and elsewhere above, that token causal relations are independent of population–level causal relations among the relevant factors. The examples of the previous section showed that an exemplification of a negative causal factor X for a factor Y may be a token cause of an exemplification of Y, and the defoliant example shows that an exemplification of a negative causal factor X for a factor Y may *fail* to be a cause of the exemplification of Y (which is perhaps less surprising). And again, our collection of examples would seem to provide reason for pessimism about the possibility of understanding token-level causal relations in terms of probability and probability change.

If our "asymmetric" intuitions about what causes what in the two kinds of examples are correct, then, given the apparent symmetry of probabilistic relations in the examples, it would seem that figuring out what is behind the asymmetry of our intuitions across the examples ("diagnosing" our intuitions) would constitute an important step in the direction of understanding (at least our concept of) token causation. In what follows in this section, I will point to a difference between the two kinds of examples. This is a difference in how I think we envision certain probabilities as changing, or *evolving in time,* in the examples, a difference that I think plausibly lies behind the difference in what we say causes what in the examples. This will suggest a characterization, a rough theory, of token causation, involving a kind of probability change in terms of which we can understand token causal relations, but which is quite inappropriate and inapplicable at the population level.

Unlike causation at the population level, what actually happens after the occurrence of a token event x is relevant to whether or not x is a token cause of a later event y. Of course, whether or not y actually occurs later on is relevant to whether or not x is a token cause of y, for "x token causes y" implies that x and y both actually occur. In addition, I believe that what happens *between* the times of x and y is *also* relevant to whether or not x is a token cause of y (each understood as of given types X and Y, as always). In particular, how the probability of y (or y's being Y) *evolves* between the times of x and y is crucially relevant.

Of course, to be precise about the idea of "the proability of token event y," we need to associate a type, Y, with y, and then consider the probability of the type Y, understood as exemplified or not at the time and place, $<t_y, s_y>$, of y – understanding this idea of a type at a time and a place as explained in Chapter 5. Also, as explained in Chapter 1, in order for $Pr(Y)$ to make sense, we need a token population as well as a kind that the token population exemplifies; the probability of a type is always relative to a given population and a kind that the population exemplifies (of course, the kind is more important than the token). I will say more about this later, but for now, we can understand the relevant population to be the singleton population of the particular case in question, and the kind to be determined by the conjunction of all factors (or all the "relevant" factors) that are actually exemplified before the time t_x of the earlier event x (this characterization of the relevant kind will be revised and clarified in the subsequent sections of this chapter).

Setting aside momentarily issues involving a type for the event x, I think that the question of whether x is a token cause of y's being Y turns significantly on how we picture $Pr(Y)$ as changing from around the time of x to the time of y. A more revealing notation, which captures the idea of $Pr(Y)$'s evolving in time is this: "$Pr_t(Y)$," where the indexed time t may vary from before the time t_x of x, to t_x, to times between t_x and the time t_y of y, to t_y, and to times after t_y. The idea is that $Pr_t(Y)$ will evolve when x occurs and when various effects of x, the ones that occur between t_x and t_y and that are causally relevant to y's being Y, fall into place. When such factors fall into place, $Pr_t(Y)$ becomes conditional on these factors; also, of course, for t at or after t_y, $Pr_t(Y) = 1$ (exactly how all this works will be explained in detail later). Differences in the evolution of $Pr_t(Y)$ mark an important difference, I think, between the cat, squirrel, and boulder cases of Section 6.1, on the one hand, and the defoliant case given in this section, on

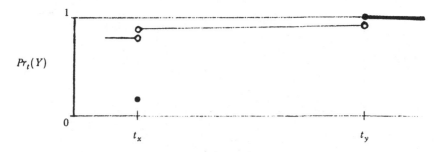

$Pr_t(Y)$

t_x t_y

Figure 6.1

the other hand. And I think this difference explains the asymmetry of our intuitions across the two kinds of case.

Let us first look at probability change in the squirrel case, and then compare this with the defoliant case. I will do this for the cat and boulder cases later in this section; the interpretation of these cases is a little more complicated. In the golf case as described (after the squirrel's kick, the ball is again on a path toward the cup), I think we envision the probability of a birdie as being high just before the kick, then falling abruptly at the time of the kick, but then just as abruptly recovering and becoming high again. See Figure 6.1. (I should note here that the many "probability-trajectory diagrams" in this chapter are not drawn "to scale.") Although *the probability that the probability of a birdie* will take this trajectory is quite low (since the squirrel's kick by hypothesis lowers the probability of a birdie), the probability of a birdie nevertheless does take this trajectory, in this particular, token case. This is because, improbably enough, the ball emerged from its encounter with the squirrel on a new trajectory into the cup. Notice that Figure 6.1 shows the probability of a birdie recovering to a higher value after t_x than it had before t_x. This, I think, is the most plausible way to enforce the intuition that the squirrel's kick *caused* the birdie. Later, I will

293

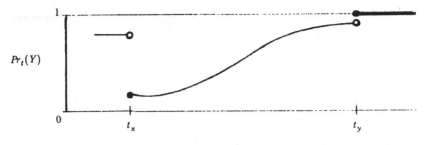

Figure 6.2

consider what we should say about the token causal significance of the kick for the birdie if the probability trajectory
were different, for example if the probability of a birdie recovered to exactly the same high value it had before the kick.[6]

In the defoliant case, I think we picture the probability of
the plant's surviving as being high at first, then abruptly
falling at the time the plant is (improbably) sprayed, but then
only slowly increasing to near 1 as the plant gradually recovers. See Figure 6.2. In this case, the application of the defoliant succeeded, for a while at least, in lowering the probability
of the plant's surviving. The spraying left that probability
low, in that after the spraying was over, the probability remained low. The plant's gradual, but uncertain, recovery is
reflected, in the diagram, by the slow increase in the probability of survival and eventual health.

For lack of better terminology that I can think of, I shall
say that in the golf case, the ball dropped in the cup (this is y
under description Y in Figure 6.1) *because* of the squirrel's
kick (for token *causation*); and I shall say that the plant sur-

[6]In Figure 6.1, as well as in other figures like this that follow, I represent the
probability as "jumping" at t_y from some value short of 1 to 1. This means that the
probability is not 1 until the event happens. However, as Tony Peressini points out,
this latter idea does not force there to be a discontinuity in the probability trajectory
at t_y. The probability could "smoothly" approach 1 consistent with the probability's
not being 1 until the event happens. However, the diagrams had to be drawn in one
of these two ways; and, fortunately, the analysis of token causal significance that
follows does not pay attention to whether or not the trajectory is smooth at t_y.

294

vived *despite* the application of the defoliant. And I think that the "probability trajectories" depicted in Figures 6.1 and 6.2 are quite close to themselves being good *characterizations* in general of the ideas of "y's being Y *because* of x" and "y's being Y *despite* x." But more adequate characterizations of these ideas will be given just below and in the following sections.

Actually, it seems that there are, basically, *three* kinds of causal significance one token event can have for another token event. The later one can exemplify a type Y *because of, despite,* or *independently of* the earlier event, where the last kind of "significance" applies to cases in which, roughly, the earlier event played no role in determining the character of the later event. I also think that these are just the extremes, and that the token causal significance of one event for another (relative to given types) can lie somewhere between these extremes (see below). Before discussing degrees, I will now explain in more detail the paradigm extremes of because, despite, and independence, and more carefully distinguish these ideas from the different ways in which one factor can be causally significant for another at the population level. What follows is a preliminary characterization of these paradigm extremes – pending qualifications, to be made in Sections 6.3 and 6.4, that are analogous to qualifications made in the population-level theory. Also, in Section 6.3, the role of the relevant *type*, X, of the earlier event x will be clarified.[7]

First, an event y exemplifies type Y *because of* event x (at the paradigm extreme) if (i) the probability of Y changes at the time of x, (ii) just after the time of x the probability of Y is high, (iii) this probability is higher than it was just before x, and (iv) this probability remains at that high value until the time of y. See Figure 6.3. Note that there is no reference here

---

[7] The reader may have noticed that the only thing about the earlier event x that explicitly enters into the probability trajectory diagrams above is its *time*, t_x. This is true also for the further diagrams that follow. The identity of the relevant type, X, of x will control certain features of the relevant background context, which in turn will control the shape of the probability trajectory, as explained in Section 6.3.

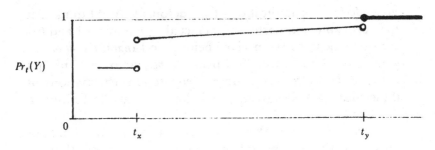

Figure 6.3

Figure 6.4

to what the probability of Y is *at* the time of x. An event y
exemplifies a type Y *despite* x (at the paradigm extreme) if (i)
the probability of Y changes at the time of x, (ii) just *after* the
time of x the probability of Y is low, and (iii) this probability
is lower than it was just before x. See Figure 6.4. Again, there
is no reference to what the probability of Y is *at* the time of x.

There are basically *two* kinds of situations in which it
would be natural to describe the causal role of x for y's being
Y by saying that y's exemplification of Y is causally independent of x. To distinguish these two kinds of situations, I will
use the terms "independent" and "autonomous." Let us say
that y is Y *independently of* x if the probability of Y is the same
just after the time of x as it is just before the time of x. See
Figure 6.5. And let us say that y is Y *autonomously of* x if: (i)
the probability of Y changes at the time of x, (ii) just after the

296

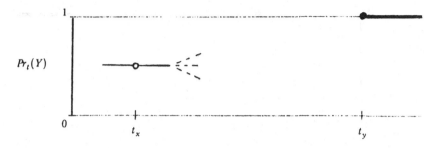

$Pr_t(Y)$

t_x　　　　　　　t_y

Figure 6.5

time of x the probability of Y is high, (iii) this probability is higher than it was just before x, but (iv) at some time after that of x and before the time of y, the probability of y drops to a low value. See Figure 6.6. As a final piece of terminology, let us say that x is *token causally relevant* to y's being Y, if y is Y either because of x or despite x; and if y is Y either token causally independently of x or autonomously of x, then x is *token causally irrelevant* to y's being Y.

There is a certain amount of organization to these four kinds of possibilities that might not be apparent at first sight. What can happen to the probability of an event's exemplifying a type Y across the time of an earlier event x? Roughly, the three possibilities are: (1) it can remain unchanged, (2) it can end up higher, and (3) it can end up lower. Possibility (2) can be split roughly into these two possibilities: (2a) it ends up higher and stays that way all the way until the time of y, and (2b) it ends up higher but comes down sometime before the time of y. (1) is token causal *independence* of y's being Y from event x; (2a) is y's being Y *because of x*; (2b) is y's being Y *autonomously of x*; and (3) is y's being Y *despite* event x.

The distinction between y's having Y because of x and y's having Y despite x is quite different from the distinction between positive and negative causal significance at the population level. And neither of the former can be characterized in terms of the latter alone (even the latter together with saying

297

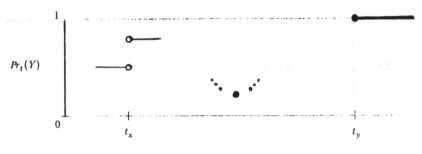

Figure 6.6

whether or not y occurred as an exemplification of type Y), nor can the latter be characterized in terms of the former. This is because what actually happens *after* the occurrence of x – namely, the actual trajectory of the probability of Y – is relevant to whether y has Y because of or despite x, but this is quite irrelevant to whether any type that x exemplifies is causally positive or negative for Y at the property level.

Also, the probability of Y at the time of x is irrelevant to whether y has Y because of or despite x. But I think it is correct to identify the probability of Y at the moment of x as the appropriate probability for assessing type level causal relations between (the relevant type for) x and Y (in the relevant context). The probability of Y at the time of x is $Pr(Y/K_a \& X)$, where X is the relevant type for x and K_a is the context actually exemplified in the token case in question. (The appropriate K_a will be characterized in detail in Sections 6.3–6.5.) The token-level theory characterizes causal relations in terms of the probabilities of Y before and after, but not *at*, the time of x, whereas the type-level theory characterizes causal relations in effect in terms of the probability of Y *at the time of x*, but not before and not after.

The only kind of support I can offer for this way of looking at probability change in the theory of token-level causation is to point out how naturally the theory arises out of the diagnosis given of the different intuitions we have in various exam-

298

ples, and to support the diagnosis given. To this end, I shall now return to the examples, consider some variations on them, and apply to them the kind of diagnosis I have suggested and the rough sketch of a theory of token causation that I have outlined above.

Let us return briefly to the squirrel example. The intuition I have about the example, and that I have diagnosed and let guide the construction of the theory, is that the squirrel's kick *caused* the birdie. But now that the idea of "despite" has surfaced, and has been distinguished from "because" (or positive token causation), it is perhaps natural to suggest instead that the birdie in this example occurred *despite* the squirrel's kick. Part of the motivation for this suggestion may stem from the perhaps unfortunate terminology I have chosen for the different kinds of token causal significance. Nevertheless, I think it is clear that in the example, as specifically formulated, the squirrel kick did token cause the birdie, the birdie "traces back to" the squirrel kick. And this kind of case should be distinguished from the class of cases I associate with the term "despite."

I have emphasized that this is a case of "because" *as the example was specifically formulated*. At the instant the ball left the squirrel's foot, *the ball was again on a trajectory into the cup,* where this is true in virtue of the contour of the golf course, the new motion of the ball, and the improbability of further interference with the motion of the ball along this trajectory. The actual effect of the squirrel's kick, the immediate result of the encounter, was that the ball was on a new course into the cup. And, as I have diagrammed the probability trajectory for this example in Figure 6.1 (which makes the example more definite), the probability of a birdie was even higher after the kick than it was before the kick. (Below I will consider a version of the example in which the squirrel kick changed the path of the ball but left the probability of a birdie at the same high value.)

Here is a variation of the golf example in which I think the

299

right answer *is* "despite." Suppose the squirrel kicked the ball, in exactly the same way as before, and that the ball came off the squirrel's foot on a path *away from the cup,* which is the more probable outcome (90 percent chance). This makes the probability of a birdie very low; something very improbable would have to happen for the ball to wind up in the cup. But a few seconds later, in this variation of the example, something very improbable does happen. A branch falls from a tree, the ball ricochets off the branch on a path into the cup, and the birdie is made. In this example, the squirrel's kick succeeded in making the probability of a birdie low for a time. For a few seconds, the probability of a birdie was very low, and something very improbable had to happen in order for the ball to return to a course into the hole. The birdie was made *despite* the squirrel kick in this example.

For a situation to qualify as a case of "despite," the earlier event must succeed in lowering the probability of the later event, for some positive amount of time. The *actual outcome,* of the earlier event must be to have left the probability of the later event lower than it was before the earlier event. That is, the probability of the later event must be lower after the earlier event has finished occurring, after it has "done its work." When the later event nevertheless does occur, then we have a situation that I think quite intuitively deserves the term "despite."

I mentioned that the token causal significances characterized above come in degrees. They do so, roughly, according to the "degree" to which the various clauses in the definitions of the different kinds of causal significance are satisfied. The idea of these degrees will be developed in detail in Section 6.5, but the idea can be illustrated here. The original golf example (Figure 6.1) is not a case of "because" of the most extreme degree. The squirrel's kick did not have the effect of raising the probability of a birdie by very much, in the example as described. The probability of a birdie was already quite high (80 percent), not leaving much room for increase. A

300

more extreme case of "because" is an example in which the ball intitially had a lower probability of falling in the cup.

Say the ball is initially rolling toward a number of obstacles between the ball and the cup, where the probability is 0.8 that the ball will be interfered with in such a way that it will miss the cup, so that at this point the probability of a birdie is 0.2. The only physical difference between this example and the original is the presence, here, of the obstacles between the ball and the cup. Then the squirrel kicks the ball in exactly the same way as in the original example, in a way that on average gives balls rolling like this one was a 90 percent chance of missing the hole. But again, the ball, improbably enough, wound up on a path into the cup, away from the obstacles. In this case, even though there is a decrease in the probability of a birdie from 0.2 to 0.1 at the time of the kick, there is an actual increase in the probability of a birdie from about 0.2, just before the kick, to, say, 0.98 just after the kick. Again, I think the intuition is clear that the squirrel's kick caused the birdie. And the original golf example is different from this one only in degree.

Now consider a version of the golf example that is a kind of limit of cases like the one just described and the original example, one in which the degree to which it is a case of "because" vanishes. In this example, the probability of a birdie is initially high, the squirrel kicks the ball as in the original example, and, after the kick, the probability of a birdie is (again improbably) very high again – but no higher, or lower, than it was initially. As in the original example, the squirrel's kick changes the path of the ball, so that again the birdie physically "traces back to" the squirrel kick; but in this new example, the kick leaves the probability of a birdie un-changed (though as before the probability of a birdie *at the instant of the kick* is low, 0.1). The probability trajectory for this example is shown in Figure 6.7; it is the same as the trajectory for the original squirrel case, shown in Figure 6.1, except that the trajectory of the probability of a birdie around

Figure 6.7

t_x forms (more or less) a straight line with one point, at t_x, missing. Note also that this trajectory conforms to the pattern shown in Figure 6.5, for *token causal independence*. At this point, it might be objected that the same intuitions that tell us that the kick token caused the birdie in the original example should tell us that this is the case in the new example as well. And this means that if the theory gives the right answer of "because" in the original example, then it is giving us the wrong answer of "token causal independence" in the new example.

Indeed, the theory does give the answer that the birdie is token causally independent of the squirrel kick, in the new example; and I think this is the correct answer. How can this be reconciled with the intuitions expressed in the objection just stated? First, we must be careful to remember that token causal significance comes in degrees. In the original golf example, the squirrel's kick did token cause the birdie, but as mentioned above, this is not a case of "because" of an extreme degree: The degree to which the kick raised the probability of a birdie is not extreme. In the new example, the kick did not raise the probability of a birdie at all. In intermediate cases, the kick would raise the probability of a birdie only a very little, and we have cases of because of very small degrees. Qualitatively, the difference between such examples

302

and the new one described above is the difference between token causal relevance and token causal irrelevance, but quantitatively, the new example is continuous with such cases.

But what about the idea that, even in the new example, the birdie "traces back," physically, to the squirrel's kick? There is, after all, a physical "process" connecting the kick and the birdie. If the theory says simply that the birdie is token causally independent of the squirrel kick, then it seems there is a causal fact that escapes the theory. But here we must be careful to remember that, according to the theory, token causal claims always specify a type for the cause event and a type for the effect event, where (again postponing discussion of types for the cause event), the cause can be causally significant in different ways for different types that the effect event exemplifies. For definiteness, let us assume that, when the golfer swings, the cup is due south of the ball; assume also that after the swing, the ball is rolling due south on level ground, and that the contour of the golf course is such that the squirrel's kick had the effect of making the ball enter the cup from the northwest (rather than from the north as it probably would have without the kick). Then the squirrel's kick is token causally irrelevant to the later event's exemplifying the type Y described by "ball enters cup," but it is token causally positive for the later event's exemplifying the type Y' described by "ball enters cup from the northwest."

The probability trajectory for assessing the causal significance of the squirrel's kick for the ball's entering the cup from the northwest (for y's being Y') is shown in Figure 6.8. Here, we trace the trajectory of $Pr_t(Y')$, rather than that of $Pr_t(Y)$. With this trajectory, the theory tells us that the ball *entered the cup from the northwest because* of the squirrel kick, where this is consistent with the fact that the ball's *simply entering the cup* is token causally *independent* of the kick. In this way, the theory is sensitive to the fact that the birdie "traces back" to the squirrel kick. The earlier event is token

303

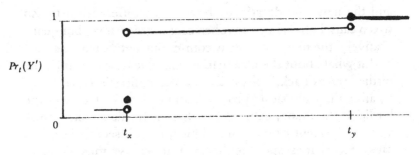

Figure 6.8

causally positive for those features of the later event that trace back to the cause.[8]

What I am calling "independence" is the first kind of token causal irrelevance explained above. A different version of the original golf example is helpful in illustrating the second kind of token causal irrelevance explained above, called "autonomy." Suppose that some time after the squirrel kicks the ball (when the ball is again on a path toward the cup), a second squirrel comes along and kicks the ball away, lowering the probability of a birdie for a time. Finally a third squirrel kicks the ball, making the ball begin to roll straight toward the cup, thus again increasing the probability of a birdie. In this case, all three kicks are, until they occur, very improbable. Here, the birdie took place more or less independently of the first squirrel's kick – or, as I shall say to distinguish this from what I have already called "independence," *autonomously* of the first squirrel's kick. The probability of a birdie was not *permanently* raised by this kick, and other, later and improbable events have to take a certain amount of the credit for the birdie. And the theory gives the answer of token causal autonomy of the birdie from the squirrel's kick in this example: it is token causal irrelevance of the second kind.

Let us now turn to the example involving Holmes, Mori-

[8]In some cases, it may be necessary to resort to "nonlocal," or relational, properties of the later event. This issue is discussed further in Section 6.6.

304

arty, Watson, and the boulder. In describing the example, I said, agreeing with Good, that Watson's push *caused* Holmes to be crushed. But again, since the idea of "despite" has surfaced, it might be suggested instead that Holmes was crushed *despite* Watson's pushing the boulder. In this case, I think the example actually has not been described definitely enough for us to choose between the "because" description and the "despite" description. What we need to know, as always, is how the probability of Holmes's being crushed changed around the time of Watson's push, and how this probability evolved thereafter. Here are two natural ways of filling in the gap.

First, it might have been that at that moment the boulder left Watson's fingertips (that is, immediately after the event of his push ended), the boulder was on a trajectory in the direction of Holmes. Of course, this is an improbable outcome of the push, given Watson's intentions. Nevertheless, in this case, the probability is high, immediately *after* the push, that Holmes will be crushed. If this probability is high enough, and remains high until the time of impact (the boulder remains on a course toward Holmes), then we have a case of positive token causation, a case of "because."

As a second way of more fully describing the example, it might have been that Watson is very skillful, that he carefully took into account the wind, the factor of air resistance, the shape of the boulder, and so on, and that he sent the boulder off the cliff *in exactly the way he should have* given his data and his intention to overshoot Holmes. However, his data only provides statistical information about what the air currents, and so on, will be like around the boulder on the way down. And on this occasion, on the boulder's way down, the air currents, and so on, combined – in a way that is very improbable given Watson's accurate data and given his accurate push – to send the boulder straight into Holmes. In this case, Watson's push succeeded in lowering the probability of Holmes's being crushed, but only for a short time. On the boulder's way down, it acquired (perhaps gradually, perhaps not) a trajectory into Holmes, thus raising the probability of Holmes's being crushed. In this case, the theory says that Holmes was crushed *despite* Watson's push, since the push succeeded in *leaving* the probability of Holmes's being crushed lower than it was before (though of course this probability did not permanently remain low).

The example about the cat and what its rescuer did is somewhat different. First let me clarify what the probabilities are and how they change at the various stages of the example. Before the rescuer does anything, the probability is 0.8 that the cat will be dead within the hour. The cat is in the pound waiting to be put to sleep, and it is very improbable that the rescuer, or anything else, will save this feline. Then (improbably enough) the rescuer takes the cat from the pound, intending and able to carry out his experiment. During the time between his getting the cat and his activating the randomizing device that controls the revolver, the probability is only (about) 10 percent that the cat will be dead within the hour, since there is only a 10 percent chance that the revolver will fire when the device is activated and since the rescuer intends to keep the cat alive if the revolver does not fire. Then the device is activated, and, improbably enough, the revolver fires and a bullet emerges from the barrel. Between the time of the emergence of the bullet and the bullet's entering the cat, the probability is extremely high that the cat will be dead within the hour. When the bullet enters, the cat dies instantly.

The probabilities in this example are depicted in Figure 6.9, where x is the event of the rescuer's taking the cat from the pound (with his intention and ability to carry out his experiment), x' is the event of the rescuer's activating the randomizing device, and y is the event of the cat's being dead within the hour. The short time between x' and y is the time it takes for the bullet to travel from the revolver into the cat (of course, this diagram, as always, is not drawn to scale).

In describing this example in the previous section, I said that the cat was dead within the hour *because of what the*

306

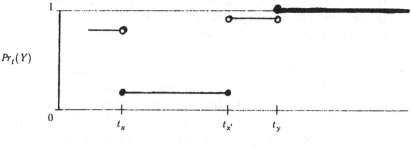

Figure 6.9

experimenter did. In describing "what the rescuer did," I included everything he did to the cat, beginning with his taking
it from the pound and ending with his activating the randomizing device – that is, everything he did between and including the times t_x and $t_{x'}$. And, indeed, Figure 6.9 shows that
immediately after the rescuer finished doing what he did (that
is, immediately after $t_{x'}$), the probability of the factor Y (the
cat's being dead within the hour) is very high, plausibly even
higher than it was (0.8) *before* the temporally extended event
of the rescuer's doing what he did (that is, before t_x). This
probability increase – from what the probability of Y is before t_x to what it is after $t_{x'}$ – is why the theory gives us the
answer that the cat was dead *because* of what the rescuer did
between t_x and $t_{x'}$.

This example illustrates how the theory applies to temporally extended events that are token causes. In general, consider Figures 6.3–6.6, which depict the four paradigm extreme kinds of token causal significance that an event x can
have for a later event y's being Y. Imagine that, for each of
these diagrams, we expand t_x into an interval, $<t_x, t_{x'}>$, and
push the parts of the trajectories shown to the right of t_x just
to the right of $t_{x'}$. That is, any point on a trajectory that was
any positive temporal distance d from t_x is now that same
distance d to the right of $t_{x'}$. Also, do not insert any trajectory
parts into the now empty spaces between t_x and $t_{x'}$. Then the

307

resulting, "stretched" diagrams depict the four paradigm extreme kinds of token causal significance that a temporally extended event (occurring between t_x and $t_{x'}$) can have for a later event at t_y.[9]

There is another causal event of interest in the example about the cat and what the rescuer did. It is the "point" event x of the rescuer's taking the cat from the pound with the intention and ability to perform his experiment. If we understand the state of the rescuer at t_x as determiing (with near certainty) both that he will have possession of the cat and that he will perform his experiment, then the event of his being in that state succeeds in *lowering* the probability of the cat's being dead within the hour. It lowers that probability from 0.8 to (about) 0.1, and this probability stays at that relatively low value all the way until $t_{x'}$, when the revolver fires. So the theory tells us that the cat was dead within the hour *despite* the earlier event x. So we have two token causal truths, according to our rough theory: The cat died *because* of the temporally extended event of what the rescuer does between t_x and $t_{x'}$, but *despite* the event x itself at t_x. And this would seem to accord with intuition. Of course, it would also be correct, according to the theory, to say that the cat died because of the event x' of the resuer's activating the randomizing device.

The sketch of a theory laid down in this section is far from complete. It merely provides an understanding of *probability change* that is appropriate for evaluating token causal relations in terms of probability change. Several important qualifications still need to be made. What we have so far, for the theory of probabilistic token causation, is analogous to having, for the probabilistic theory of causation at the population level, just the "basic idea" of type-level probability increase. In the next two sections, I will explain why we have to include cer-

[9]What about temporally extended token *effects*, things' being certain ways between two times t_y and $t_{y'}$? Here, it seems that the analysis should be the same, except that the trajectories in the diagrams may not achieve the value 1 until the later endpoint, $t_{y'}$, of an interval $<t_y, t_{y'}>$.

tain qualifications in this probability-increase theory, qualifications similar to those involving independent causes and independent interacting factors that we had to make in the population-level probability-increase theory. When this has been done, it will be possible to explain the role of the *type, X,* of the earlier event *x,* which has not been taken account of in the analysis so far.

6.3 THE CAUSAL BACKGROUND CONTEXT AND SEPARATE CAUSES

This section begins with a description of several (genuine) counterexamples to the rough theory presented in the previous section. This shows not exactly the *inadequacy* of the basic probability-increase approach described there, but rather its *incompleteness.* The counterexamples show the need for a certain qualification of the basic probability-increase idea, a qualification involving holding fixed certain items when making probability assessments for evaluating token causal relations. I then formulate the qualification and incorporate it into the theory. In the next section, further counterexamples will show the need for yet another qualification.

In the rough theory given in the previous section, the *probability* of the relevant effect factor, exemplified by the token effect, played a crucial role: this is the probability whose trajectory over time is central to the theory. As to the token cause, only its *time* entered into the analysis: We looked at changes in the trajectory of the *probability of Y* across the *time of x* (as well as later on), and this is the only way in which the earlier event *x* entered the picture. So if we considered any other token event, say *z,* that takes place at the same time as *x,* the theory would give the same answer about *z*'s causal role for *y*'s being *Y* as it does for *x*'s causal role for *y*'s being *Y*. The qualification of the theory introduced in this section will allow us to distinguish the causal role of *x* from the causal roles of other events simultaneous with *x.*

The first counterexample is directed specifically at the characterization of "despite" above, but, as we shall see, it also has force against the characterization of "because," in a couple of ways.[10] Recall that all "y's being Y *despite x*" means, according to the theory as developed so far, is that the probability of Y fell at the time of x and remained low for some time. This gave us the intuitively correct answer in the defoliant case described above. But consider this modification of the defoliant case. The plant was poisoned as before, say on January 1, 1990. And as before, the probability of the plant's being alive in a year fell sharply. Say that by late April, the plant is well on its way to recovery, and the probability of its being alive and healthy at the end of the year is also "recovering." By the end of June, the plant is fully recovered, and the probability of its being alive and healthy at the end of the year is as high as it ever was.

Here is the twist in the example. On July 1, 1990, something very improbable happened. The plant received a second, more severe, shock, from *twice* the amount of the *same* defoliant as was administered on January 1. Again, the probability of the plant's being alive and healthy at the end of the year abruptly falls. However, in recovering from the first shock, the plant acquired a certain amount of resistance to this kind of defoliant. Due to the resistance the plant has acquired, it is able to recover from the second shock, though again slowly. By the end of the year, the plant is again alive and healthy. See Figure 6.10, in which x is the event of the first poisoning, z is the event of the second poisoning, and y is the event of the plant's being alive and healthy at the end of the year (under that description, Y).

It seems clear that, even though the first shock caused the probability of survival in a year to fall and remain low for some time, we should nevertheless say that the plant's even-

---

[10] The counterexample was explained to me by Bill Tolson. It took me some time to appreciate that the counterexample was genuine, and to see the natural kinds of qualifications the token level theory required.

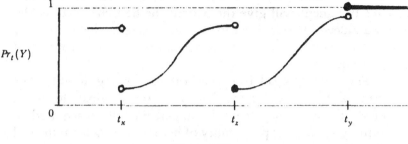

Figure 6.10

tual survival is *because* of the first shock, and the first shock token caused the plant's eventual survival and health. Without the first shock, the plant would not have acquired the resistance needed to recover from the second, more severe, shock. The first shock caused the resistance, which is what enabled the plant to survive the second, doubly severe, shock. Of course, this is more or less what happens when we receive inoculations, which at first weaken us but eventually enable us to resist a disease when exposed.

If our intuitions about what caused what in this example are correct, then the example also is straightforwardly a counterexample to my earlier characterization of "because," for it *is* a case of "because," but it *does not* have a probability trajectory of a shape characteristic of because, on the definition of "because" given in the previous section. There is also the following kind of counterexample to the way the theory of the previous section characterizes "because."

A patient is very ill on January 1, 1990. The disease gives her very little chance to survive the year. But she has an excellent chance (as good as anybody else) to survive at least six months. Improbably enough, a treatment that is 90 percent effective in curing the disease, and that takes only one day to administer, is both discovered and administered on January 1. However, this treatment has very severe side effects that themselves have a very good chance of killing the

311

patient within the year. On January 1, however, it seems that this treatment will give the patient the best chance possible, and indeed the treatment in fact does increase the probability of surviving the year, but not by very much – say from 2 to 4 percent.

The patient does fairly well in the following months. The probability of her recovering from the original disease is gradually increasing (from the 90 percent value it acquired on January 1), and the probability of her recovering from the bad side effects is also gradually increasing (from the low value, just above 4 percent, that it achieved on January 1). After six months, the probability of this patient's surviving to the end of the year has increased from 4 percent to 50 percent, where, of course, both these probabilities take account of both the (relatively small) threat from the original disease and the (relatively large) threat from the side effects of the treatment.

On July 1, 1990, another very improbable medical event occurs. A new treatment for the patient's original disease is discovered. This treatment is better than the one that was administered on January 1: it is nearly 100 percent effective in curing the disease, it has no bad side effects, and, like the first treatment, it also takes only one day to administer. Since the patient still has *some* chance of dying from the original disease (indeed still within the year), the new treatment is administered on July 1. Of course, this raises the probability of the patient's surviving the year, since this treatment completely eliminates the original disease without itself causing any bad side effects. But the new treatment raises the probability of surviving the year only by a little, since the patient still has to contend with the more serious problem of the side effects of the first treatment. Let us suppose that on July 1 the new treatment raised the probability of surviving the year from 50 percent to 52 percent. During the next six months, the patient continues gradually to recover, and she makes it to the end of the year.

Figure 6.11 depicts our patient's progress, where x is the

312

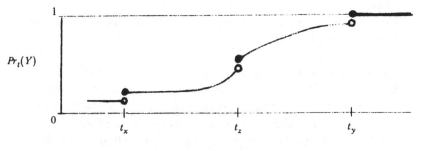

Figure 6.11

administration of the first treatment on January 1, z is the administration of the new treatment on July 1, and y is the event of the patient's being alive at the end of the year (under that description, Y).

According to the theory of the previous section, this is a case of x's being a token cause of y's being Y: It is a case of the "because" variety, though not a very extreme case. The probability of survival for a year (that is, of y's being Y) increased at the time of the first treatment (t_x), and never fell. However, intuitively, it would seem that the patient survived *despite* the first treatment, given that the better treatment in fact was to become available, and given the severe side effects of the first, crude, and soon to become obsolete, treatment. Had the patient not been administered the first treatment, the following would have happened (almost certainly): the patient would have survived until July 1, she would have been given the better cure, she would have recovered from the disease, and she would not have suffered from the deadly side effects of the first treatment. Thus, intuitively, it seems that we should say that the patient survived *despite* the first, crude, and soon obsolete, treatment.

This is, I think, a genuine counterexample to the characterization of "because" given in the previous section, since the rough theory classifies it as a case of because, whereas in fact it is a case of despite. Also, of course, it is a counterexample

313

to the theory's characterization of "despite," since the trajectory for the example is not of the kind of shape the theory associates with "despite."

Once it is understood how these counterexamples to the theory's characterizations of "despite" and "because" work, it is not difficult to think up counterexamples to the theory's characterization of token causal *independence* as well. Slight modifications of the above counterexamples will do.

Modify the two-shock version of the defoliant example as follows: the first event, x, is not really a shock at all, but rather just a mild inoculation of the plant against a not so severe application of the defoliant, where the expected level of benefit from the inoculation (deriving from the low probability of encountering the defoliant within a year) exactly offsets the expected level of detriment from side effects. In this case, the probability of survival in a year does not change across the time of x, so that the theory tells us that the plant survived token causally *independently* of the inoculation. But, since the plant in fact *is* (improbably enough) exposed to the poison, the plant in fact survives *because* of the inoculation. Of course, in addition to being a counterexample to the theory's definition of "independence," this "because" example also is a counterexample to the theory's definition of "because," since the probability trajectory is not of the kind of shape that the theory associates with "because."

For another counterexample to the theory's characterization of token causal independence, modify the example about the severely ill patient as follows: Suppose the first treatment simply evenly trades the threat of death from the disease for an equal threat of death from the side effects of the treatment. Then the probability of surviving the year does not change around the time of x, so that theory says the patient's survival was token causally *independent* of the treatment. But as in the example above, the causal truth is still that the patient survived *despite* the first treatment, since, given that the better treatment in fact was soon in the offing, the first treatment in fact unnecessarily put the patient at risk from the severe side effects (which the second treatment does not cure). Of course, in addition to being a counterexample to the theory's definition of "independence," this "despite" example also is a counterexample to the theory's definition of "despite," since the probability trajectory is not of the kind of shape that the theory associates with despite.

We now have several counterexamples each to the characterizations, given by the rough theory of the previous section, of the token causal ideas of "because," "despite," and "independence." (I invite the reader to concoct counterexamples to the way "token causal autonomy" was characterized, for example by finding a case in which the truth is "because" or "despite" but the theory says "autonomy.") Each of these counterexamples involves a *temporally intermediate* event. The defoliant counterexamples involve the temporally intermediate event of the second shock. And the examples about the severely ill patient involve the temporally intermediate event of the administration of the second, better treatment. *There are two important things to notice about these temporally intermediate events.*

First, in each case, the temporally intermediate event was *itself causally relevant to the effect in question* (namely, y's being Y) – indeed, both at the token level (both intuitively and according to the definitions of the previous section) and at the type level (for the type level, consider the temporally intermediate event, z, to be of the type Z given by the description of z in the examples). In the defoliant counterexamples, the plant survived *token-despite* the second shock (both intuitively and on the rough theory of the previous section); and, in the relevant (actual) context, the factor of being administered that second shock is a *negative causal factor* for survival (even given the resistance the plant has acquired). In the examples about the severely ill patient, the patient survived *token-because* of being administered the better treatment (both intuitively and on the rough theory of the previous section); and,

315

in the relevant (actual) context, being administered the better treatment is a *positive causal factor* for survival.[11]

The second important thing to notice about the temporally intermediate events is that it is crucial to the counterexamples that the earlier event, x, in question *not be a token-level cause of the temporally intermediate event, z: z must not occur because of x*, for the counterexamples to be genuine. That is, x must not be a token cause of z's being Z, where Z is the type given by the description of z in the examples.[12] Consider the first two-shock version of the defoliant example (Figure 6.10, the counterexample directed against the rough theory's definition of "despite," but also a counterexample to the definition of "because"). Here, x is the first shock, z's being Z is the plant's suffering the second shock, and y's being Y is this plant's survival. If x token caused z's being Z (if z's being Z occurred because of x), then, both intuitively and according to the basic theory, the plant survived the year (y was Y) *despite* x. Supposing that x caused z, then x is responsible not only for the weakness of the plant that immediately follows *it*, and for the plant's resistance later on, but also (to some degree) for the weakness of the plant directly following the *second* shock. In this case, x is responsible (to some degrees) for *both* injuries, so that intuitively the plant survived *despite x*, and the initial drop in the probability of Y correctly marks a case of "despite," on the rough theory of the previous section.

Note that when x causes z then the first drop in the probability of Y (at t_x) should be greater, and the second drop (at t_z) should be less, than in the case in which z is not caused by x. This is because, for x to token cause z's being Z, x must raise

[11] This is a feature of all the counterexamples we have looked at so far. In the next section, I will discuss several counterexamples that do not have this feature, and these will require a somewhat different kind of qualification of the basic theory from what is required in light of the kind of counterexamples considered so far. Both qualifications will be natural in light of what we learned about background contexts in the type level theory, about holding fixed both independent causes and independent interactive factors.

[12] This point will have to be qualified below; the qualification has to do with degrees of token causal significance.

the probability of Z (in the sense explained in the previous section, of course), so that some of z's negative impact on y's being Y is "absorbed" by x, "from" z. In this case in which x is a token cause of z, how steep the *second* drop, at t_z, of the probability of Y will be, and whether or not there even is a second drop in the probability of Y, will depend on how strongly and how quickly the event x plays out its causal role for z's being Z. But whether or not there is such a second drop, there will remain the first drop, so that the rough theory gives the correct answer of "despite." Only the *magnitude* of the first drop will depend on the strength of x's impact on z's being Z.

I should add here that the question of the *degree* of x's token causal significance for z's being Z has an important bearing on the degree to which y occurs despite or because x. This issue will be explained further just below, and then fully addressed in Section 6.5, after the basic approach to the counterexamples has been explained and the question of degrees of token causal significance has been explored.

Now consider again the first example about the severely ill patient (Figure 6.11, the counterexample directed at the rough theory's definition of "because," but also a counterexample to the theory's definition of "despite"). If the first treatment somehow caused the second treatment (perhaps by inspiring the discovery of the second as a result in part of the bad reactions to the first), then, both intuitively and according to the rough theory of the previous section, the first treatment *is* a positive token cause of the patient's survival. In this case, the first treatment had, all things considered, an enhanced positive causal impact on survival. The first treatment is responsible not only for the benefits directly attributable to it, and for the negative side effects, but also (to some degree) for the benefits derived from the second treatment.

In this case in which the first treatment somehow caused the second, the first increase in the probability of survival (at the time of the first treatment) will be greater, and the second increase (at the time of the second treatment) will be less, than in the case in which the second treatment is not caused by the first. This is because, for the first treatment to token cause the second, the first must increase the probability of the second (in the sense explained in the previous section, of course), so that some of the second treatment's positive causal impact on surviving the year is "absorbed" by the first treatment, "from" the second treatment. The magnitude of the second jump in the probability of survival, at the time of the second treatment, and whether or not there is a second jump, will depend on how strong a token causal impact the first treatment has on the second, and how quickly the first treatment plays out its causal role for the second. The first jump will nevertheless remain, though its magnitude will depend on the degree of the first treatment's impact on the second. The question of degree, to be explored in Section 6.5, has an important bearing here, as noted above for the example considered previously.

The situation is similar for the modifications of these examples that were counterexamples to the rough theory's characterization of token causal independence. A difference, however, is this. In the previous two cases, when we change the example so that the earlier event x is a cause of the intermediate event z, the *magnitude* of the decrease or the increase in the probability of Y, at the time of x, is increased. In the two counterexamples to the rough theory's characterization of token causal independence, however, there is no change in the probability of Y across the time of x. I leave it to the reader to see that if the only change in the two counterexamples is that we now let x *cause* z, then there will be a decrease or an increase in the probability of Y across the time of x, in each case in the direction that makes the rough theory give the intuitively right answer ("despite" in the defoliant variation, and "because" in the severely ill patient variation).

I should finally describe the importance of the idea of *degrees* of token causal significance in the evaluation of these

318

examples, though this matter cannot be finally resolved at this point. I have suggested that for the counterexamples to be genuine, the earlier events x cannot be token causes of the intermediate events z being Z. The suggestion is that it cannot be that z is Z *because of x*. Let us say that z's being Z is *token uncaused by* event x if the only kinds of token causal significance x has for z's being Z are of the "despite," "independence," or "autonomy" varieties: There is no "because" component to the causal significance of x for z's being Z.[13] Then for the counterexamples to the rough theory of the previous section to be genuine, it is not necessary for the temporally intermediate event z's being Z to be *strictly* token uncaused by the earlier event x. The counterexamples may still be genuine even if the earlier event x is, *to a sufficiently small degree*, positively token causally relevant to the intermediate event z's being Z.

For example, in the first two-shock version of the defoliant example, if the first shock were, to an extremely minute degree, positively token causally relevant to (a very weak token cause of) the second shock – and the second shock occurred *largely*, but not *strictly, independently* of the first shock – then we should still say that the plant survived because of the first shock, even though the first shock lowered the probability of survival and the rough theory says "despite."[14]

However, we have not yet discussed degrees of token causal significance in detail. And a detailed understanding of degrees of token causal significance will be required to handle

[13] In Section 6.5, I define natural quantitative measures of the four components of the causal significance of one event for another (the "because," "despite," "independence," and "autonomy" components). It will turn out that $B + D + A = 0$ and $I = 1 - |D|$ (where "B" stands for the "*because*" component, "D" for the "*despite*" component, and so on). Then z's being Z is *token uncaused by* x if (to put it roughly, suppressing some notation) $B \leq 0$. As we will see, the way in which this can happen is this: At some time after x and before z, the probability of z's being Z is not higher than it was before the time of x.

[14] Also, of course, the counterexamples would still be genuine if the intermediate event occurs despite, or autonomously of the earlier one, instead of independently of it. This is not much to the point, though, since in such cases the intermediate event cannot at all occur because of the earlier one.

319

the kind of case just described. Thus, for now, let us understand the counterexamples in such a way that the intermediate event occurs *strictly* token uncaused by the earlier event. Other versions of the counterexamples, in which there is sufficiently small positive token causal relevance of the earlier event to the intermediate event for the counterexamples to remain genuine, will be handled in Section 6.5.

So each of the counterexamples involves an event z that is *both* token uncaused by the earlier event x *and* causally relevant to the later event y's being of type Y. In the examples, z is a temporally intermediate event, of a type Z, such that (1) z's being Z is token uncaused by x, (2) z is token causally relevant to y's being Y, and (3) Z is, on the type level, causally relevant to Y. (As will be made clear below, it is really just (1) and (3) that are crucial here.) Of course, the idea of independent causes is familiar from the discussion of spurious correlation in Chapter 2. There, we found it necessary to include in the type-level theory of probabilistic causation the requirement that we must hold fixed factors that are *causally independent of the causal factor in question* and *causally relevant to the candidate effect factor in question*. We must do the analogous thing in the theory of token-level causation, except that in the theory of token-level causation, there is only one context, and the "confounding" event must be held fixed in just one way, the way it actually happens.

Consider again the first two-shock version of the defoliant example. Various events occur between the time t_y of the first shock and the time t_x of the event of the plant's being alive and healthy at the end of the year. Some of these temporally intermediate events owe their occurrence, to some extent, to the occurrence of x, the first shock. For example, the event of the plant's being weakened just after t_x and the event of the plant's developing a resistance to the defoliant are such events. Other events that occur between t_x and t_y occur token uncaused by x; they do not owe their occurrence to the occurrence of x. Some of these, such as the second shock in our example, are themselves causally relevant to y. In parallel to a

320

requirement in the theory of type-level causation, when we assess the probabilistic or causal impact that x *alone actually* has on y's being Y, these causes of y's being Y that are token uncaused by x must, somehow, be held fixed. Doing this affects the relevant probability trajectory.

When we assess the probability of Y and trace its evolution between the time of x and the time of y, we do so relative to the way things are in the particular case in question. $Pr_t(Y)$, the probability of Y at time t, is the probability of y's being Y, given the way things are in the actual context, at time t. This is the same as $Pr_t(Y/K_a)$, where K_a is a causal background context, as in the probabilistic theory of a type-level causation. Then $Pr_t(Y/K_a)$ varies as t varies, and as the actual consequences of x itself, which are causally relevant to y's being Y, unfold. (Think of $Pr_t(Y/K_a)$ as the same as $Pr(Y/K_a \& W_t)$, where W_t is the conjunction of all factors that have fallen into place by time t and that are consequences of x that are causally relevant to y's being Y; this will be further clarified below.) In parallel to the type level theory, K_a must hold fixed, positively or negatively, all factors whose exemplifications are token uncaused by x and causally relevant to y's being Y. But there are three important differences between the K_a's used in the assessment of token-level causal significance and the K_i's used in the assessment of property-level causal significance.

First, of course, if a factor must be held fixed in a K_a, then it must be held fixed positively or negatively *according to whether it was actually exemplified or not*. Suppose a factor Z is causally relevant to Y, and that whether or not it is exemplified is token uncaused by x. Then if Z was actually exemplified in the particular token case in question, then Z should be held fixed only positively; and if Z was not actually exemplified, then it should be held fixed only negatively. In the theory of type-level probabilistic causation, on the other hand, we hold the relevant factors fixed in all possible combinations, *uncontrolled by what actually happens*. The second difference, then, is that for token causation, we have only *one* causal background context.

321

The third difference concerns how we should understand "causally independent" and "causally relevant," if we formulate the theories' requirements in terms of causally independent (of x or X) factors that are causally relevant to Y. This is a little tricky. First, of course, this terminology is not quite right, when used to characterize what to hold fixed in the theory of token causation. For the type-level theory, one factor's being causally independent of a second factor is the same as the second's being causally neutral for the first, and causal relevance means "nonindependence" (that is, positive, negative, or mixed causal factorhood). On this understanding of the terms, it is correct to say, for the type level theory, that we hold fixed factors causally *independent of X* and causally *relevant to Y*. However, in the token-level theory, the terms have quite different meanings, and putting the requirement as formulated above is incorrect. The correct way to put the requirement for the token-level theory is this: Hold fixed all factors that are such that, (1) they are actually *exemplified*, (2) their exemplifications are *token uncaused* by x and (3) they are *type-level causally relevant* to y's being Y.

To assess the causal impact of x on y's being Y, we must look at the probabilistic impact x had on y's being Y, *given the way things actually are in respects x actually had nothing to do with*. When we hold fixed factors whose exemplifications x actually (token causally) had nothing to do with – that is, factors whose exemplifications are *token uncaused* by x – we isolate the impact that x itself actually had on the probability of y's being Y – beyond the impact that other events, uncaused by x, had on y's being Y. That is why the causal background context within which an event x operates, and within which we should assess its token causal significance for a later event y's being Y, should be understood as including the fact of the exemplification of all factors that are relevant to y's being Y and that are actually uncaused by x. We do not hold fixed factors that are exemplified, after x, *because of x*, for these may be causally intermediate; we only hold fixed factors that, regardless of when they occur, are exemplified either *independently of x, autonomously of x*, or *despite x*.

Must these factors also be, at the *type* level, independent of the factor X (the relevant type of x, whose role is explained below)? No. The idea is that we hold fixed all those factors whose exemplifications x *actually* is not causally responsible for, in the particular case in question. As will be clear when this approach is applied to the counterexamples described above, if we require type-level causal independence for the factors we hold fixed, then the procedure of holding fixed other causes will not in all cases allow the token causal significance of the earlier event for the later one to show up in the relevant probability changes.

In what way must a factor be causally relevant to y's being Y for it to be appropriate to require that it be held fixed (assuming, of course, that it is exemplified and its exemplification is token uncaused by x)? As long as we are only holding fixed factors whose exemplifications are token uncaused by x, it cannot hurt to hold fixed factors that, strictly speaking, need not be held fixed. In fact, it would not be inappropriate to hold fixed *all* exemplified factors whose exemplifications are token uncaused by x. However, for the kinds of counterexamples described above, I suggest that it will suffice just to require holding fixed factors that are causally relevant to Y, at the *type* level – where, of course, this is qualified by the first part of the requirement, which says we only hold fixed factors whose exemplifications are token uncaused by the earlier event x.

Of course, type-level causal relevance is relative to a token population and a kind that the token population exemplifies. So, if we require holding fixed factors Z that are type-level causally relevant to Y, then we must specify a population and a kind.[15] Here is what seems to be the most natural sugges-

---

[15] This would not be necessary if we took the easier approach of holding fixed *all* factors whose exemplifications are token uncaused by x. The easier approach is fine, but the kind of approach being developed here points to just what is going on in the counterexamples already described and in others to be described later.

323

tion for this. Let the population P be the singleton population of the actual case in question. The kind Q that P exemplifies can be determined by the conjunction of all factors that were actually exemplified at some time before the time t_z of the exemplification of Z.[16] Then we evaluate the type level causal significance of Z for Y relative to population P, considered as of kind Q. If Z is, relative to this population and kind, type-level causally relevant to Y (and if Z is actually exemplified and its exemplification is token uncaused by x), then hold Z fixed (positively, of course) in the causal background context K_a. Note that the kind Q will hold fixed the relevant type, X, of the earlier event x: X will be a conjunct of the conjunction that determines Q.

One may wonder why the requirement about what to hold fixed focuses on the *type*-level causal significance, for Y, of factors Z whose exemplifications are token uncaused by x. In saying how x may be causally significant for the exemplification of factors Z that we may hold fixed, we focused on the *token* level; the exemplification of Z must be token uncaused by x, if it is to be allowable to hold Z fixed. But, as the requirement was formulated, it is at the type level that Z must be causally relevant to Y, if it is to be required that we hold Z fixed – why at the *type* level rather than at the *token* level? That is, why not instead require that Z be held fixed if its exemplification is causally relevant to Y at the token level (and, of course, if the exemplification of Z is token uncaused by x)?

While I am unsure about the merits of this alternative proposal, the reason I have focused on the type level here is to connect more directly with the fact that, in the probability trajectories we examine, it is the probability of Y, conditional

[16] Another natural suggestion would have the kind Q be determined by the conjunction that includes as a conjunct any factor W such that (1) W is exemplified before t_z in the token instance in question and (2) there is some population relative to which W is either an independent (of Z) cause of Y or interactive for Y (relative to Z). This suggestion would seem to include in the characterization of Q just those factors that matter.

on all factors that are held fixed or that have fallen into place at successive times, that is traced. And it is the type-level causal significance of the constellations of conditioning factors that determines the successive conditional probabilities of Y. The idea is to hold fixed factors whose exemplifications are token uncaused by x and *that affect the value of $Pr_t(Y)$* – and, among factors whose exemplifications are token uncaused by x, it is just those that are type-level causally relevant to Y that affect the position of the trajectory of $Pr_t(Y)$.[17]

One might worry that there can be factors Z that are causally neutral for Y at the type level and yet whose exemplifications are, for example, positive token causes of y's being Y. In this case, our requirement does not force us to hold Z fixed, even if the exemplification of Z is token uncaused by x, and even though the Z's exemplification resulted in an increase in the probability y's being Y, *this increase being token uncaused by x*. The worry is that failing to hold Z fixed (or failing to hold fixed Z and its exemplification's token causal impact on Y) would distort the trajectory of $Pr_t(Y)$, making it too low. Of course, there can be such factors. However, the increase in the probability of y's being Y must be because of the exemplification, *after the time of z*, of factors that are causally positive for Y at the type level. If the probability of Y is higher after the time of z than it is before the time of z, this means that the probability of Y conditional on factors that have fallen into place by the later time is greater than the probability of Y conditional on all factors that have fallen into place by the earlier time. For example, it could be that a factor W fell into place after the time of z, where it is being conditional on W that makes the probability of Y greater after z than it was before z. Then, in the relevant context, W is a positive causal factor for Y, and *it* must be held fixed.

---

[17] At least this is true for the examples we have considered so far; later we will see that interactive factors, understood in a certain way, whose exemplifications are token uncaused by x, can also affect the position of the trajectory, so that they also must be held fixed.

Note that this kind of possibility actually points to an advantage of focusing on the type-level causal relevance, to Y, of factors like Z and W, rather than on their exemplifications' token-level impact on Y. Holding fixed z's being Z will not itself bring the probability trajectory up to where it should be, for by hypothesis, Z does not increase the probability of Y in the way in which it must in order to force the whole trajectory up to where it should be: Z is causally neutral for Y at the type level. Also, the exemplification of W may not be a token cause of Y.

On the other hand, it cannot hurt to hold fixed, in addition, factors whose exemplifications are token uncaused by x and that are *type-level neutral* for y's being Y yet *token causally relevant* to y's being Y. Because of their type causal neutrality for y's being Y, holding them fixed will not affect any of the probabilities. Again, however, the exemplification of such factors Z will be token causes of the exemplification of factors W that are themselves type-level causally relevant for the type of the later event y. And these factors W *will* be held fixed.

Another possible worry is that it can happen that a factor Z is, for example, causally positive for Y at the type level, yet at the token level Y occurs token causally independently of the exemplification of Z, all while the exemplification of Z is token uncaused by x. In this case, the probability of Y is the same after z as it is before, but at the time of z, the probability of Y is higher. Does holding Z fixed distort the trajectory of $Pr_t(Y)$, making it too high? This worry can be set aside in the same way as the first. Again, there will be factors W in place just after z that, relative to a context in which z's being Z is held fixed, are negative causal factors for Y at the type level; these are the factors that bring the probability trajectory back down, and they will be held fixed.

Let us now finally see how our new understanding of the causal background context K_a handles the counterexamples that we have seen so far. Consider first the first two-shock version of the defoliant case, in which the rough theory of the previous section gave the answer that the plant survived *despite* the first shock, but in which the truth is *because*. In this example, we must hold fixed (positively) the factor of the plant's being shocked at the later time, the factor of the second shock. This is because the second shock was *token uncaused* by the first shock, and because the factor of being shocked (more severely) at the later time is *causally relevant at the type level* to survival.

When we take into account, by holding fixed, the fact of the second, more severe shock, it is clear that, before the first shock (before t_x), the probability of surviving a year is already quite low. Also, the first, less severe shock should *increase* the probability of the plant's surviving the year, since it gives the plant a chance to develop resistance to the defoliant, so that it might survive the second, more severe shock. Of course, the first shock will not raise the probability of survival by very much; after all, the can of defoliant said that it is 90 percent effective in killing plants of this kind. Furthermore, there will be no change of the probability of survival across the time t_z of the second shock, since the factor of the second shock, Z, is held fixed all along.

Holding fixed the fact of the second shock, the probability trajectory for survival in this example is depicted in Figure 6.12. The differences between the trajectory shown in Figure 6.12 and the trajectory shown in Figure 6.10 result simply from the fact that the second shock is held fixed for Figure 6.12 but not for Figure 6.10. The revised theory of probabilistic token causation (the rough theory of the previous section together with the qualification about holding fixed factors that are type-level causally relevant to Y but whose exemplifications are token uncaused by x) now gives us the correct answer that the plant survived *because of* the first shock (the first shock *token caused* the survival of the plant). At t_x, the probability of Y goes up, and, in the example, it does not fall.

Incidentally, a slight variation of this example shows that

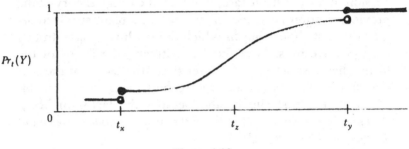

$Pr_t(Y)$

t_x t_z t_y

Figure 6.12

we should not require that the factors we hold fixed be *type-level* causally independent of the relevant type of the earlier event *x*. We should only require that *their exemplifications* be *token* uncaused by the earlier event. Suppose that, for some reason, the factor of the plant's being sprayed by the defoliant a first time is a negative causal factor for its being sprayed again at a later time; this may be because the gardener only sprays plants that need to be gotten rid of, and usually one application of the defoliant does it. The token causal facts in this version are the same in the original two-shock example: The plant survived *because of*, and *not despite*, the first shock. But in this case, the factor of being sprayed the second time is not, at the type level, causally independent of the factor of being sprayed earlier; yet the exemplification of the factor of the second spraying is still token uncaused by the first spraying (indeed, most likely the second spraying is *despite* the first). So, if we required that the factors we hold fixed be, at the type level, causally independent of the relevant type of the earlier event, then we would not be able to hold fixed the factor of the second shock, and we would get the wrong answer that the plant survived despite the first shock.

Let us turn now turn to the example about the severely ill patient, in which the rough theory of the previous section says that the patient survived *because of* the first medical treatment, but in which the truth is *despite*. In this example, we

328

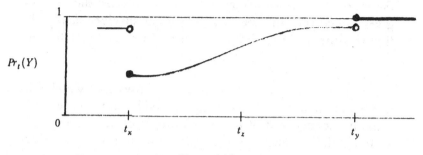

Figure 6.13

must hold fixed the factor of the second, better treatment. This is because the event of the second treatment is *token uncaused* by the first treatment, and because the factor of being administered the second treatment is a *type-level positive causal factor* for survival. When we hold this factor fixed (positively), the probability of survival is pretty good even before the first treatment. This is because the second treatment is (nearly) 100 percent effective in curing the patient's disease, so that if the patient fails to survive the year, her death would have to result from some other cause (which, in the example, is improbable). At the time the first treatment is administered, however, the probability of survival *decreases*. This is because the first treatment has little or no effect on the chances of the patient's recovering from the disease (which are already 100 percent, or nearly so, taking the fact of the second treatment into account), and because it causes the severe side affects associated with the treatment. Also, the probability of survival should not change at time t_z, the time of the second treatment, since the factor of the second treatment, Z, is held fixed, positively, all along.

Holding fixed the fact of the second treatment, the probability trajectory for survival in this example is depicted in Figure 6.13. The differences between the trajectory shown in Figure 6.13 and the trajectory shown in Figure 6.11 result simply from the fact that the independent cause is held fixed for

329

Figure 6.13 but not for Figure 6.11. The revised theory now gives the correct answer that the patient survived *despite* the first treatment. At t_x, the probability of y's being Y falls; and that is what it means, according to the theory, for y to be Y despite x.

Again, a variation of this example shows that we should not require that the factors we hold fixed be, at the type level, causally independent of the relevant type of the earlier event x. If we did require this, and if, for some reason, the first treatment were, say, a negative causal factor for the second treatment, then the token causal facts may remain the same, but we would not be able to hold fixed the factor of the second treatment, and we would get the wrong answer that the patient survived because of the first treatment.

The other two counterexamples discussed in this section, which were modifications of the first two, and counterexamples to the rough theory's characterization of token causal independence, are also taken care of by the requirement that we must hold fixed factors that are causally relevant to Y at the type level and whose exemplifications are token uncaused by x. In each case, holding fixed the appropriate intermediate factor results in a probability change at t_x, and in the right direction, as the reader can easily verify.

Thus, all of the counterexamples to the basic probability-increase idea for token causation discussed so far can be dealt with by including, along with the basic probability-increase idea, a qualification analogous to the important qualification made in Chapter 2 for the basic probability-increase idea for type-level causation: the qualification, roughly, that independent causes must be held fixed. This is an important qualification, both of the basic probability-increase idea for type-level causation and of the (different) basic probability-increase idea for token-level causation. Of course, the formulation of the qualification is somewhat different for the two levels. For the type level, we hold fixed factors that are themselves type-level causally independent of the causal factor in question,

330

while for the token level, we hold fixed factors whose exemplifications are token uncaused by the causal token event in question, where for each level, the factors that must be held fixed are type-level causally relevant to the effect type (or the relevant type of the effect) in question.

As noted at the beginning of this section, one problem with the theory as developed prior to this section is that the only aspect of the token cause that entered into the analysis was its *time*. By contrast, the *probability* of the effect has all along been crucial. The only notation used in connection with the earlier event has been "x" and "t_x," and only "t_x" appears explicitly in our probability trajectory diagrams. This means, for one thing, that (at least prior to this section) the analysis gives the same answer about the causal role of x for y as it gives about the causal role of any other token event z for y, as long as z occurs at the same time as x. And this is clearly wrong. Also, since no "relevant" type for x has explicitly entered into the analysis (it does not show up in the probability trajectory diagrams, for example), it seems that the theory cannot distinguish the causal role of x's being X for y's being Y from the causal role of x's being X' for y's being Y, for types X and X' both of which x exemplifies. And this also is clearly a defect of the theory, but only as developed prior to this section. I will show now how the qualification introduced in this section solves these difficulties.

For an example that illustrates the first problem, about events simultaneous with x, consider again the case of the squirrel's kicking the golf ball. The probability trajectory for this example is given in Figure 6.1, and the analysis says the squirrel's kick is a token cause of the birdie. The answer given is based on the shape of the probability trajectory, and this trajectory just traces how the probability of y's being Y (the ball's falling in the cup) evolves from t_x to t_y. The broken curve in Figure 6.1 depicts the evolution of $Pr_t(Y)$ – and this expression, "$Pr_t(Y)$," does not explicitly involve anything about x. Now let z be any event that is simultaneous with x

and that plays no causal role for y's being Y. For example, z may be the event of some *other* squirrel kicking a *tree* on some *other* golf course. Since, as far as x goes, it is only its *time* that enters into the analysis, and since the time of z is the same as the time of x ($t_z = t_x$), the probability trajectory used to assess z's token causal role for y would seem to have to be the same as the one used to assess x's token causal role for y. So it seems that the theory (at least prior to this section) would have to give the same answer about z's causal role as it gives about x's causal role: that z also is a token cause of y's being Y, which is false by hypothesis.[18]

However, the new requirement introduced in this section, that we must hold fixed type-level causal factors for y's being Y whose exemplifications are token uncaused by x, involves the event x in a fuller way, in a way that allows the theory to distinguish x's causal role from the causal roles of other events z that are simultaneous with x. For the example just described, the new requirement tells us that in assessing z's token causal role for y's being Y, we must hold fixed all factors that are type-level causal factors for y's being Y and whose exemplifications are token uncaused by z. In this example, this includes all factors exemplified in the causal chain from x to y, including the factor of the ball's being kicked by the first squirrel, the ball's emerging from the collision on a path into the cup, the ball's nearing the cup, and so on. This is because these events occur, by hypothesis, token uncaused by z, and because the factors exemplified there are type-level

[18] I thank Igal Kvart for making this point, with an example much like this one. His suggestion was to take care of examples such as this one by using counterfactual conditionals: At least a necessary condition for an event x to be the cause is that the probability trajectory would have been different had token event x not occurred (or not exemplified the relevant type X). The problem will be dealt with here in a different way that does not involve any further revisions of the theory. I note also that if we insist that appropriate factors X for the event x be spatially local to x (see Section 6.6), then we can solve the problem for the particular example in question. However, the problem is more general: As well as factors exemplified spatially removed from x, there will in general be multiple factors all exemplified (all locally) by x and that have different token causal roles for a later event y's exemplifying a factor Y, as explained more fully below.

causally relevant to y's being Y (in the relevant contexts, specified in the requirement). Clearly, when all this is held fixed, there will be no change in $Pr_t(Y)$ across the time of z (in fact, the probability will be high all along), so that the theory now gives the right answer that the ball fell into the cup token causally independently of z, the other squirrel's kicking a tree on the other golf course.

The second difficulty mentioned above is similar to the first one but involves not the possibility of identifying the *wrong event* as the cause but rather the possibility of identifying the *wrong type* of the right event. Recall from the beginning of this chapter that the causal relation I wish to analyze can be expressed as follows: things' being X at $<t_x,s_x>$ caused things to be Y at $<t_y,s_y>$ (where x is what takes place at time and place $<t_x,s_x>$ and y is what takes place at time and place $<t_y,s_y>$). It could be that things are many ways at $<t_x,s_x>$ – X, X', X''', and so on – where it is things' being X – and not their being X' or X'', and so on – that is the cause of things' being Y at $<t_y,s_y>$. To put it somewhat differently (suppressing the times and places), it could be that x is X, X', X'', and so on, and that x's being X caused y to be Y, and that x's being X', and x's being X'', and so on, are token causally irrelevant to y's being Y; or it could be, for example, that y is Y *because of* x's being X, *despite* x's being X', *independently of* x's being X'', and *autonomously of* x's being X'''. So the question that arises is this: How does the theory distinguish the token causal role of x's being X for y's being Y from the token causal roles of x's being X', X'', and so on, for y's being Y?

Here is an example that illustrates this kind of possibility. Again, let us consider the original squirrel case (Figure 6.1), and suppose that in addition to the squirrel's kicking the ball in the way it did, X, at $<t_x,s_x>$, another thing, X', that was true at $<t_x,s_x>$ is that the squirrel blinked. Also, say, the ball was white then and there (X''), and so on. It is intended, as part of the example, that y's being Y is token causally independent of x's being X' and of x's being X''. Then *it is x's being X* (the squirrel's kicking the ball in the way it did) – and *not x's being X'* (the squirrel's blinking then and there, another aspect of the same event), and *also not x's being X''* (the ball's being white then and there, another aspect of the same event) – that is the token cause of y's being Y. The question, then, is this: Since the probability trajectory ($Pr_t(Y)$, as t varies) simply traces the evolution of the probability *of Y*, how does the theory distinguish between the causal roles, for y's being Y, of *x's being X, x's being X'*, and *x's being X''*? The problem is to show that the probability trajectories should differ, on the theory, according to whether we think of x as an X, an X', or an X''.

Again, the new requirement of this section enables the theory to separate the different token causal roles, for y's being Y, of x's being of the different types it exemplifies. Now that we are explicitly considering types for x, and since the theory is about the causal role of one event's exemplifying one type for another event's exemplifying another type, we should state the new requirement, more fully, as follows: When assessing the token causal role of x's being X for y's being Y (in terms of probability trajectories) we must hold fixed factors Z such that Z is exemplified token un-caused by x's *being X* and Z is, at the property level, causally relevant to y's being Y. Until now, the difficulties we have encountered for the basic probability–increase idea of token probabilistic causation, and the qualification introduced to handle some of them, have been independent of the role of the relevant type of the causal event x. So it was convenient to ignore this factor. But now we must explicitly include this type in the formulation of the requirement.[19] Of course, as already seen in Sections 6.1 and 6.2, specifying a type Y

[19] Note, then, that the causal role of x's being X for y's being Y is explicated in terms of the token causal role of x's being X for *other* events, z, being of *other* types, Z (and in terms of the property-level causal role z's being Z for y's being Y). So there is a circularity in the token-level theory that is somewhat parallel to the circularity found in the property-level theory; this will be discussed further below.

for the *later* event is important, and the earlier event can have different token causal significances for the later event's exemplifying the various types it exemplifies – and we have seen how the theory accommodates this fact. It is time now to appreciate, and to see how the theory accommodates, the fact that it is *also* true that the *first* event's exemplifications of the various types it exemplifies can have different token causal significances for the later event's exemplifying a given type that it exemplifies.

Thus, in the example just described, in assessing the token causal role of x's being X' (or its being X'') for y's being Y, we must hold fixed factors that are causally relevant to Y at the type level and whose exemplifications are token causally independent of x's *being* X' (or its *being* X''). For example, we must hold fixed the factor of the ball's being on a trajectory straight into the cup after the time of x, since the exemplification of this factor is token uncaused by x's being X' (and of x's being X''), and since this factor is type-level causally relevant to y's being Y. When we do this, when considering x as of type X' (or the type X''), then obviously there will be no change in $Pr_t(Y)$ across the time of x.

Thus, the requirement of holding fixed factors that are causally relevant to the later event y's being Y, and that are exemplified token uncaused by the earlier event x's being X, has the result not only that the theory can distinguish the causal roles of different simultaneous events, but also that the theory can distinguish the causal roles of the earlier event's exemplifying the different types it exemplifies. In the examples I have used to illustrate this, the later event y's being Y was *token because* of the earlier event x's being X, and *token independently* of other earlier events' exemplifying the other types considered, or of the same earlier event's being of the other types, different from X, considered. It is easy to see that the same issues can arise in terms of other combinations of token causal significances.

In general, when y is Y, it can be that there are four different events, x, z, w, and v, of types X, Z, W, and V, respectively, such that y is Y token because of x's being X, token despite z's being Z, token independently of w's being W, and token autonomously of v's being V. And, of course, this can happen even if x, z, w, and v are simultaneous. Also, when y is Y, it can be that an earlier event x is of four types, X, X', X'', and X''', such that y is Y token because of x's being X, token despite x's being X', token independently of x's being X'', and token autonomously of x's being X'''. Subject to one more qualification, in each of these cases (and I leave it to the reader to ponder this), the requirement of holding fixed type-level causes of the later event, whose exemplifications are token uncaused by the earlier event, has the effect of enabling the theory to correctly identify the token causal roles of the various events, considered as of the various types they exemplify, for the later event.

The qualification just mentioned has to do with another kind of factor that must be held fixed. This is the topic of the next section. Also, of course, we must still deal with versions of the examples discussed here that involve the idea (to put it roughly) of the temporally intermediate event being *only to a sufficiently large degree* token uncaused by the earlier event; this will be done in Section 6.5.

6.4 THE CAUSAL BACKGROUND CONTEXT AND INTERACTION

In Chapter 3, we found that for the type-level theory it does not suffice to hold fixed just the independent causes of the effect factor in question. We found that we must, in a separate qualification, explicitly require that independent interactive factors also be held fixed. We saw examples in which, in order for the type-level theory to deliver the correct assessment of property-level causal roles, we must hold fixed independent factors F that the causal factor X interacts with, with respect to Y, where F may not itself be an independent cause

336

of Y. And it was because such interactive factors F need not be causes of Y that we needed a separate qualification for interactive factors. Just as the theory of token causation requires, parallel to the type-level theory, a qualification about causal factors whose exemplifications are token uncaused by the causal event in question, so also the theory of token causation requires, again parallel to the type-level theory, a qualification about interactive factors whose exemplifications are token uncaused by the causal event in question. The following examples show the need for this.

These examples will be genuine counterexamples to the theory of token-level causation as developed so far, which includes the basic probability increase idea for token-level causation (the rough theory of Section 6.2) and the qualification about holding fixed factors that are token uncaused by the earlier event and causally relevant at the property level to the relevant type of the later event (the qualification from Section 6.3). These new counterexamples will be dealt with by including, as in the type-level theory, a further qualification about interactive factors whose exemplifications are token uncaused by the causal event in question.

The first example is a version of the defoliant case, which may at first sight seem parallel to the two-shock version discussed above, but which in fact is not.[20] There is a healthy tree that is sprayed with defoliant; this is token event x, of type X (as described), which takes place at time t_x. One of the effects of the defoliant is that the tree quickly loses its leaves. As in the original example, the occurrence of event x reduces the probability of the tree's being alive and healthy in a year. Also as in the original example, however, the tree gradually recovers and it is alive and healthy in a year; the plant's being alive and healthy after a year has passed is token event y, of type Y (as described), at time t_y. The probability trajectory for the example is of the same shape as the one for the original

[20] This example was posed by Malcolm Forster. It was this example that inspired the natural qualification of holding fixed appropriate interactive factors.

defoliant example, as shown in Figure 6.2, so that the theory tells us that y was Y despite x.

Here is the wrinkle in this example, a wrinkle that will change our assessment of token causal roles but will not change the shape of the probability trajectory for the example (given the theory as developed so far). At some time between t_x and t_y, the entire area where this tree lives was invaded by parasites. Call this event z, of type Z (as described). These parasites eat the leaves of trees and leave a tree with a disease that makes it almost inevitable that the tree will die soon; the parasites are even more deadly than the defoliant. All the trees around the one that was defoliated die; but our defoliated tree has no leaves, and the parasites are not interested in trees with no leaves. Because of the application of the defoliant, the parasites leave our tree alone. In this case, we are forced to say that the tree survived *because of* the application of the defoliant: event x *caused* event y (to be of type Y). But the probability trajectory for the example is still of the shape that the theory associates with y's being Y *despite* x. The theory, as developed so far, delivers the wrong answer.

The remedy of holding fixed factors causally relevant to Y, and whose exemplifications are token uncaused by x, will not work for this counterexample, as it did for the previous ones. It is a natural guess that our qualification about holding fixed such factors should force us to hold fixed (positively) the factor Z of the parasite invasion. Just as in the previous examples, doing this would result in the theory's giving the right answer of "because." Of course, in order for the counterexample to work, the invasion of the parasites must be (more or less) token uncaused by x.[21] But although the exemplification of Z is token uncaused by x, the factor Z is *not* a property-level causal factor for our plant's survival, in the

---

[21] This is for the same reason that the intermediate events had to be (more or less) token uncaused by the earlier events in the counterexamples discussed previously; again, the idea of *degrees* of token causal significance is relevant, and this will be taken up in Section 6.5.

338

relevant context, as it must be for our qualification to force us to hold Z fixed. As explained above, the appropriate population (or context) for assessing the type-level causal role of such a factor Z, in applying our qualification, holds fixed, positively, the earlier factor X. In this example, therefore, we must hold fixed the fact that our tree has been sprayed with defoliant, which makes the tree lose its leaves. Indeed, given the way the qualification was explained, we must even hold fixed the fact that our tree has already lost its leaves by the time of the invasion. But for leafless trees, invasions of parasites of the kind in the example are *causally neutral* for survival. So our qualification about holding fixed type-level causes of Y whose exemplifications are token uncaused by x does not apply to this example.

As another example of this, take the first two-shock version of the defoliant case, and change it so that, by the time of the second, more severe shock, the plant has developed complete immunity to the defoliant, as a result of the first shock. Again, the intermediate event of the second shock is type-level causally neutral for the effect event in question, in the relevant population (or context). Holding fixed the plant's complete immunity, the factor of being shocked by the second application of the same defoliant is causally neutral for the factor of survival. Again, the qualification introduced in the previous section does not apply.

In these examples, the truth is that the earlier event x is a *token cause* of the later event y's being Y, while the theory as developed so far says that the later event happens *token causally despite* the earlier event. An example in which the truth is "*despite*," while the theory says "*because*," is easily constructed from the example, discussed above, about the severely ill patient (Figure 6.11). Simply change the example so that, as a result of the first treatment, the second, improved, treatment is completely inefficacious in curing the patient's disease: the first and second treatments *interact* in such a way that, if the first treatment has been administered, the second is inefficacious in treating the disease, while if the first treatment is not administered, the second treatment gives the best possible results. The probability trajectory for this version of the severely ill patient example is like the one for the original example (Figure 6.11), except that the trajectory is unbroken and smooth through the time, t_z, of the second treatment. (Another fact about interaction in this example is that, as explained below, assuming the second treatment will be given, the first treatment is highly negative for survival, but assuming the second treatment is not given, the first is slightly positive for survival.)

Again, in the relevant population (or context) in which the first treatment *has* been administered, the factor of the second treatment is, at the property level, causally neutral for survival. So again, our qualification that requires holding fixed property-level *causes* of Y, whose exemplifications are token uncaused by x, does not apply. As in the original example, the truth is still that the patient survived *token causally despite* the first treatment, while the theory as developed so far says that the first treatment is a *token cause* of survival.

The outline of a remedy to these counterexamples should be fairly clear by now. In each case, there is an interaction between the relevant type X of the earlier event x and the type Z of the intermediate event z, with respect to the type Y of the later event y – in exactly the sense of interaction among factors that was explained in Chapter 3. By analogy with the theory of type-level causation, with its qualification about holding fixed independent interactive factors, this suggests a qualification for the theory of token causation that will force us to hold Z fixed in the three examples discussed above.

First, let me explain why it is that, for each of the three examples, the type X of the earlier event interacts with the type Z of the intermediate event, with respect to the type Y of the later event. Of course, just like causation at the type level, interaction at the token level is relative to a token popu-

340

lation and a kind that the population exemplifies. In each case, we consider a population of individuals (trees, plants, severely ill people) just like the one in the example, with respect to all relevant features of the individual in question. This can be the singleton population consisting of just the particular case in question. The kind that we associate with the population can be determined by the conjunction of all factors that were in place in the particular case in question before the time, t_x, of the earlier event x. Of course, we cannot hold X or Z fixed in the description of the population. The reason is that, in order for an interaction to show up, we must compare the probabilistic impact of X (versus $\sim X$) on Y in the presence of Z with the probabilistic impact of X (versus $\sim X$) on Y in the absence of Z – and this cannot be done if either of X and Z is held fixed.

In the example about the tree, the defoliant (X), and the parasites (Z), we have, where Y is survival:

$$Pr(Y/Z\&X) > Pr(Y/Z\&\sim X)$$

and

$$Pr(Y/\sim Z\&X) < Pr(Y/\sim Z\&\sim X).$$

Given that the parasites will invade the area where the tree lives, the defoliant is causally and probabilistically positive for survival, since the defoliant causes the tree to lose its leaves, thus causing the parasites to be uninterested in the tree. And recall that the parasites were even deadlier than the defoliant, in the example. Given that the parasites will not invade, the defoliant is, of course, causally and probabilistically negative for survival.

The very same probabilistic relations obtain in the example about the plant, in which the plant acquires total immunity as a result of the first shock (where X and Z are the factors of the first and second shocks, respectively, and Y is survival). Given that there will be a second, more severe shock, the first shock increases the probability of survival (because the sec-

341

ond shock is more severe and the first shock is positive for developing immunity). And given that there will be no second shock, then, of course, the first shock decreases the probability of survival.

For the version of the case of the severely ill patient described just above, the factor of the first treatment (X) interacts with the factor of the second treatment (Z), with respect to survival (Y), where for this example the inequalities go in the opposite direction from those that characterize the first two examples described. We have

$$Pr(Y/Z\&X) < Pr(Y/Z\&\sim X)$$

and

$$Pr(Y/\sim Z\&X) > Pr(Y/\sim Z\&\sim X).$$

Assuming that the second, highly effective treatment, which has no bad side effects, will be administered, the probability of survival is very high if the first treatment is *not* administered. But, still assuming the second treatment, the probability of survival is lowered by the first treatment, due to the first treatment's serious side effects and its making the second treatment inefficacious in curing the disease. But if we assume that there will be no second treatment, the first treatment increases the probability of survival somewhat, since the side effects of the first treatment are somewhat less serious than the patient's disease.

The requirement of holding fixed interactive factors in the theory of token causation is as follows. In assessing the token causal role of x's being X for y's being Y, we must hold fixed, positively, factors Z that are such that: (1) Z is exemplified in the case in question, (2) the exemplification of Z in the case in question is token uncaused by x, and (3) X interacts with Z with respect to Y (in the relevant population, or context, as described above). So, when a factor Z is exemplified token uncaused by the earlier event x, we require that it be held fixed if it is either a property-level *causal* factor for the later

342

factor Y or *interactive'* for Y relative to X. Whether Z is a property-level cause of Y or interactive for Y, its presence or absence makes a contextual difference that affects the earlier event's causal role for the later event; and it is for this reason that it must be held fixed in the context K_a.

Thus, in the example involving the tree, the defoliant, and the parasites, we must hold fixed, positively, the factor Z of the parasite invasion. Taking the parasite invasion into account, the probability trajectory for the example is just like the one for the first two-shock version of the defoliant example, holding fixed the second shock, Figure 6.12. So now the theory gives the correct answer that the tree survived *because of* the defoliant: the defoliant *token caused* the survival of the tree. The version of the two-shock example in which the first shock gave the plant complete immunity is also characterized by Figure 6.12, when we hold fixed the interactive factor Z of the second shock. And again we get the right answer that the plant survived because of the first shock, that the first shock token caused the survival. And in the example about the severely ill patient, when we hold fixed the interactive factor Z of the second treatment, the probability trajectory is just like that shown in Figure 6.13. Again, we get the right answer that the patient survived token despite the first treatment.

It is worth noting that in all of the examples discussed in this and the previous section – in all of these counterexamples to the rough theory formulated in Section 6.2 – there is interaction of the type X of the earlier event with the type Z of the intermediate event, with respect to the type Y of the later event. For example, in the original two-shock version of the defoliant example, presented in Section 6.3, the factor of the first spraying interacts with the second spraying with respect to survival. Again, assuming that there will be a second, more severe, application, the first spraying increases the probability of survival, and assuming that there will not be a second, more severe, application, the first spraying decreases the probability of survival. And in the original example about

343

the severely ill patient, presented in Section 6.3, the factor of the first treatment interacts with the second treatment with respect to survival. Again, assuming that the patient will be given the second, better, treatment, the first treatment lowers the probability of survival, and assuming that there will be no second treatment, the first treatment increases the probability of survival.

This suggests that it may be the idea of *interaction* that should be fundamental in the evaluation of all these examples, and that the idea of the temporally intermediate factor's being a type-level *causal* factor for the later factor (in the relevant population, or context) need not play a (separate) role. This may be correct when the question at issue is just the qualitative question about the kind of causal significance that a character of one event has for a character of another. However, when it comes to degree of the various kinds of causal significance, it is clearly necessary to control for all causes of the later event (when token uncaused by the earlier event), whether there is interaction or not. For example, if we failed to control for a property-level cause of the later factor (where the exemplification of the property-level cause is token uncaused by the earlier event), the relevant probability trajectory may be too high or too low; also, the earlier event may be more, or less, efficacious for the later event in the presence of a separate cause than in its absence (degrees of causal significance are not always simply additive). On the other hand, of course, a suitably general understanding of interaction, as formulated in Chapter 3, will include all relevant cases of separate causes. But, as before, it is nevertheless of some value, conceptually, to separate the ideas of interaction and separate causation.

I should now summarize the characterization of the background context K_a used to assess the causal relevance of x's being X for y's being Y, and used in particular for the assessment of the values $Pr_t(Y)$ traced in the probability trajectory diagrams. K_a can also be thought of as the relevant *kind of*

344

population for assessing the relevant probability values, where the relevant token population can be just the singular case in question. K_a is the causal background context obtained by holding fixed, positively, all factors of the following two kinds:

(1) any factor Z such that (a) Z is exemplified in the case in question, (b) Z's exemplification is token uncaused by x's being X, and (c) Z is type level causally relevant to y's being Y in the context (or population plus kind) determined by the way things are before t_z (holding fixed all actually exemplified pre-t_z factors);

and,

(2) any factor Z such that (a) Z is exemplified in the case in question, (b) Z's exemplification is token uncaused by x's being X, and (c) X interacts with Z with respect to Y in the context (or population plus kind) determined by the way things are before t_x (holding fixed all actually exemplified pre-t_x factors).

Note that it is not specified in (1) or (2) at what *time* the factor Z must be exemplified for it to be required that Z be held fixed. The time of Z's exemplification does not matter: it may be before t_x, the same as t_x, or between t_x and t_y (of course, it will not be at or after t_y).

Now, where t is any time at all (either before, the same as, or after t_x, and either before, the same as, or after t_y), let W_t be the conjunction of all factors $F_{t'}$ such that (i) $F_{t'}$ is not included in K_a, (ii) $F_{t'}$ has fallen into place at or before time t, and (iii) $F_{t'}$ is type-level causally relevant to (positive, negative, or mixed for) Y, in the context K_a – with the stipulation that X be included in any W_t for times t the same as or after t_x, and Y be included in any W_t for times t the same as or after t_y. The idea is that for a time t between t_x and t_y, W_t specifies all exemplified factors that are causally relevant to y's being Y and whose exemplification, within K_a, can be traced back to the exemplification of X at t_x – and Y is specified for times t at or after t_y whether or not y's being Y traces back to x's being X, and X is specified just at and after t_x.[22] Then, for all times t,

$$Pr_t(Y) = Pr(Y/K_a \& W_t).$$

And token causal significances are defined, as in Section 6.2, in terms of the trajectory of $Pr_t(Y)$, as just defined. This just summarizes the theory of token causation as developed so far in this and the previous two sections.

Before refining the theory, in the next section, to take account of *degrees* of the various kinds of token causal significance, there are two issues that should be briefly addressed in closing this section.

First, there is the possibility, mentioned in Section 6.1, that events prior to t_x may already make the event of x's being X (at t_x) very probable; and these events may exemplify factors that have to be held fixed in K_a and in W_t's for t's before t_x. In this case, there may be little or no change in the probability of Y across the time t_x of x's being X, this being because x's being X was already so probable. And this would seem to be possible even if x's being X is token causally significant for y's being Y. However, the kind of causal significance x's being X has for y's being Y is defined (in part) in terms of *the kind of change there is in $Pr_t(Y)$ across the time t_x*. In particular, it cannot be that y is Y *because of* or *despite* x's being X unless there is a change in $Pr_t(Y)$ across t_x.

This kind of possibility simply points to a fundamental limitation of this theory, analogous to a limitation of the type-level theory. It applies only to cases in which constellations of causes and interactive factors fail to confer extreme probabilities on their effects. As mentioned at the beginning of this chapter, the theory assumes that determinism is false. And as long as events prior to t_x fail to confer probability 1 on

[22] Another possibility would be to include in W_t the specification of *all* factors that have fallen into place by time t. This would be fine, I think, but what is important is just those factors that have fallen into place by t and that are causally relevant to y's being Y.

346

x's being X, there will still be room for X's falling into place at t_x to affect the probability of Y – as it will, according to the theory, if y is Y token causally because of, or despite, x's being X.

For the theory of property-level probabilistic causation, the issue of transitivity of causation was discussed after the chapters on separate causes and interactive factors. It is time now to address, if only briefly, this question for the token-level of probabilistic causation. I will give an example that shows that token-level probabilistic causal chains are not in general transitive, on the theory developed here. And I will suggest a sufficient, but nonnecessary, condition for transitivity that is similar to the sufficient, and nonnecessary, condition for transitivity of property level causal chains that was given in Chapter 4.

The example involves three events, x, z, and y, exhibiting factors X, Z, and Y, respectively, where x occurs first, z next, and y last.[23] The event z will be Z *because of* x's being X; y will be Y *because of* z's being Z; but y will be Y *despite* x's being X. There is a windowless building that is equipped with an emergency backup electrical generator. For some reason, it is important that the lights inside this building never go out for more than a minute or so. The backup generator is designed to start supplying electricity at one minute after any power failure from the electrical utility. However, this backup generator is not very reliable. The probability of its supplying electricity, if needed, is only 0.1. And let us suppose that the unreliable component of the generator is one that comes into play just at the time the generator is supposed to start supplying power, so that, once called upon, the probability that the generator will supply power remains at 0.1 for a full minute. After this minute, the generator will either supply power (so that the probability jumps from 0.1 to 1) or not (the probability drops from 0.1 to 0). On the

[23] This example is patterned after one originally described to me by Aladdin Yaqub.

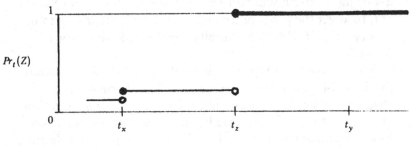

$Pr_t(Z)$

0 t_x t_z t_y

Figure 6.14

other hand, if there is a power failure from the electrical utility, and if the generator succeeds in supplying power, then it will thereafter be reliable throughout the emergency. Luckily, the electrical utility is very reliable: the probability of its failing is extremely small.

On one occasion, however, there was a power failure from the electrical utility – say an electrical surge caused a power line to snap. This is event x's being X. Given this power failure, the generator now has a 10 percent chance of supplying electricity in a minute (without a power failure, the chance of the generator's kicking in is 0). Fortunately, the generator worked this time; in one minute, the generator began supplying electricity. This is event z's being Z. At this time, the lights in the building turn back on, and they remain on. The event y's being Y is the lights' being on after another minute (*two* minutes after the power failure from the utility).

It is clear that the generator supplied electricity because of the power line's snapping: z is Z *because of* x's being X. Before t_x, the chance of a power failure was very small, so that the chance of the generator's being called upon was also very small. At the time of the power failure, the generator is called upon, so that, with its 10 percent reliability, the chance of its supplying electricity in a minute, at t_z, jumps up to 10 percent. The probability trajectory for assessing the significance of x's being X for z's being Z is shown in Figure 6.14.

348

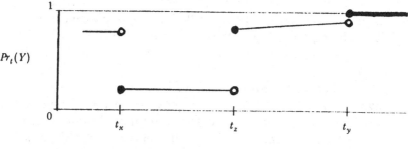

$Pr_t(Y)$

1

0

t_x t_z t_y

Figure 6.15

It is also clear that, given that there was a power failure, the lights were on at the later time t_y because of the success of backup generator: y was Y *because of* z's being Z. At the time the generator began supplying electricity, the probability of the lights being on a minute later jumped from 10 percent to nearly 100 percent. Figure 6.15 shows the probability trajectory for assessing the significance of z's being Z for y's being Y. Finally, it is also clear that we should say that the lights were on at the later time t_y despite the event of the power failure: y was Y *despite* x's being X. Figure 6.15 shows the relevant probability trajectory, in which there is a sharp drop in $Pr_t(Y)$ across the time of x.

The snapping of the power line was a positive token cause of the generator's supplying electricity, and the latter was a positive token cause of the lights' being on later; but the lights were on later despite the snapping of the power line. In this example, the earlier event is a positive token cause of an intermediate positive token cause of the later event. The reason why we say that the earlier event is not a positive token cause of the later event is that the earlier event also was a positive token cause of intermediate events' exemplifying factors that are negative causal factors for the relevant type of the later event (in the context determined by the way things were just before the earlier event). In this case, such an intermediate event is the power line's lying on the ground discon-

349

nected; and this is more negative for the lights being on at t_y than the generator's being called upon is positive for the lights being on at t_y (and the surge was a stronger cause of the line's being disconnected than it is of the generator's supplying electricity).

This suggests the following sufficient condition for transitivity of token causal chains. Let us say that x's being X is *unanimous* for y's being Y if, for all temporally intermediate events z and all factors Z that such events z exemplify, z is Z because of x's being X and y is Y because of z's being Z. Then it seems to be a good bet that this kind of unanimity is a sufficient condition for transitivity of token causal chains. Of course, it is not necessary. To see this, we can change the example above, so that the surge was anticipated, and the generator was disassembled and its parts used to strengthen the power line. In the new example, the power line withstood the surge, and the action of strengthening the power line caused the lights to remain on and to be on at t_y. But this action is not unanimous: It resulted in there being no backup generator, which is negative for the lights' being on at the later time.

In this and the previous section, two requirements were added to the basic probability-increase idea for token causation. The basic probability-increase idea, described in Section 6.2, is analogous in the development of the theory of token causation to the basic probability-increase idea for type-level causation, in that it stood in need of qualifications. And the two qualifications made in this and the previous section are analogous to the qualifications made in Chapters 2 and 3 for the type-level theory. One involves holding fixed separate causes and the other involves holding fixed interactive factors. A further analogy between the theories of type- and token-level probabilistic causation is worth noting: Both theories are circular in that each makes use of the idea of items that are both *causally relevant* to the candidate effect item and *caus-*

350

ally independent of (or, in the case of the token-level theory, *token uncaused* by) the causal item in question.[24]

A disanalogy, however, is this: In the type-level theory, to assess the causal role of one factor X for a second factor Y, it was only necessary to appeal to other *type-level* causal relations (between X and certain factors Z that are causally independent of X, and between such factors Z and Y). In the token-level theory, however, in assessing the token causal role of an event x's being X for a second event y's being Y, it is necessary to appeal *both* to the *token causal role* of x's being X for certain events z being Z *and* to the *type-level causal role* of factors Z (whose exemplifications are token uncaused by x's being X) for y's being Y. But this disanalogy is natural when one contemplates both the *analogy of circularity* between the two theories and the fact that, in this book at least, *probability attaches to types, to factors,* and not directly to token events themselves. A consequence of this disanalogy is that while type-level causal relations are autonomous from token-level causal facts, the token-level causal facts depend, in a subtle way involving the appropriate background context, on what the type-level causal relations are.

This consequence is in conflict with the dependence relations recently urged by Cartwright (1988a, 1989). She argues that (i) type-level causal facts depend both on other type-level facts and on token-level facts (her argument for this was examined in Chapter 4), and (ii) token-level facts are "basic" in that they *do not* depend on type-level facts (see especially Cartwright 1989). Thesis (i) implies the thesis, (iii), that token causal facts are not reducible to type-level facts. I agree with (iii), of course; but, of course, this is for reasons other than (i). It has been a conviction in (iii) that has motivated the

[24] However, neither theory is circular in the sense of characterizing the causal role of one item for another in terms of the causal role of the first item itself for the second item itself. In each theory, the causal role of a first item for a second is characterized in terms of causal roles of the first for third items and causal roles of such third items for the second.

development of a separate theory of token causation in this chapter. The theories of this book are in conflict with (i) and (ii). As to (i), I argued in Chapter 4 that Cartwright's examples in fact do not show that we must bring token-level causal facts into the explication of type-level causal relations. And as to (ii), the theory of this chapter appeals both to token-level causal facts and to type-level causal facts in the explication of token-level causal facts.

The theory of probabilistic token causation is not complete, however. As mentioned earlier, in order to handle certain variations of the counterexamples dealt with in this and the previous section, it is necessary to understand the idea of *degrees* of the various kinds of token causal significance that have been discussed. This is the topic of the next section.

6.5 DEGREES OF TOKEN CAUSAL SIGNIFICANCE

The idea of *degrees* of the various kinds of token causal significance is of interest in itself, of course. One token event can be *more or less* causally significant for another, in the various ways in which one event can be token causally significant for another. For example, an event x's exhibiting a type X can be a more or less strong, or a more or less weak, token cause of a later event y's exhibiting a type Y. This is how the *because* variety of token causal significance can come in degrees. Also, the character of a later event can be more or less *despite* the character of an earlier event. And the same goes for the other two kinds of token causal significance, *independence* and *autonomy,* first described in Section 6.2.

In addition to the idea's being of interest in itself, an understanding of the idea of degrees of token causal significance is crucial to the understanding of certain remaining counterexamples to the theory of probabilistic token causation, as the theory has been developed so far. And the development of a "calculus" of degrees of token causal significance is crucial for seeing how the theory should be adjusted in light of the possibility of the natural kinds of situations that remain as counterexamples to the theory as developed so far.

In the previous two sections, two kinds of counterexamples to the basic probability-increase idea for probabilistic token causation were handled. The counterexamples were handled by including two qualifications in the theory, analogous to two qualifications in the theory of property-level probabilistic causation (about separate causes and interactive factors). The qualifications involved holding fixed certain factors whose exemplifications are *strictly* token uncaused by the earlier event in question being of its character in question. And, as explained above, I assumed that, in all the counterexamples, the relevant "confounding" events were of a character such that these events' being of that character was, strictly, token uncaused by the earlier event's being of its character in question. But versions of these counterexamples were described in which the relevant confounding events were not strictly, but only largely (or to a large degree), token uncaused by the earlier event's being of its character in question. These versions remain genuine counterexamples to the theory as developed so far. An understanding of degrees of token causal significance will help us to better understand these counterexamples, and to see how the theory should be adjusted in a natural way.

In this section, I develop a quantitative calculus of degrees of the various kinds of token causal significance that have been described, mainly qualitatively, in Sections 6.2–6.4. I then revise the theory of probabilistic token causation in such a way that, for the remaining counterexamples, we must hold fixed the relevant confounding factors in a "partial way" – that is, to a degree that is appropriate given the degree to which the earlier event is token causally significant for the confounding event. With this adjustment, the theory will give the right answers for the examples in question, and the theory of probabilistic token causation will be complete.

Let x and y be the relevant earlier and later events, where their characters in question are X and Y, respectively. Then the problem is to characterize the degrees to which y is Y *because of, despite, independently of,* and *autonomously of,* x's being X. When this is done, we can understand the idea of x's being, to a degree, token causally significant for the confounding events z being Z in the remaining counterexamples, and finally we will be in a position to hold such factors Z fixed in an appropriate partial way.

We begin with the probability trajectory in terms of which we assess, as explained above, the qualitative token causal significance of x's being X for y's being Y. There are *three* crucial points on this trajectory, in terms of which degrees of token causal significance will be defined. First, there is the probability of Y immediately *before* the time t_x of the event x. This is $Pr_t(Y)$ for t immediately before t_x. This probability is the "last" value $Pr_t(Y)$ takes on before time t_x. More precisely, we can say that this probability is the limit of $Pr_t(Y)$ as t approaches t_x from the past, and it is an *assumption* in this analysis that this limit exists. I will denote this value by "P^-." Second, there is the probability of Y immediately *after* the time t_x of the event x. This is $Pr_t(Y)$ for t immediately after t_x. This probability is the "first" value $Pr_t(Y)$ takes on after time t_x. More precisely, we can understand this value to be the limit of $Pr_t(Y)$ as t approaches t_x from the future, and again it is an *assumption* in this analysis that this limit exists. I will denote this value by "P^+." And third, there is the smallest value that $Pr_t(Y)$ takes on in the open interval (t_x, t_y). This is $\min\{Pr_t(Y) : t \text{ in } (t_x, t_y)\}$. Of course, there need not be a least value in this set, but we could instead define this value to be the greatest lower bound of the set, which may or may not be in the set. It is, intuitively, the lowest point in the part of the probability trajectory between t_x and t_y. I will denote it by "M."

In what follows, it will be simpler to understand P^-, P^+, and M, intuitively as the "last" value of $Pr_t(Y)$ before t_x, the "first" value of $Pr_t(Y)$ after t_x, and as the "lowest" value of Pr_t strictly between t_x and t_y, respectively. Of course, it is an assumption that this is appropriate. In the case of P^- and P^+, the assumption would be that there are t' and t'' flanking t_x such that $Pr_t(Y)$ is constant both in (t',t_x) and in (t_x,t''). And in the case of M, the assumption would be that the greatest lower bound of the probability trajectory strictly between t_x and t_y is in the trajectory. Understanding things in this way makes all three values defined strictly between 0 and 1 (as explained further below). It is perhaps also worth noting that a more complete notation for the three crucial values of $Pr_t(Y)$ would be "$P^-(x,X,y,Y)$," "$P^+(x,X,y,Y)$," and "$M(x,X,y,Y)$," but I will for the most part suppress the part of the notation for the events and their types, since, unless noted otherwise, we will always be concerned with events denoted by "x" and "y" being of types denoted by "X" and "Y."

The degrees to which y is Y because of, despite, independently of, and autonomously of x's being X will be denoted by "B," "D," "I," and "A," respectively. A more complete notation would be "$B(x,X,y,Y)$," "$D(x,X,y,Y)$," "$I(x,X,y,Y)$," and "$A(x,X,y,Y)$," but again I will for the most part suppress the part of the notation for the events and their types, since, until noted otherwise, we will always be concerned with the events denoted by "x" and "y" being of types denoted by "X" and "Y." Then these degrees can be defined as follows:

$$B = M - P^-;$$
$$D = P^- - P^+;$$
$$A = P^+ - M; \text{ and,}$$
$$I = 1 - |P^- - P^+|.$$

Before explaining these choices of definitions, I will point out some properties of these degrees and some relations among the four kinds of degrees.

First, notice that M is always less than or equal to P^+: $M \le P^+$. This is because M is the *smallest* value of $Pr_t(Y)$ after t_x and before t_y; since P^+ is a value of $Pr_t(Y)$ after t_x and before t_y, P^+ cannot be less than M. This means that A, or $P^+ - M$, cannot be negative; in fact,

$$0 \leq A < 1.$$

The relation $M \leq P^+$ is the *only* constraint on the three crucial points of $Pr_t(Y)$, except that they are all probabilities, and hence between 0 and 1; and it is natural to assume that they are all *strictly* between 0 and 1, since y in fact *is* Y at t_y (so that $Pr_t(Y)$ should never be 0) and since it is only at and after t_y that y's being Y is certain (so that $Pr_t(Y)$ should not be 1 at any of the crucial points, which are all before t_y). This is why A is always strictly less than 1. For the other degrees, we have,

$$-1 < B < 1;$$
$$-1 < D < 1; \text{ and,}$$
$$0 < I \leq 1.$$

The most interesting relation among the three kinds of degrees is this: $B + D + A = 0$. This means that each the three degrees involved in the equality is the negative sum of the other two: $B = -(D + A)$; $D = -(B + A)$; and $A = -(B + D)$. Also, if $A = 0$ (so that $M = P^+$), then $B = -D$. Finally, notice that $I = 1 - |D|$. Of course, it is not possible for all of B, D, and A to be greater than 0, since they add to 0. Nor can they all be less than 0. Furthermore, it is not possible for B and D each to be greater than 0. This is because $A \geq 0$ and $A = -(B + D)$, so that $-(B + D) \geq 0$. When $A = 0$, we have $-1 < B = -D < 1$. When A approaches 1, B and D will both be less than 0, and B approaches $1 - D$. Figure 6.16 depicts the possible relations among B, D, and A. The antidiagonal lines are lines of constant A; the highest, and longest, antidiagonal line corresponds to $A = 0$, and proceeding to lower and lower, and shorter and shorter, antidiagonals corresponds to A's becoming larger and larger, approaching 1. The horizontal D-axis and the vertical B-axis are interpreted in the standard way, and the antidiagonal A-axis is interpreted in the way just explained.

The definitions of B, D, I, and A capture, in a natural way, the idea that token causal significances come in degrees, in the

356

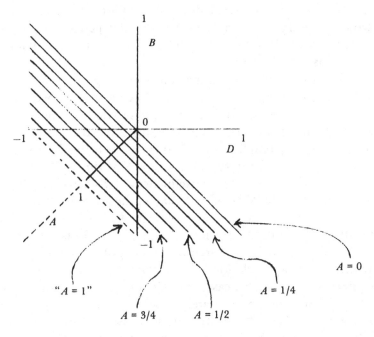

Figure 6.16

way briefly motivated and explained in Section 6.2. Recall the characterization of the paradigm extreme of "because," given in Section 6.2 (where this is now taken to be relative to the relevant type, X, of the earlier event x): (i) $Pr_t(Y)$ changes at t_x, (ii) this probability is high just after t_x, (iii) this probability is higher just after t_x than it is just before t_x, and (iv) this probability remains at its high value until t_y (at which point it becomes 1). According to the brief explanation of degrees in Section 6.2, the degree to which a token causal significance is of the because variety is "the degree to which the four clauses in this characterization of because are satisfied." It is determined by *how much* the probability of Y changes at t_x, by *how high* this probability is just after t_x, by *how much* higher it is after t_x than before t_x, and by *how high* it remains after t_x until

357

t_y. Of course, with intermediate probabilities, the degrees to which these clauses are satisfied can vary independently. However, the value $B = M - P^-$ will be close to 1 if and only if all four clauses are satisfied to high degrees. And whenever B is positive, they will all be satisfied to some degree.

The reason for choosing the difference $M - P^-$, instead of $P^+ - P^-$, as our measure of because, has to do with the fourth clause of the characterization of because, and the motivation for it as explained in Section 6.2. If event x's being X succeeds in raising the probability of y's being Y at time t_x, but then later events – improbable consequences of x's being X – later lowered the probability of y's being Y, then the occurrence of those later events can rob x's being X of some, or all, of its positive token causal significance for y's being Y, as explained in Section 6.2 with the variation of the golf ball example in which other squirrels kicked the rolling ball after the first squirrel kick. In such a case, $P^+ - P^-$ can be very high, while $M - P^-$ will be appropriately low, or even negative. The value $B = M - P^-$ appropriately measures the degree to which x's being X succeeded in permanently increasing the probability of y's being Y. If it is positive, then the probability of y's being Y is always higher after t_x than it was just before t_x. Thus, I take B's being positive to mean that y's being Y was, to some degree (namely, B), because of x's being X. When B is negative or 0, then y's being Y was not at all because of x's being X.

The rationales for the definitions of D, A, and I are parallel to the rationale given for the definition of B. In each case, the defined measure is one way of making precise the idea of satisfying, to a degree, the clauses of the corresponding definition of a kind of token causal significance.

I say that the defined measures give *one* way of making the idea of degrees precise, for each kind of token causal significance. There are, of course, many others. For example, an alternative measure of "degree of becauseness" would be $B^\star = \max\{0, M - P^-\}$. This would capture the intuition, which one might have, that whenever $M - P^-$ is negative, at any value, then there is simply no because component at all. However, although this will not figure in the analysis of the remaining counterexamples below, I do not see any problem with the idea of negative degrees of the because variety of causal significance (which is not to be confused with positive degrees $D, = P^- - P^+$, of the despite kind of causal significance). Thus, I will stick with the definition of B given in the text, which in any case is simpler. Corresponding to $B^\star$, which is always greater than or equal to 0, there is this alternative understanding of "degree of despiteness": $D^\star = \max\{0, P^- - P^+\}$, which also is always greater than or equal to 0. Again, however, I will stick with the definition of D given above, and the same goes for A and I.

Having now developed the idea of degrees of the four kinds of token causal significance, we can turn finally to the versions of the examples described previously that remain as genuine counterexamples to the theory as developed so far. I will only discuss the relevant versions of the two-shock defoliant example and the example involving the severely ill patient, described in Section 6.3. However, the very same considerations apply to the other examples discussed in Sections 6.3 and 6.4, and it will be clear how to apply the ideas developed below to these examples as well. As we will see, it is the idea of the degrees to which one event can be *because of* an earlier event that is crucial to dealing with all these examples.

Let us turn first to the two-shock defoliant example. Recall that in this example, the plant actually survived *because* of the first shock (because the first shock enabled the plant to develop immunity to the defoliant), though the simple probability increase idea developed in Section 6.2 gave the answer that the plant survived *despite* the first shock (Figure 6.10). However, if the second shock is strictly token uncaused by the first shock, then the qualification introduced in Section 6.3 ("hold fixed factors whose exemplifications are *strictly* token uncaused by the earlier event and that are causally rele-

359

vant to the later event") tells us to hold fixed the factor of the second shock; in this case, the theory gives us the correct answer that the plant survived because of the first shock (Figure 6.12). The qualification applies because the second shock is, in this version of the example, strictly token uncaused by the first shock. In terms of the degrees characterized above, the second shock's being strictly token uncaused by the first shock means: $B(x,X,z,Z) \leq 0$ (where the first shock is x's being X and the second shock is z's being Z).

Now suppose that the second shock occurred only *largely*, and *not strictly*, token uncaused by the first shock. The first shock was a cause, but only a very, very weak cause, of the second shock. This means that the first shock raised the probability of the second shock, but only by a little: $B(x,X,z,Z)$ is greater than 0, but it is very small. If $B(x,X,z,Z)$ is small enough, we should still say that the plant survived because of the first shock. However, the qualification introduced in Section 6.3 no longer applies, since that qualification only requires holding fixed factors whose exemplifications are *strictly* token uncaused by the earlier event. Thus, if $B(x,X,z,Z)$ is greater than 0, but small enough, then the counterexample remains genuine.

The solution to the problem is to describe a way of holding fixed the factor of the second shock "*to the appropriate degree*," which "degree" will be a function of the degree, $B(x,X,z,Z)$, to which the second shock was *not token uncaused by* the first shock. Figure 6.17 gives the probability trajectory for assessing the token causal significance of x's being X for z's being Z; the trajectory must have (roughly) this shape for this token causal significance to be of the because variety.[25] (In this figure, P^- is an abbreviation for $P^-(x,X,z,Z)$, and similarly for P^+, M, and B). At t_z, the probability of z's being Z attains

[25] I assume that all appropriate factors have been held fixed in drawing this trajectory. This includes holding fixed factors that, in the previous sections, we have learned must be held fixed. This also includes holding fixed factors that we will see in this section need to be held fixed, in the way in which we will see they should be held fixed. This is the same kind of circularity noted at the end of the previous section.

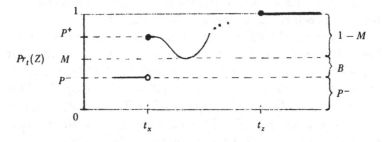

Figure 6.17

the value 1. We can ask, intuitively, "How did that probability get there?" Since z's being Z is partly due to x's being X, we must apportion part of the "responsibility" for z's being Z to x's being X. The value $B(x,X,z,Z)$ is a measure of the part of the unitary probability of z's being Z that x's being X is responsible for. The other components of the unitary probability of z's being Z are: $P^-(x,X,z,Z)$ (the probability of z's being Z just before t_x) and $1 - M(x,X,z,Z)$ (what the probability of z's being Z had to "make up," after the positive impact of x's being X on z's being Z finished playing itself out).

Note that $P^-(x,X,z,Z) + B(x,X,z,Z) + (1 - M(x,X,z,Z))$ $= 1, = Pr_t(Z)$ for t at and after t_z. Event x's being X is "not responsible" for the first and last components of this sum, which adds up to the unitary probability (at and after t_z) of z's being Z. The value $P^-(x,X,z,Z)$ can be thought of as measuring "the amount of probability" that z's being Z already had before x's being X came into play. And $1 - M(x,X,z,Z)$ is the amount that had to be made up after x's being X left $Pr_t(Z)$ as high as it could, in the end, leave it. And we can say that the probability of z's being Z has these two components strictly token uncaused by x's being X, where the only component of the eventually unitary probability of z's being Z that x's being X is token causally responsible for is $B(x,X,z,Z)$. So the idea is to hold fixed, all along, the fact that the eventually unitary

361

probability of z's being Z has those two components that x's being X is not token causally responsible for.

In assessing the causal significance of x's being X for y's being Y, we should, to put it roughly, hold fixed the fact that z's being Z has the "amount of probability," $P^-(x,X,z,Z)$ + $(1 - M(x,X,z,Z))$, token uncaused by x's being X. This "amount of probability" is equal to $1 - B(x,X,z,Z)$, and it is very large when $B(x,X,z,Z)$ is very small. How do we hold this fixed? I suggest that the following is the natural way. For readability, let b abbreviate $B(x,X,z,Z)$. Then, roughly and intuitively, we want to hold fixed this factor: $Pr(Z) = 1 - b$. Call this factor V. Then "holding fixed V" (or "holding fixed Z to degree $1 - b$") amounts to this: Instead of letting $Pr_t(Y)$ $= Pr(Y/K_a\&W_t)$ (as near the end of the previous section), we set

$$Pr_t(Y) = (1 - b)Pr(Y/K_a\&W_t\&Z) + bPr(Y/K_a\&W_t\&\sim Z).$$

Thus, the context K_a will "fully" hold fixed causal and interactive factors whose exemplifications are *strictly* token uncaused by x's being X. And for causal and interactive factors whose exemplifications are *not strictly* token uncaused by x's being X (those such factors Z for which $B(x,X,z,Z)$ is greater than 0), we hold them fixed in the "partial" way just described. Generalization of this way of partially holding fixed *one* factor Z to the case of *more than one* such factors can proceed by iteration.[26]

When we hold fixed, in this partial way, the factor of the second shock in the two-shock defoliant example, we get the probability trajectory shown in Figure 6.18, when $B(x,X,z,Z)$ is very small. All the way up until t_z, $Pr_t(Z)$ is very high. So the trajectory resembles the one shown in Figure 6.12, in which $Pr_t(Z)$ is 1 at all times. A difference is at time t_z. In Figure 6.18, there is a small drop in $Pr_t(Y)$ at this

---

[26] Whether simple iteration is appropriate here is an issue worth exploring in detail, but I will not pursue that here. Note the parallel between what I have called "partially holding fixed a factor" and Jeffrey's idea of "probability kinematics" (1983).

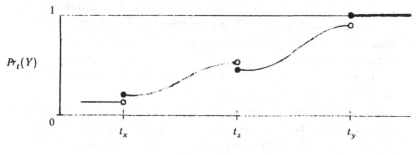

Figure 6.18

time. This is because, at this time, $Pr_t(Z)$, the probability of the second shock, jumps from its very high value to 1, which decreases, but only slightly, the probability of survival. In Figure 6.12, there is no change in $Pr_t(Y)$ at this time, since $Pr_t(Z)$ does not change at this time, being 1 throughout. The difference between Figures 6.12 and 6.18 is just this: At times before t_z, the probability of Y is shown slightly higher in Figure 6.18 than it is in Figure 6.12 (since Z is not yet quite certain in the Figure 6.18 case, and x's being X only slightly increases the probability of Z, whereas Z is always certain in the Figure 6.12 case); otherwise (after t_z), the trajectories are identical.

For the trajectory shown in Figure 6.18, I have assumed that $B(x,X,z,Z)$ is very, very small (so that $Pr_t(Z)$ is always very high). That is why, as in the example depicted in Figure 6.12, there is an increase in $Pr_t(Y)$ across the time of x, so that we again get the right answer that the plant survived because of the first shock. The increase is not as great in Figure 6.18 as it is in Figure 6.12, since in the Figure 6.18 case, x's being X is a token cause, though a very weak one, of the second shock. In other versions of the example, where $B(x,X,z,Z)$ is larger, there will not be an increase at all; in some such versions, there will be no change, and in others there will be a decrease. As $B(x,X,z,Z)$ becomes larger and larger, $B(x,X,y,Y)$ will become smaller and smaller, and even drop below 0 (and

363

$D(x,X,y,Y)$ will become larger and larger). The extremes are represented in Figure 6.12 (in which $B(x,X,z,Z) \leq 0$) and a figure like Figure 6.2 (with the probability drop even more extreme, to represent $B(x,X,z,Z)$'s being, or being very close to, 1); and Figure 6.18 depicts the case of a very, very small, but nonzero, value of $B(x,X,z,Z)$.

Now recall the example about the severely ill patient. In this example, the patient survived *despite* the first medical treatment, at the beginning of the year. Although the first treatment is pretty effective against the patient's disease, it causes serious side effects, and a better treatment, better against the disease and without bad side effects, was soon to be discovered and administered to the patient. The simple probability-increase idea of Section 6.2 gave the answer that the patient survived because of the first treatment (Figure 6.11), but the qualification introduced in Section 6.3 forced us to hold fixed the factor of the second treatment, giving the right answer that the patient survived despite the first treatment (Figure 6.13).

But holding fixed the factor of the second treatment was forced, in Section 6.3, only because of the assumption that the second treatment was *strictly* token uncaused by the first treatment: $B(x,X,z,Z) \leq 0$, where x's being X is the first treatment and z's being Z is the second. Now suppose that the second treatment is only largely, and not strictly, token uncaused by the first treatment: $B(x,X,z,Z)$ is greater than 0, but is very, very small. The first treatment *only slightly* raised the probability that the superior treatment would be discovered and administered. Still, we should say that the patient survived despite the first treatment, but the qualification introduced in Section 6.3 no longer applies.

But when we apply the qualification introduced in this section, and hold fixed the factor of the second treatment in the partial way described above, we get the probability trajectory depicted in Figure 6.19 for this version of the example. Assuming that $B(x,X,z,Z)$ is very, very small, the probabil-

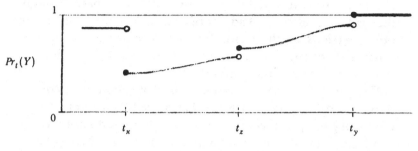

Figure 6.19

ity of Z will always be very, very high – all the way up to t_z, at which time the probability of Z becomes 1. So Figure 6.19 resembles Figure 6.13, in which the probability is Z is always 1. The difference is that, at times before t_z, the probability of survival is shown slightly lower in Figure 6.19 than it is in Figure 6.13. This is because, in the Figure 6.19 case, the second treatment is not certain until t_z (and the first treatment only very, very slightly raises the probability of the second), whereas in the Figure 6.13 case, the second treatment is certain all along. And as in Figure 6.13, we have a drop in the probability of survival at t_x, which in this case is because the second, superior, treatment is *almost* certain. The drop is slightly smaller in Figure 6.19 than it is in Figure 6.13, because in the Figure 6.19 case, the first treatment is a slight cause of the second, superior, treatment.

As in the two-shock defoliant example, we can consider other versions of the example of the severely ill patient, where $B(x,X,z,Z)$ varies from less than or equal to 0, to larger and larger positive values. Figure 6.13 represents the extreme where $B(x,X,z,Z) \leq 0$. In Figure 6.19, $B(x,X,z,Z)$ is only slightly positive. When $B(x,X,z,Z)$ becomes larger, we will find a value at which there is no change, across time t_x, in the probability of survival. This is the point at which the first treatment becomes just sufficiently causally positive for the second treatment that this positive effect of the first treat-

365

ment on survival exactly offsets its negative impact through the bad side effects. And as $B(x,X,z,Z)$ approaches 1, in other versions of the example, the trajectory depicting the causal role of the first treatment for survival approaches the shape for the paradigm extreme of because.

We began this chapter by noticing that the kind of conditional probability comparisons that played the central role in the theory of type-level probabilistic causation do not always coincide, in the natural way, with the token causal role of one event for another. We saw in Section 6.1 that an event x's being of a type X can be a positive token cause of another event y's being of a type Y, even if the probability of Y conditional on X is lower than the probability of Y conditional on $\sim X$. Token causation does not coincide with probability increase, in the sense of probability increase central to type-level probabilistic causation. However, in Section 6.2, a different conception of probability change was described, a conception of probability change around which the theory of probabilistic token causation was built.

This alternative way of understanding probability change, in terms of *probability trajectories,* provided the theory of probabilistic token causation with a basic probability-increase idea. This basic idea stands in the same relation to token causation as the idea of probability increase, understood in terms of *conditional probability comparisons,* stood to property-level causation. And both of the basic ideas stood in need of qualifications. In Sections 6.3 and 6.4, the qualifications, about separate causes and interactive factors, were incorporated into the theory of probabilistic token causation.

However, these qualifications of the basic idea did not deal with quite all the kinds of cases that motivated them. They required holding fixed causal and interactive factors whose exemplifications are *strictly* token uncaused by the token causal event in question. But the kinds of examples that motivated the qualifications retain their point if the exemplifications of the separate causes and interactive factors are not strictly, but *only largely* (or *almost*), token uncaused by the token causal event in question. To deal with this kind of possibility, a calculus of *degrees* of token causal significance was developed. And the theory was again qualified, by incorporating a final requirement, instructing us how and when to hold fixed the relevant factors *"to the appropriate degrees."*

6.6 COMPARISONS, AND EXPLANATION OF PARTICULAR EVENTS

In this section, I will briefly, and somewhat vaguely, compare the theory developed above with theories proposed by I. J. Good and Wesley Salmon.[27] I will also briefly and tentatively suggest how the theory of token *causation* developed above may figure in a theory of the *explanation* of particular events, here suggesting some comparisons and contrasts between the idea of probability trajectories developed above and Salmon's (1984) idea of causal processes and probability distributions "carried by" causal processes.

I. J. Good (1961–2) was, as far as I know, the first to recognize the distinction between, as he puts it, "the tendency of [X] to cause [Y]" and "the degree to which [X] actually caused [Y]". And he gives analyses of both ideas. As described near the end of Chapter 5, for the tendency of X to cause Y, Good proposes, basically, a measure of statistical relevance. Here, I will describe Good's analysis of the degree to which X actually caused Y.

Recall Good's example about Holmes, Moriarty, Watson, the cliff, and the boulder. In this example, Watson's pushing the boulder off the cliff had a *negative tendency to cause* Holmes's death by crushing, but nevertheless *actually caused* Holmes's death by crushing. Good says, "F [Watson's push] had a tendency to prevent E [Holmes's death by crushing] and yet caused it. We say that F was a cause of E because there

---

[27] This part of this section leans on my (1988b) "Probabilistic Causal Levels."

was a chain of events connecting F to E, each of which was strongly caused by the preceding one" (1983, pp. 216–17). It is natural to interpret Good's "degree to which an X actually caused a Y" as a measure of strength of token causation. This degree is high to the extent to which links of causal chains of actually occurring events, from the X to the Y, are, basically, highly probabilistically relevant to their successors (or, better: events of each of the successive *kinds* that were actually exemplified have a high *tendency to cause* – are highly causally positive for at the property level – events of the *kinds* of their successors).

The essence of Good's theory can, I think, be expressed as follows (I will not present the formal details). An X *token causes* (*or caused*) a Y if there is a chain of events (actually, event types) – X, Z_1, Z_2, . . . , Z_n, Y – such that (1) X, Y, and the Z_i's actually occurred (these factors were actually exemplified); (2) they (their exemplifications) are all spatiotemporally "adjacent" without "overlapping" very much; and (3) each has a strong *tendency* to cause its successor (at the property level, each is a strong positive causal factor for its successor). And the *degree* to which the X actually causes the Y (a number ≥ 0) is a function of the strengths of the individual links in the chain.[28]

Good's theory and the one developed in this chapter can give different answers to the question of the token causal significance of one event for another. Consider again the first example described above involving the sprayed plant. Recall that the application of the defoliant reduces the probability of survival from some high value to the low value of 0.1, but the plant gradually recovers and survives, and the probability of survival gradually recovers in time. The theory of this chapter says that the plant survived *despite* the application of the defoliant (since the probability of survival became low at

---

[28] Actually, Good's chains are a special case of his "nets." And the degree to which X actually causes Y is actually defined as the *limit* of the strength of nets connecting X to Y when the nets are analyzed in more and more detail, more and more "fine grained."

the time of spraying and stayed low for some time). However, it seems likely that there exists a chain of events connecting the application of the defoliant to the plant's eventual good health, such that each event in the chain is strongly causally positive for its successor. The spraying has a strong tendency to cause the plant to be unhealthy in some particular way; this in turn has a strong tendency to cause the plant to react in a certain way, setting into motion a process of combatting the effects of the poison; and so on. (Whether or not this is biologically quite accurate is not, of course, to the point; surely there are biologically correct examples that illustrate the point, perhaps involving a body's immune system's reaction to a typically deadly disease.) So Good's analysis can imply that the spraying actually caused the plant's survival, indeed to a degree that is high to the extent that the connections between successive members of the relevant chain are strong.

There are two key features of the difference between Good's theory and the one developed in this chapter that allow the theories to give conflicting answers. First, the theory developed in this chapter focuses at all times on the probability of Y, rather than on the relevance of events Z_i in a chain from X to Y to succeeding events Z_{i+1} in the chain. It is possible, of course, to put a lot of strong connections together in a chain and still wind up with a rather weak connection between the first and last elements of the chain. The probability trajectory model developed in this chapter reflects this feature of the defoliant case in the *gradual* increase in the probability of survival: As time goes on and more and more of the strong connections in the chain are successfully made, the higher the probability of survival (gradually) becomes, because there are fewer and fewer of the strong connections left that have to be made in order for the plant to survive.

Although Good's theory is sensitive to the fact that a series of strong connections may itself be a weak connection, it is important to note that the strength of a chain from an event X

369

to an event Y is, on Good's analysis, *independent of the actual initial probability of Y* (that is, at the time of X), so that, for example, this strength (the degree to which X actually caused Y) can be the same whether the trajectory of the probability of Y is high all the way from the time of X to the time of Y, or gradually recovers from a low value at the time of X to a high value at the time of Y.[29]

A second key difference between Good's theory and the one advocated here is that the latter includes the causal category of "despite," while the former does not. In Good's theory, the strength of a causal connection between an event X and a later event Y is a number greater than or equal to 0. So, intuitively, and in terms of the terminology of this chapter, Good's theory allows for the extreme of Y's happening *independently of X* and for degrees to which Y may happen *because of X*. To capture the token causal truth in cases like the defoliant example, however, it seems that a third dimension is needed. Surely the plant did not survive because it was sprayed; and surely the spraying had *some* causal role with respect to the survival of the plant (so that the plant's survival was not causally independent of the spraying). This is part of the reason why the theory of this chapter allows for the separate category of the "despite" kind of token causal significance. Given the way "despite" has been explicated above, we can say that one way in which the first difference between the two theories noted just above shows up, is in this second difference.

The two theories also seem to give different answers in the case of the golf birdie, described above. Just after the squirrel's kick, recall, the ball's momentum, the contour of the golf course, the positions of obstacles on the course, and so on, were, improbably enough, such that the probability of making the shot became very high again. But that was improbable; there was a very strong tendency of the squirrel's kick to *pre-*

<hr>

[29] Incidentally, this feature of Good's theory is illustrated by Wesley Salmon's example (1980, pp. 54–5) involving Joe Doakes and Jane Bloggs.

vent this. Thus, Good's theory says that the degree to which the squirrel kick actually caused the ball to fall into the cup was very low. (I note that, *on Good's theory, a causal chain is no stronger than its weakest link.*) The theory developed in this chapter, however, classifies the causal significance of the kick for the ball's falling into the cup as more or less of the "because" kind. The difference between the two theories that makes the difference in this case is that, while Good's theory uses just *type-level conditional probability comparisons,* the theory of this chapter uses the *actual trajectory of the probability of Y,* regardless of what the causal role or tendency might have been of the cause *X,* or of intermediate events, to *produce* just the trajectory that the probability of *Y* actually took.

On the other hand, it may seem possible to describe a sequence of factors, all exemplified between the kick and the ball's falling into the cup, such that each factor increases the probability of its successor. For example, the factor of the squirrel kick may increase the probability of the ball's entering the cup *from the particular angle* that the kick in fact caused the ball to enter the cup from. However, it is possible to modify the example in such a way that an analogue of the squirrel kick event confers an overwhelming probability on the ball's just disappearing: The event, in the modified example, makes the probability of the ball's disappearing so high that the probabilities of each of the ways in which the ball might enter the cup (or even survive the event) are all lower than they would have been in the absence of the event. That is, the conditional probability of the ball's entering the cup from any angle at all, conditional on the event, is lower than the conditional probability of the ball's entering the cup conditional on the absence of the event. Consistent with this, it is still possible that, in the particular case in question, the ball emerged from the event with a higher probability of entering the cup than it had before. This kind of *actual probability trajectory* is still possible, consistent with the *conditional probability comparisons* specified.

Wesley Salmon has called the approach of invoking a causal chain, and requiring just positive probabilistic relations between successive elements of it, "the method of successive reconditionalization" (1984).[30] Salmon (1980, 1984) has objected that there *may not be* any intermediate factors between a cause X and an effect Y that are all strongly probabilistically relevant to their immediate successors and predecessors. The kind of modification of the squirrel's kick example described just above is like this. Salmon describes a different, simpler, kind of example like this. He considers a fictitious case of an atom that can be in one of (at least) four states, labeled "1," "2," "3," and "4." The atom starts in state 4. The probabilities of going from state 4 to states 3 and 2 are, respectively, 0.75 and 0.25. If it goes to state 3, then the probability of its going into state 1 is 0.75 (otherwise, I suppose, the atom disappears, or goes into a state "0"). If it goes to state 2, then the probability of its going into state 1 is 0.25 (otherwise, I suppose, the atom disappears, or goes into a state "0"). See Figure 6.20. The most probable result is that the atom ends up in state 1 (probability = $(\frac{3}{4})(\frac{3}{4}) + (\frac{1}{4})(\frac{1}{4}) = \frac{10}{16}$). The most probable *way* for the atom to do this is to go from 4 to 3 to 1: the probability of the 4–3–1 chain is $(\frac{3}{4})(\frac{3}{4}) = \frac{9}{16}$. What actually happens, however, is that the atom goes from 4 to 2 to 1, improbably enough: the probability of the 4–2–1 chain is only $(\frac{1}{4})(\frac{1}{4}) = \frac{1}{16}$. The actual causal chain is 4–2–1; being in state 2 is negatively probabilistically relevant to going into state 1; and it seems that there is no way to fill in intermediate links, each positively probabilistically relevant to its successor, to "mend" the simple 4–2–1 chain.

I agree, of course, that there are cases in which an X token

[30] For a related proposal, see David Lewis's 1973 paper and 1986 book. Causation is characterized there in terms of causal chains, as the transitive closure of the relation of "causal dependence," which idea in turn is characterized in terms of counterfactual conditional relations. The proposal in his 1986 book (see especially p. 179), about what I have called "token-level probabilistic causation," is similar to "the method of successive reconditionalization" (except for Lewis's use counterfactual conditional relations).

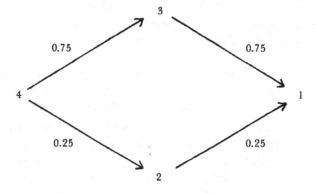

Figure 6.20

causes a Y but in which X may be negatively probabilistically relevant to Y (in the actual or in all appropriate causal background contexts), and in which there are no immediate events that can be interpolated to "mend the chain" in such a way that all events in the chain are probabilistically positive for their successors. However, I don't think that the example just described is most naturally construed as such a case. I am not sure whether or not Salmon would say that the atom's being in state 4 *token caused* the atom's being in state 2, or whether or not the idea is that the atom's being in state 2 *token caused* its being in state 1; though he does write that the 4–2–1 sequence is a "causal chain." In any case, it seems most natural to me, in light of the theory of this chapter, to say that, in the relevant context (the atom's originally being in state 4), the atom occupied state 1 *despite* its occupying state 2.[31] The natural probability trajectory diagram for this example – in which the probability of the atom's being in state 1 (at an appropriate later time) drops, from $^{10}\!/_{16}$ just before the time of the atom's actually entering state 2, to $^{1}\!/_{4}$ just after the time of

[31] Salmon notes (does not endorse) the possibility of an approach *somewhat* similar to this: namely, that "the transitions 'just happen by chance'; nothing *causes* them" (1980, note 24).

the atom's actually entering state 2 – gives exactly this diagnosis, given the theory elaborated in this chapter.

Salmon (1984) discussed "the method of successive reconditionalization" in the context of criticizing defenses of the idea that causes raise the probability of their effects, in the sense of conditional probability comparisons. This method was a last resort – after dismissing "the method of more precise specification of events" (discussed in Section 6.1 above) and "the method of interpolated causal links" (a sort of intermediate strategy between the above mentioned methods) – for saving a probability increase conception of probabilistic causation. If a cause X does not, strictly speaking, raise the probability of its effect Y, then at least, according to the successive reconditionalization idea, X raises the probability of a "direct" effect of it, which in turn raises the probability of a "direct" effect of it, and so on, until the next to the last member of the causal chain raises the probability of the final effect Y, a "direct" effect of the next to the last member of the chain. (Of course, this is best understood as an approach to token causal relations.)

As we have seen, Salmon rejects this approach. And, of course, I concur. And Salmon rejects altogether the idea of probability increase, understood in terms of conditional probability comparisons, as a mark of (token) causation. I concur with this as well. However, I have suggested an actual probability trajectory conception of probability change for assessing token-level probabilistic causal relations, as opposed to the conditional probability comparison conception that is appropriate for assessing property-level causal relations. Also, it seems that as a result of his rejection of probabilistic relevance relations as the mark of causation, Salmon further *rejects the idea of classifying token causal significances,* as (for example) positive, independent, or whatever – favoring instead his now well-known idea of a *causal process:* "The basic causal mechanism, in my opinion, is a causal process that carries with it probability distributions for various types of interactions" (1984, p. 203). Although Salmon discusses various possible structures of interactions, he does not attempt to classify different kinds of token causal significance in terms of the character of the probability distributions that a causal process carries with it, or in terms of the evolution of such distributions.

Roughly, a "causal process" is, for Salmon, a "self-determined" preservation of some kind of physical structure from one point in space and time to another. (My description of Salmon's important ideas about processes and related aspects of causality and explanation will be very rough here; see his 1984 for details on processes, interactions, events, "mark transmission," and so on, and how these ideas are characterized in terms of principles that rely on counterfactual conditionals). Physical objects, and propagations of waves, energy, and information are typical kinds of causal processes. Causal processes can intersect in space and time and as a result "causally interact" with one another. The causal interaction is called an "event." For example, the intersection of a defoliant spray process with a plant process results in an interaction, when the plant is poisoned. The plant process carries with it a propensity to change in character upon interaction with a defoliant spray process; and the defoliant spray process carries with it a propensity to undergo certain changes in structure upon interaction with a plant process. The intersection/interaction is the event of the poisoning of the plant. The idea that causal processes *carry with them these propensities,* which may change or evolve over time, seems closely related to the idea of *probability trajectories,* described in this chapter.

One of Salmon's primary concerns (1984) was with *the explanation of particular events* (explanations of other kinds of things, such as regularities, is also discussed in his 1984 book). The leading idea is that to have an explanation of an event is to understand how the event fits in, causally, with other events in the rough proximity of the event in question, and this in turn is given in terms of the characters of processes connecting other events to the event in question. Salmon, however, did not attempt a *classification* of the different kinds of token causal significance that one event can have for another, in terms of the characters of the causal processes that connect the one to the other. The primary concern of this chapter, on the other hand, has been with the *kinds* of token causal significance one particular event can have for another. And this study has taken as basic the probability trajectory that, to put it in Salmonian terms, is carried by the processes connecting the earlier event to the later event. I have called these kinds of causal significance "because," "despite," "independence," and "autonomy." Although the choice of these terms was *somewhat* arbitrary (as admitted above), I note that some of them carry some explanatory, as well as some causal, connotations.

In any case, consider: (1) Salmon's emphasis on the idea (which I endorse) of explaining an event in terms of token causal relations between the event in question and other events; (2) Salmon's emphasis on the idea of causal processes, that carry with them propensities for different kinds of interactions, as a basic aspect of causality as a relation between token events; (3) the rough idea that probability trajectories (as developed in this chapter) can model the "propensity-carrying" aspect of causal processes; and (4) the way in which different kinds of token causal significances, as well as their degrees, can be characterized in terms of the profile of probability trajectories (as described in this chapter). All this suggests, as a very rough characterization of the form of the explanation of a particular event, this: token event y has character Y *because of* events $x_{1,1}$, $x_{1,2}$, . . . having characters $X_{1,1}$, $X_{1,2}$, . . . , *despite* events $x_{2,1}$, $x_{2,2}$, . . . having characters $X_{2,1}$, $X_{2,2}$, . . . , *independently of* events $x_{3,1}$, $x_{3,2}$, . . . having characters $X_{3,1}$, $X_{3,2}$, . . . , and *autonomously of* events $x_{4,1}$, $x_{4,2}$, . . . having characters $X_{4,1}$, $X_{4,2}$, ... And degrees can be attached to the explanatory significance of such and such events' having such and such characters in the ways described above for causal significances.[32]

Of course, this sketch of the form of a theory of probabilistic causal explanation of particular events, based on the theory of probabilistic token causation developed in this chapter, is only rough and vague. And there are issues that confront a theory of probabilistic causal explanation of particular events that do not confront, and were not addressed in, the theory of probabilistic token causation developed in this chapter.

For example, in developing the theory of probabilistic token causation, I assumed that the "causal question" we wanted to answer *came with* an earlier event, a later event, and the "relevant" types for each event; I have tried to give forms and explications of answers to questions of the form: "What is the causal significance of event x's being of type X for event y's being of type Y?" Thus, I assumed that the relation under scrutiny includes *both* one earlier event, and one type for it, *and* one later event, and one type for it. As to the question of *explanation* of particular events, this does not seem very plausible. Although we may ask about the explanatory significance of one earlier event's exemplifying some one factor for a later event's exemplifying some one factor, it seems that a more "canonical form" for an explanation-seeking question, about a particular event, would be more like this: "Why was event y of character Y?" Thus, the first main difference between the analysis of token probabilistic causation (as developed in this chapter) and an analysis token probabilistic explanation, is

---

[32] Compare Humphreys (1981, 1983), on "aleatory explanations." Humphreys suggests that the canonical form of probabilistic causal explanations is this: "*A* because *phi*, despite *psi*." Here, *phi* is a nonempty set of contributing causes and *psi* is a (possibly empty) set of counteracting causes. Positive and negative causes are distinguished in terms of conditional probability comparisons, and an explanation need not cite particular probability values. Salmon (1984) cites this approach with general approval, and notes that, "If Humphreys' theory has a serious shortcoming, it would be . . . that he places insufficient emphasis on the actual [causal] mechanisms" (p. 266). Perhaps Salmon would pass the same judgment on the approach sketched here, which focusses on the *form* of probabilistic causal explanations, and is intended to isolate the formal, probabilistic features of processes that have explanatory significance.

that questions about the explanation of particular events will not generally "come with" an earlier "relevant" event, or a type for it, whereas questions about the causation of an event do (as such questions were understood in this chapter).

A second difference between the theory of probabilistic token causation proposed here and a development of it into a theory of probabilistic causal explanation of particular events is related to the point just noted. Not only did the "causal questions" addressed above come with an earlier event and a type for it, these question came with *just one* earlier event and *just one* type for it. In explaining events, however, it is typical or common to refer to *multiple* earlier events – and one can imagine that *multiple* types of *multiple* earlier events can be relevant to the explanation of a later event. This has already been indicated in the suggestion, above, for a "canonical form" for the structure of an explanation of a particular event. The suggestion involves possibly multiple explanatory events and types.

I have suggested some comparisons and contrasts between, on the one hand, some of Salmon's ideas on causation, processes, and explanation, and, on the other hand, some of the ideas developed here on causation, probability trajectories, and, very roughly just above, on explanation. Some further points of comparison between these approaches are worth noting.

First, in Salmon's (1984) theory of explanation, there are two levels: the "S-R (statistical relevance) tier" and the "causal tier." This bifurcation of levels of explanation parallels, in a way, the distinction between the property level and the token level of causation described in this book. Salmon's S-R basis of an explanation specifies, roughly, all factors that are statistically relevant to the character of the event to be explained, and just how these factors are statistically relevant to the character of the event to be explained. Further, there are principles about homogeneity and "maximal" homogeneity that are supposed to capture, as well as can be done in

378

terms of statistics, the idea that the factors cited are all and only the explanatorily relevant factors for the character of the event to be explained. (Again, I refer the reader to Salmon 1984 for details.) At an earlier time in the development of Salmon's theory of explanation (for example, in his 1971 book), it seemed that Salmon believed that explanatory relevance could be captured entirely in terms of statistical relevance relations among factors.

In his 1984 book, however, Salmon introduced the causal tier when he came to believe that statistical relations among factors can only be *evidence for* the true explanatory, causal relations among the event to be explained and other events. The causal level of explanation involves Salmon's ideas about causal processes, interactions, events, and so on (as very briefly described above). In his 1984 book, it appears that the causal level is more central to the explanation of events than is the S-R, statistical, level. In fact, Salmon seems tempted by the idea that once the causal level is adequately developed, the S-R basis level of an explanation may be discarded (pp. 265 ff.). Ultimately, the ideally complete S-R basis functions just as *evidence* for the causal relations described at the causal level, which in turn give the real explanatory information.

Likewise, given the way type and token causation have been explicated in this book, type-level causal relations function as *evidence* for token-level causal relations, in two ways. First, the probability trajectories used in the theory of token-level probabilistic causation give the probability of the *effect type* given the appropriate *types* that have fallen into place at the relevant times. Throughout, probability has always attached to factors (or properties, or types). And type-level statistical relations (as well as the falling into place of possibly improbable occurrences) ground the vertical placement of values in the probability trajectory diagrams that are central to the analysis of this chapter. And second, when assessing a potential token-level causal connection, we must figure out which types to hold fixed when drawing the probability tra-

379

jectory. These are the types such that (1) their exemplifications are, roughly, token uncaused by the earlier token event, and (2) they are, *at the type level,* causally relevant to the relevant type of the later token event. Thus, assessments of type-level probabilistic and causal relevance are, in this way as well, evidence for, but not constitutive of, token-level probabilistic causal relations, in the theory of this chapter.

A second point of comparison involves some issues of "locality" in the characterization of causal processes and in the factors relevant to the contour of probability trajectories as described in this chapter. Salmon distinguishes genuine causal processes from what he calls "pseudo processes" in terms of principles involving counterfactual conditionals and the idea of local interactions with the process in question. Roughly (again, see Salmon 1984 for details), a pseudoprocess maintains a continuity of structure, but the continuity of its structure is dependent on the continuity of genuine causal processes. Causal processes, in contrast, are "self-determined." For example, the now well-known pseudoprocess of the traveling spot of white light on the interior wall of the astrodome (Salmon 1984) is dependent on the genuine causal processes of the rotating beacon in the center of the dome, and on the propagations of white light from the beacon to the wall. The idea of locality, as well as the idea of counterfactual dependence, in Salmon's analysis, is easily explained in terms of this example.

Salmon points out that a local interference with the spot of white light on the interior wall of the astrodome – for example, by placing a red filter near the wall when a light beam from the beacon is about to reach the wall – *would not* affect the succeeding structure (color) of the spot of light. The spot of light cannot be "marked" by a local interaction. This fact, involving both *locality* and *a counterfactual element,* rules out the spot of light from counting as a *causal* process. Let us set aside the counterfactual element; I would like to concentrate on the locality feature here. An event that is *nonlocal* to the

380

spot of light, such as attaching a red filter to the front of the beacon light, *would* affect the succeeding structure of the spot of light (turning it from white to red). However, being a nonlocal interference with the spot of light, the effect of this kind of interference with the spot does not make the spot of light a causal process. This illustrates, in a rough way, how this locality idea figures in Salmon's way of distinguishing causal processes from pseudoprocesses.[33]

The issue of locality also confronts the theory of token-level probabilistic causation developed in this chapter – as well, therefore, as the sketch of a theory of explanation briefly suggested above. (I have, however, avoided the use of counterfactuals here, except for in the discussion of interaction and disjunctive causal factors, mainly in Chapter 3.) First, in thinking about the idea of probability trajectories in the characterization of token-level probabilistic causation, it is natural to think that the probability values traced in the diagrams should take into account (conditionalize on) only the appropriate factors that are *local* to the relevant process. Second, it is also natural to think that factors X and Y – exemplified by the token cause event and the token effect event, respectively – in the relation of token-level probabilistic causation should be restricted to factors that are *local* to the cause and effect events x and y. However, intuitions may vary in connection with these two ways in which the locality issue confronts the theory of this chapter.

Consider first the idea that we hold fixed only factors local to the relevant process, at the relevant times, when assessing probabilities for probability trajectories. In the original golf example of the ball kicked by the squirrel, the relevant "causal process" would seem to be the golf ball. The way in which the locality idea applies here may be illustrated as fol-

[33] See also Salmon's discussion of Richard Otte's *prima facie* counterexample to Salmon's principle "CI" (for "Causal Interaction") in Salmon's (1984, see pp. 172–4), and Philip Kitcher's (1985), as well as Elliott Sober's (1987b, see especially pp. 253–4).

lows. As the time of the squirrel's kick approaches, the squirrel is approaching the golf ball, the squirrel has a certain momentum, and the squirrel's collision with the ball becomes more and more inevitable. If one examines the (not so local) environment of the golf ball (which includes the approaching squirrel), the collision between the golf ball and the squirrel's foot becomes more and more probable. And given the way the example was described above, this would seem to make the probability of the golfer's making the shot improbable *already before the time of the kick,* which conflicts with the way the probability trajectory for this example was drawn. In the description of the example, however, I assumed that the squirrel's kick was improbable all the way until the time of the kick.

The correctness of the probability trajectory, as drawn, can be secured in a different way, however: We can stipulate that the factors conditionalized on, as they fall into place at the times against which the probability of making the shot is plotted, are all local to the golf ball. In this case, we do not conditionalize on anything having to do with the squirrel until the collision actually happens, when the squirrel's paw makes contact with (becomes local to) the ball. When I originally described the example, the point I wished to illustrate was just that (a type exemplified by) a token cause can be, at the property level, a negative causal factor for (a type exemplified by) a token effect of it. Thus, it was not inappropriate to assume that the squirrel's kick was improbable until the very moment of the kick. I don't think this assumption about the example should affect our verdict about the token causal role of the kick for the birdie – and the locality idea, about what kinds of factors to conditionalize on in the assessments of probabilities diagrammed in the probability trajectories, makes this assumption about the example superfluous.

Other examples considered above illustrate the same point. Consider again the plant-defoliant example. According to the way the probabilities are diagrammed in the relevant probability-trajectory diagrams, it was improbable until the very moment the defoliant reached the surface of the plant that the plant would be sprayed with defoliant. Again, this was an assumption of the example. But the locality stipulation again makes the assumption superfluous: We do not conditionalize on anything about the defoliant until the time at which the defoliant makes contact with (becomes local to) the plant process; at the appropriate time, we conditionalize on the factor of the defoliant's being on the surface of the plant. In other cases, the identification of the relevant *process* is more intricate – in particular, we must decide "how local" (just "how spatiotemporally confined") the process should be considered to be. In the Holmes–Moriarty–Watson case, for example, it seems natural to think of the Holmes–Moriarty–cliff–boulder system as the relevant process, a process that Watson interferes with by intervening and pushing the boulder of the cliff himself.

Note also that it is spatial locality, and not temporal locality, that is the point here. We have seen examples in which we must conditionalize all along on factors temporally intermediate between the cause x and the effect y. In Tolson's two-shock version of the original defoliant example, we conditionalize all along on the factor, eventually exemplified, of the stronger dose of the defoliant arriving on the surface of the plant in question at the relevant intermediate time. And in Forster's example about the defoliated tree and a subsequent invasion by parasites (that eat leaves and leave behind a deadly poison), we condition all along on the factor, eventually exemplified, of the parasites being in the vicinity of the tree at the relevant later time.

The second way, mentioned above, in which the idea of locality affects the theory of this chapter, has to do with the causal and effect factors, X and Y, that enter into the relation of token-level probabilistic causation. I have suggested that these also be restricted to factors local to the cause and effect events x and y. Recall that, until Sections 6.3 and 6.4, the only way in which the candidate token cause, x, entered into the analysis was by way of its *time*, t_x. This gave any other event happening simultaneously with x the same causal role as x, with respect to the later event y's exemplifying Y. (Recall the example involving a second squirrel doing something entirely irrelevant on a distant golf course.) However, in Sections 6.3 and 6.4, we saw that the relevant *type* X and the relevant *event* x have roles (beyond the relevant time t_x) in determining which factors Z, possibly temporally intermediate between t_x and t_y, should be held fixed all along in assessing the appropriate probabilities of y's being Y. This gave the relevant type X of x a role in the analysis. Now, the idea that the relevant factors X must be *local* to the relevant candidate token cause event x further enforces our theoretical identification of the intuitively relevant types X exemplified by a candidate token cause x.

Even more important in this connection, and for the intuitive appraisal of the theory of this chapter, is the "relevant" type Y of the candidate token effect y. Recall that an earlier event x can exemplify various (local) factors, X, X', X'', and so on, and a later event y can exemplify various (local) factors Y, Y', Y'', and so on, and the various factors that x exemplifies can have different token causal significances for the different types that y exemplifies. This idea was used above to diagnose certain intuitions when the probability-trajectory analysis seemed to give an intuitively incorrect verdict.

Consider again the possibility, in the original golf example, that, just after the kick, the golf ball emerged from the squirrel's foot on a different path toward the cup, but which conferred *exactly the same probability* on making the shot as the ball had just before the kick. In this case, I said that the ball's (simply) falling into the cup was token causally independent of the squirrel kick, but that the intuition that the birdie causally traced back to the squirrel kick could be captured, in the probability trajectory theory, by the idea that the golf ball's exemplifying the *local* factor of entering the cup from

384

the angle it entered the cup from was token causally *because* of the kick (where the probability trajectories vindicate this suggestion). But what if the ball entered the cup, after the squirrel kick, from the exact same angle (and in all other ways in the same way) as it (probably) would have without the kick?

Also, in the Holmes–Moriarty–Watson example, it is possible that the boulder emerged, improbably enough, from Watson's fingertips with the very same probability of crushing Holmes as it had before Watson intervened. In this case, it may well be that the effect (Holmes's death) exemplified the local factor of there being a boulder with Watson's fingerprints on it there. This may accommodate, within the theory developed in this chapter, the intuition that Holmes's being crushed traces back to Watson's action. But what if Watson wore gloves, and left no physical trace on the boulder of his interaction with the Holmes–Moriarty–cliff–boulder process. That is, what if the effect, Watson's death, happened in a way that is locally indistinguishable from the way it (probably) would have happened were it not for his intervention?

In cases of this kind, the idea was that, where x and y are the earlier and later events, even when certain "natural" factors Y that y exemplified had their probabilities unchanged by the exemplification of X by x (when seemingly clearly these features of y "causally traced back to" to the occurrence of x), there nevertheless *could be found* some factors Y', exemplified by y, whose exemplification by y was, on the theory, token because of x's exemplifying some appropriate factor X or X'. If we were to drop the locality ideas just described, then it would be easy to find such factors Y'. In the golf example, we could easily say, and it would fit the theory, that the birdie exemplified the factor, "falling into the cup *after being kicked by a squirrel*" token because of the squirrel kick. And in the Holmes case, we could say that the event of Holmes's death exemplified the factor "crushed by a boulder *pushed off a cliff by Watson*" token because of what Watson did.

What should we say about this? I can only say that I have

385

intuitions two ways. On the one hand, I have sympathy with the "physically tracing back to" idea. Dropping the locality stipulations (or being less restrictive about what is to count as locality to a process) gives us one way of directly accommodating the relevant intuitions in the examples just described, by using the factors described in the previous paragraph. On the other hand, I see the possibility that, in the examples just described, the earlier event has no actual effect on the (probability trajectory) chance of the later event, however the later event may be (locally) described. I think it is not implausible to say that in the version of the golf example just considered, the ball entered the cup token causally independently of the kick, and that in the version of the Holmes et al. case just considered, Holmes was crushed to death token causally independently of Watson's action. In neither case did the earlier event actually succeed in making a (local) difference in the character of the later event.

I favor the locality stipulations. And in the examples just considered, if the later event *does* physically trace back to the earlier event, then the earlier event must have made some difference for events causally intermediate between the earlier and later events. By looking at these causally intermediate events, at the earlier events' causal roles for them, and at the causal roles of the intermediate events for the later event, it seems that we can, within the probability trajectory theory, identify the fact that the later event traces back to the earlier one, where, on the concept of cause developed in this chapter, the later event is nevertheless called "token causally independent of" the earlier event.

A final point of comparison between Salmon's ideas on causality and the ideas of this chapter is worth noting. This concerns what might be called the "laws of motion" of probability trajectories, or of the probability distributions "carried by" causal processes.[34] We can ask, "How, if at all, these

---

[34] I thank Elliott Sober for raising this issue with me.

probabilities can change in the absence of any interactions between the process in question and other processes?"

I begin by noting that Salmon (1984) characterizes causal interactions beteen processes (in his principle "CI," p. 171) in terms of processes P_1 and P_2, of characters Q and R respectively, that *would not* have undergone changes in their characteristics Q and R *were it not* for the intersection of the two processes P_1 and P_2 in question, but which processes nevertheless do undergo changes, from Q to Q' in the case of P_1 and from R to R' in the case of P_2, at the time and place of the intersection of the processes. Later (p. 175), the idea is extended to not only a sufficient but also a necessary condition: Roughly, two processes causally interact with one another *if and only if* there are *some* characteristics possessed by the processes that *would not have changed were it not for the intersection between the two processes,* but nevertheless do change, at the time of the intersection of the two processes. This suggests that, for Salmon, processes do not change unless interacted with by other processes. And if we include, among characteristics Q, R, and so on, that processes can exhibit, also *the probability distributions that the processes carry with them,* then all this suggests that the "laws of motion," for the probability distributions carried by causal processes, are what may be described as "Newtonian" in character: the probabilities do not change unless the relevant process is acted upon.

I agree with the idea that causal processes should be thought of as having a kind of stability of character in the absence of external interventions, or interactions. In this connection, however, Salmon also discusses Russell's idea of a "causal line," according to which, in part, "Throughout a given causal line, there *may be* constancy of quality, constancy of structure, *or gradual change in either, but not sudden change of any considerable magnitude* (Russell 1948, p. 459; quoted from Salmon 1984, p. 144; emphasis added here). Russell's description of a causal line is somewhat similar to Salmon's description of a causal process, though there are differences.[35] Here, I want to draw attention to the idea of the possibility of a "gradual change" in a causal line, or process, the idea I emphasized in the quotation from Russell. This possibility raises an interesting question about the "laws of motion" of probability trajectories and of the probability distributions "carried by" causal processes. Salmon's characterizaton of a causal process seems to suggest no change in processes without interaction with another process, and Russell's characterization of a causal line suggests the possibility of a "gradual change."

I have four points to make myself in this connection. First, it seems that causal processes are hardly ever isolated from other ones. This fact may account not only for "discontinuities" in a probability trajectory, but also for gradual changes over time in the evolution of a probability traced by a probability trajectory, and the same goes for the probability distributions that a causal process carries with it. For example, Salmon (1984) several times makes use of the case in which the process of a kid's baseball is making its way toward a neighbor's window. Salmon says, at one point, "As the ball travels . . . it loses energy, and its propensity to shatter glass changes along its path" (pp. 203–4). Of course, this loss of propensity to shatter a glass pane is naturally describable as due to the interaction between the ball and the surrounding air (air resistance), as a result of which the ball gradually loses momentum, or its glass-shattering propensity. Whatever the physical details, this example may serve to illustrate the idea that, in the first place, causal processes seldom, if ever, actually occur without interactions with other causal processes,

[35] Salmon (1984, pp. 144–5) points out several differences. For example, Salmon explicitly characterizes and distinguishes between *causal* and *pseudo* processes, and he discusses the effects that nonlocal events can have on the structure of pseudo-processes. Also, Russell mentions, in the characterization of causal lines, the idea, about the "events" in a causal line, that "given some of them, something can be inferred about the others" (Russell 1948, quoted from Salmon 1984, as just above), while Salmon on the other hand emphasizes that his idea of causal processes is characterized in terms of objective, ontic, ideas, rather than in epistemic terms.

and gradual changes in probability trajectories, or in causal processes or lines, may in fact be attributable to interactions with other causal processes after all.

But second, in the analysis of probability trajectories above, I suggested that we hold fixed certain factors whose exemplifications are token uncaused by an earlier event, when assessing the token causal significance of the earlier event for a later event. Consider again the event of setting a baseball into motion in the direction of a glass pane. In this case, *the presence of air* between the place of the setting into motion of the baseball and the place of the glass pane, as well as the air's capacity to offer resistance against the motion of the ball, are factors whose exemplifications are token uncaused by the setting into motion of the ball; and these factors are causally relevant to whether or not a ball will shatter a glass. So it seems that we should hold these factors fixed, all along, in plotting the probability trajectory for assessing the causal significance of setting the ball into motion for the shattering of the glass. In plotting probability trajectories, all the relevant factors that have fallen into place, as well as certain factors whose exemplifications are token uncaused by the relevant earlier event, should already be taken into account at each time. The impact of this is to restore some stability to the evolution of the probability distributions traced in probability trajectories, or carried by the relevant causal process. This again would seem to reinforce the idea that the "laws of motion" for probability trajectories, and for the distributions carried by causal processes, are basically "Newtonian" in character.

Third, some processes are more complex than others, consisting themselves of more basic, or simpler, processes. For example, a plant process consists of numerous internal, sometimes interacting, biological processes. And when a plant is recovering from being sprayed with defoliant, there are various processes internal to the plant that, when working together, can gradually effect the recovery of the plant. The

389

processes internal to a complex process can interact with each other, and thus, perhaps typically gradually, affect the probability distributions carried by the larger, more complex process constituted of its simpler, interacting, component processes. It was with this kind of idea in mind that I have drawn the probability trajectories for the defoliated plant examples (and others) above: in each case, the probability of eventual survival, in the correct diagram for each case, was drawn as *gradually* increasing. I don't know whether the internal interactions in such a complex process result in discrete or continuous changes in the probabilities carried by the larger process, but in any case I have drawn the trajectories as continuous.

Finally, there is the factor (so to speak) of time's passing that can affect the slope of a probability trajectory. As time goes on between the time of a token cause and the time of a token effect of it, if all is "going well" for the occurrence of the token effect, then less and less is left to "go wrong" as far as the occurrence of the effect is concerned. Consider the original golf example. After the squirrel kicks the golf ball, the probability of a birdie is high again, but not 1. The reason that it is not 1 is that various things can go wrong between just after the kick and the time of the birdie: The ball can collide with another squirrel's foot, with a small new weed, with a sudden unpredictable gust of wind, and so on. But, as time goes on, such possibilities gradually disappear, and, intuitively, less and less is left to go wrong. This is reflected, in the probability trajectories I have drawn, in a *small positive slope* (southwest to northeast, but still pretty flat) between the time of the squirrel's kick and the time of the ball's falling into the cup. Of course, this "factor" of the passing of time also should contribute some "positive component" to the slope of the probability trajectories in the defoliant examples – and the same goes, of course, for the other examples. Also, I do not mean to say here that the very passing of time is itself a causal factor. Rather, the relevant causal factors here are the various genuine causal factors that are negative for the relevant factor exemplified by the token effect event, and that happen to fail to be exemplified at the relevant (earlier) times.

In this section, I have compared the theory of token-level probabilistic causation developed in this chapter with theories advanced by I. J. Good and by Wesley Salmon. And I have suggested some comparisons between the theory of this chapter and the way Salmon applies his ideas about causation to the question of the explanation of particular events. I have also suggested a sketch of the form of probabilistic explanations of particular events (for a particular event's exemplifying some property).

As to the comparison between the ideas on causation advanced in this chapter, on the one hand, and Salmon's ideas on causation and explanation, on the other hand, I think that perhaps the closest parallel involves the idea that causal processes – or "extended" events, objects, or whatever is involved in token probabilistic causal relations – have or carry with them probabilities that can evolve on their own or as a result of interactions with other such items. And perhaps the thing that distinguishes the two approaches the most is that while I have attempted to classify different kinds of token probabilistic causal significance (which may also be kinds of explanatory significance), it seems that, as a result of his dismissal of positive statistical relevance (a conditional probability comparison idea) as the mark of causation, Salmon has rejected the very idea of such a classification.

I think that one way of looking at the theory of probabilistic token causation developed in this chapter is as an attempt to identify and analyze, in terms of probability trajectories, the salient features of the evolution of the probability distributions that Salmon says are "carried by" causal processses. Whether or not these comparisons are valid – and in any case, much of what I have offered in this last section is somewhat tentative, vague, and suggestive – I am hopeful that the distinctions involving the various kinds of token *causal* significance may be helpful in elaborating a formally and intuitively acceptable theory of the *explanation* of particular events. As indicated above, however, this would seem to involve formal and intuitive intricacies perhaps suitable for a separate project.

Appendix 1: Logic

The only symbolic logic used in this book is a small part of *propositional logic,* also called *sentential logic* or *Boolean logic.* In this appendix, I review the relevant part of this simple area of logic and clarify some notation and terminology. This appendix is not an introduction to logic; various important fine points and distinctions are not be mentioned. But I hope this will suffice as an introduction to the basic ideas in the elementary part of logic used in this book, and as a clarification of the logical terminology and symbols used in this book.

The basic entities of the formal propositional calculus are usually called the *propositions* and the propositional *connectives* (and the language of propositional logic usually includes *punctuation marks,* usually parentheses, that are used to avoid ambiguity of grouping when "propositions" are "connected" in complex ways).

In this book, it is *factors* (or *properties,* or *types*) that play the role of the so-called *propositions* of propositional logic. The abstract and formal propositional calculus can be *interpreted* as applying to propositions in a number of ways in which the term "proposition" could be understood. For example, we could think of propositions as sentences (which may be understood as concrete linguistic entities such as utterances or inscriptions). Or we could think of them as statements (understood in such a way that many sentences can all be used to "make" the same statement, and the same sentence, if used in different contexts, would make different statements). Or we can think of propositions as *sets of "possible worlds"* (as many philosophers and logicians have done in studies of modalities, such as possibility and necessity, and various kinds of logical relations, such as counterfactual conditional connections). And the propositions can also be understood as factors (or properties, or types).

In interpreting the abstract and formal propositional calculus, we can think of propositions *either* as something like sentences or statements, in which case they are, roughly speaking, *true or false in a given situation, or* as factors (properties or types), in which case they are *exemplified or not exemplified in a given instance.* In this appendix, which is a general review of some *formal* aspects of *abstract* propositional logic, I will use the term "proposition" to refer to the entities that play this role in the calculus. And I will generally speak of propositions as being either true or false, rather than of factors as being either exemplified or not exemplified. Formally, the idea of the truth or falsity of a given sentence or statement in a situation (or a possible world) is quite parallel to the idea of the presence or absence of a given factor (or property or type) in an instance. For example, a factor is exemplified or not exemplified in a situation, according to whether the statement asserting it is exemplified is true or false in the situation. And the formal parallelism remains intact with the introduction of the propositional connectives.

The *connectives* of propositional logic, sometimes called *sentential connectives* or *truth-functional connectives* or *Boolean connectives,* can be used to form "new" propositions from "old" ones. The "new" propositions are called *truth-functional compounds,* or *Boolean compounds,* of the "old" propositions. The most usual connectives (and the only propositional connectives used in this book) are the ones for which I use the following three symbols (as is most standard): "~", "&", and "$\lor$". These represent negation ("not"), conjunction ("and"), and disjunction "or, or both").

Let X and Y be any propositions. Then, understood as sentences or statements, X and Y are entities that are either true or false, in any given situation. (Alternatively, X and Y

can be factors, which are entities that are, in any given instance, either exemplified or not exemplified.) Then: (1) $\sim X$, the negation of X, is the proposition that is true in a situation just in case (if and only if) X is false in the situation (or it is the factor that is exemplified in an instance just in case X is not exemplified in the instance); (2) $X\&Y$, the *conjunction* of X and Y, is the proposition that is true in a situation just in case both X is true in the situation and Y is true in the situation (or it is the factor that is exemplified in an instance just in case both X is exemplified in the instance and Y is exemplified in the instance); and (3) $X\lor Y$, the *disjunction of* X and Y, is the proposition that is true in a situation just in case at least one of X and Y is true in the situation (or it is the factor that is exemplified in an instance just in case at least one of X and Y is exemplified in the instance).

X is sometimes called the *negatum* of $\sim X$. X and Y are called the *conjuncts* of $X\&Y$. And X and Y are called the *disjuncts* of $X\lor Y$. More than two propositions can be *conjoined:* Their conjunction is true (exemplified) if all the conjuncts are true (exemplified). And more than two propositions can be *disjoined:* Their disjunction is true (exemplified) if at least one of the disjuncts is true (exemplified).

Three important kinds of propositions are the tautologies, the contradictions, and the contingent propositions. A proposition is a *tautology* (or *logically true*) if it cannot be false (it cannot not be exemplified). A proposition is a *contradiction* (or *logically false*) if it cannot be true (it cannot be exemplified). And a proposition is *contingent* (or *logically indeterminate*) if it is both possible for it to be true (exemplified) and possible for it to be false (be not exemplified). If X is any proposition, then examples of tautologies are $X\lor\sim X$ and $\sim(X\&\sim X)$, and examples of contradictions are $X\&\sim X$ and $\sim(X\lor\sim X)$. (Hereafter, I will not add the parenthetical "factor *exemplified* in an instance" after "proposition *true* in a situation", or "factor *not exemplified* in an instance" after "proposition *false* in a situation" – the parallel is always the same.)

Three important relations that can obtain between two propositions are those of implication, equivalence, and independence. A proposition X *implies* (or *logically implies*) a proposition Y if it is not possible for X to be true while Y is false (in the same situation). Propositions X and Y are *equivalent* (or *logically equivalent*) if they must be either both true or both false (they cannot differ from each other with respect to truth and falsity in a given situation). Another way of putting this is to say that X and Y are equivalent if each implies the other. Finally, two propositions are *independent* (*logically independent*) if all four combinations of the truth and falsity of X and the truth and falsity of Y are possible. Another way of putting this is to say that X and Y are independent if neither of X and $\sim X$ implies either of Y or $\sim Y$ (and vice versa, to be redundant).

A *set of propositions* is said to be *closed* under the propositional connectives ($\sim$, &, $\vee$) if it contains all Boolean compounds ($\sim X$, $X \& Y$, $X \vee Y$) of propositions (X and Y) that it contains. And the *closure* of a set of propositions is the smallest set that contains the given set and is closed under the propositional connectives.

Propositions in a set of propositions are *mutually exclusive* if at most one of them could be true: *The conjunction of any two or more of them is a contradiction (or logically false)*. Propositions in a set of propositions are *collectively exhaustive* if they cannot all be false: *The disjunction of all of them is a tautology (or logically true)*. A *partition* is a set of mutually exclusive *and* collectively exhaustive propositions. For any partition, exactly one proposition in it is true (in a given situation).

Relative to a set S of propositions that is closed under the usual connectives, a proposition X is *maximally specific* if X is a member of S and there is no proposition Y in S such that both (1) Y implies X and (2) Y is not equivalent to X – that is, any proposition Y in S that implies X is equivalent to X. If we assume an interpretation of propositions under which *logically equivalent* propositions are *identical* propositions (which

396

is plausible for propositions understood as statements, as sets of possible worlds, or as factors, or properties or types, but *not* for propositions understood as sentences), then we could say that X is maximally specific relative to a closed set S if no proposition in S, other than X, implies X. *Relative to any set T of propositions, X is maximally specific* if X is maximally specific relative to the closure of T under the usual connectives.

Any element of a set of propositions is equivalent to a disjunction whose disjuncts are propositions that are maximally specific relative to the set. And a maximally specific proposition relative to a set is always equivalent to a conjunction whose conjuncts are all either members of the set or negations of members of the set. The set of propositions maximally specific relative to a set is always (*modulo equivalence* – that is, treating equivalent propositions as identical) a partition. If $S_1, \ldots, S_n$ are all partitions, then a proposition is maximally specific relative to the union of the S_i's if and only if it is equivalent to a conjunction $X_1 \& \ldots \& X_n$, where each X_i is a member of S_i.

Not every set of propositions is such that, relative to it, there exist maximally specific propositions; there are the "atomless Boolean algebras." A set of propositions that is closed under the usual connectives, together with the relation of implication, is one kind of Boolean algebra. A *Boolean algebra, B*, is simply any structure, $B = \langle S, \sim, \&, \vee, \Rightarrow, 0, 1 \rangle$, in which S plays the formally analogous role of a set of propositions that is closed under the operations formally analogous to the ways "$\sim$", "$\&$", and "$\vee$" were described above, where $\Rightarrow$ corresponds to implication, and where for any element X of S, 0 is equivalent (and identical) to $X \& \sim X$ and 1 is equivalent (and identical) to $X \vee \sim X$.

A Boolean algebra is called *atomless* if, for every element X of it, there is another element Y of it that is "strictly less than" X. Understanding the algebra as a set of propositions, this means that Y implies X but X does not imply Y (so that $X \neq Y$). Of course, all atomless Boolean algebras are infinite. And atomless Boolean algebras do not have any maximally specific elements. All finite Boolean algebras have maximally specific elements, called *atoms,* such that every member of the algebra is a disjunction of these atoms. A Boolean algebra is *complete* (or *sigma-additive*) if it contains all infinite conjunctions and disjunctions of its members, as well as the finite conjunctions and disjunctions of its members. All complete Boolean algebras have atoms such that every member of the algebra is a disjunction of atoms; not all infinite Boolean algebras are atomless.

Appendix 2: Probability

In this appendix, I will present some of the basic ideas of the mathematical theory of probability. As in the case of Appendix 1, this will not be a comprehensive or detailed survey; it is only intended to introduce the basic formal probability concepts and rules used in this book, and to clarify the terminology and notation used in this book. Here I will discuss only the *abstract and formal* calculus of probability; in Chapter 1, the question of *interpretation* is addressed.

A probability function, *Pr*, is any function (or rule of association) that assigns to (or associates with) each element X of some Boolean algebra B (see Appendix 1) a real number, $Pr(X)$, in accordance with the following three conditions:

For all X and Y in B,

(1) $Pr(X) \geq 0$;
(2) $Pr(X) = 1$, if X is a tautology (that is, if X is logically true, or $X = 1$ in B);
(3) $Pr(X \vee Y) = Pr(X) + Pr(Y)$, if $X \& Y$ is a contradiction (that is, if $X \& Y$ is logically false, or $X \& Y = 0$ in B).

These three conditions are the *probability axioms,* also called "the Kolmogorov axioms" (for Kolmogorov 1933). A function *Pr* that satisfies the axioms, relative to an algebra B, is said to be a *probability function on B* – that is, with "domain" B (that is, the set of propositions of B) and range the closed interval $[0,1]$. In what follows, reference to an assumed algebra B will be implicit.

In Appendix 1, I explained how the *propositional* calculus is applicable to "propositions" understood as sentences or statements as well as to "propositions" understood as factors or properties – and the same goes for the *probability* calculus. Roughly speaking, "$Pr(X) = r$" can be understood either as asserting that a *sentence or statement* X has a probability of r of being *true* (in a given situation), or as asserting that a *factor or property* has a probability of r of being *exemplified* (in a given instance or population). Specifying an interpretation of the propositions is part what must be done to "interpret" a probability function on an algebra; the other part is interpreting "*Pr*." Various interpretations of probability (such as frequency, degree of belief, and partial logical entailment interpretations) are discussed in Chapter 1; here, the focus is on the formal calculus.

Here are some easy consequences of the probability axioms.

(4)　$Pr(\sim X) = 1 - Pr(X)$, for all X.
　　 Proof:　By (1), $Pr(X \lor \sim X) = 1$; and by (3), $Pr(X \lor \sim X)$
　　　　　　$= Pr(X) + Pr(\sim X)$. So, $1 = Pr(X) + Pr(\sim X)$,
　　　　　　and thus $Pr(\sim X) = 1 - Pr(X)$.

(5)　$Pr(X) = 0$, if X is a contradiction.
　　 Proof:　$\sim X$ is a tautology, so by (2), $Pr(\sim X) = 1$. By
　　　　　　(4), $Pr(\sim X) = 1 - Pr(X)$. So, $1 = 1 - Pr(X)$,
　　　　　　and thus $Pr(X) = 0$.

(6)　$Pr(X) = Pr(Y)$, if X and Y are logically equivalent.
　　 Proof:　X and $\sim Y$ are mutually exclusive and $X \lor \sim Y$ is
　　　　　　a tautology. So by (2), (3), and (4), $1 =$
　　　　　　$Pr(X \lor \sim Y) = P(X) + Pr(\sim Y) = Pr(X) + 1 -$
　　　　　　$Pr(Y)$. So, $1 = Pr(X) + 1 - Pr(Y)$, and $0 =$
　　　　　　$Pr(X) - Pr(Y)$, and thus $Pr(X) = Pr(Y)$.

(7)　$Pr(X) \le Pr(Y)$, if X logically implies Y.
(8)　$0 \le Pr(X) \le 1$, for all X.
(9)　$Pr(X \lor Y) = Pr(X) + Pr(Y) - Pr(X \& Y)$, for all X and Y.

The probability of Y *conditional on (or given)* X, written $Pr(Y/X)$, is defined to be equal to $Pr(X \& Y)/Pr(X)$. Note that $Pr(Y/X)$ is defined only when $Pr(X) > 0$. Since for any X and Y, $Pr(X \& Y) = Pr(Y \& X)$ (by (6) above), an immediate consequence of the definition of conditional probability is what is often called the *multiplication rule:*

(9) $Pr(X \& Y) = Pr(X)Pr(Y/X) = Pr(Y)Pr(X/Y)$, for all X and Y.

From (9) follows this simple version of *Bayes's theorem*:

$Pr(Y/X) = Pr(X/Y)Pr(Y)/Pr(X)$, for all X and Y.

A proposition Y is said to be *probabilistically* (or *statistically*) *independent* of a proposition X if $Pr(Y/X) = Pr(Y)$. Alternatively, and equivalently, Y's being probabilistically independent of X can be defined as $Pr(X \& Y) = Pr(X)Pr(Y)$. Thus, probabilistic independence is symmetric: If Y is probabilistically independent of X, then X is probabilistically independent of Y, for all X and Y.

If propositions X and Y are not probabilistically independent, then there is said to be a *probabilistic* (or *statistical*) *correlation* (or *dependence*) between X and Y. The correlation is called *positive* or *negative* according to whether $Pr(Y/X)$ is greater or less than $Pr(Y)$. This is sometimes described by saying that X is *positively or negatively probabilistically relevant to Y*, or that X has *positive or negative probabilistic significance for Y*. It is easy to see that the following six probabilistic relations are equivalent:

$$Pr(Y/X) > Pr(Y);$$
$$Pr(X/Y) > Pr(X);$$
$$Pr(Y) > Pr(Y/\sim X);$$
$$Pr(X) > Pr(X/\sim Y);$$
$$Pr(Y/X) > Pr(Y/\sim X);$$
$$Pr(X/Y) > Pr(X/\sim Y).$$

Also, these six relations would remain equivalent if the ">" 's were all replaced with "<" 's, or with "=" 's. Thus, the two kinds of probabilistic correlation (positive and negative), as well as probabilistic independence, are symmetric. If $Pr(Y/Z \& X) = Pr(Y/Z \& \sim X)$, then Z is said to *screen off* Y from X.

Two propositions X and Y are called *probabilistically equivalent* if $Pr((X\&Y) \lor (\sim X\&\sim Y)) = 1$. Another way of putting this is as follows. A common propositional connective, not mentioned in Appendix 1, is the *biconditional connective*, "$\leftrightarrow$". The biconditional of two propositions X and Y is the proposition that is true just in case X and Y have the same *truth value* – that is, either they are both true or they are both false. The *biconditional* of X and Y is often expressed as "X if and only if Y," or, for short, "X iff Y" (*X if Y, and X only if Y*). Then X and Y are probabilistically equivalent just when $Pr(X\leftrightarrow Y) = 1$. When two propositions X and Y are probabilistically equivalent, then they are "interchangeable in all probabilistic contexts." That is, given that X and Y are probabilistically equivalent, if (possibly truth–functionally complex) propositions $Z(X, Y)$ and $W(X, Y)$ result from any (possibly truth-functionally complex) propositions Z and W, respectively, by changing X's to Y's or Y's to X's, in any way, then $Pr(Z/W) = Pr(Z(X,Y)/(W(X,Y))$.

A generalization of the common idea of an average is the statistical idea of expectation, or expected value. Given a variable N that can take on the possible values $n_1, \ldots, n_r$, and a probability Pr on propositions of the form "$N = n_i$", the *expectation*, or *expected value*, of N (calculated in terms of the probability Pr) is

$$\Sigma_{i=1}^r Pr(N = n_i)n_i.$$

If the probabilities in terms of which an expectation is calculated are conditional probabilities, then the expectation is a *conditional expectation*, or *conditional expected value*. For example, if R is a proposition that may be relevant to the value of N, then

$$\Sigma_{i=1}^r Pr(N = n_i/R)n_i$$

is a conditional expectation.

Bibliography

Anderson, J. (1938), "The Problem of Causality," *Australasian Journal of Psychology and Philosophy 16:* 127–42.

Asquith, P. D., and Giere, R. N. (eds.) (1980), *PSA 1980,* vol. 1. Philosophy of Science Association, East Lansing, MI.

Asquith, P. D., and Kitcher, P. (eds.) (1985), *PSA 1984,* vol. 2. Philosophy of Science Association, East Lansing, MI.

Asquith, P. D., and Nickles, T. (eds.) (1983), *PSA 1982,* vol. 2. Philosophy of Science Association, East Lansing, MI.

Bell, J. S. (1964), "On the Einstein Podolsky Rosen Paradox," *Physics 1:* 195–200.

(1971), "Introduction to the Hidden Variable Question," in *Foundations of Quantum Mechanics,* edited by B. d'Espagnat. Academic Press, New York.

Benenson, F. C. (1984), *Probability, Objectivity and Evidence.* Routledge and Kegan Paul, London.

Bickel, P. J., Hammel, E. A., and O'Connell, J. W. (1977), "Sex Bias in Graduate Admissions: Data from Berkeley," in Fairley and Mosteller (1977).

Buck, R. C., and Cohen, R. S. (eds.) (1971), *Boston Studies in the Philosophy of Science,* vol. 8. Reidel, Dordrecht.

Carnap, R. (1945), "The Two Concepts of Probability," *Philosophy and Phenomenological Research 5:* 513–32.

(1950), *Logical Foundations of Probability* (2nd edition, 1962). University of Chicago Press, Chicago and London.

(1952), *The Continuum of Inductive Methods.* University of Chicago Press, Chicago and London.

Carnap, R., and Jeffrey, R. C. (eds.) (1971), *Studies in Inductive Logic, I.* University of California Press, Berkeley and Los Angeles.

Cartwright, N. (1979), "Causal Laws and Effective Strategies," *Nous 13:* 419–37.

(1988a), "Regular Associations and Singular Causes," in Skyrms and Harper (1988), pp. 79–97.

(1988b), "Reply to Ellery Eells," in Skyrms and Harper (1988), pp. 105–8.

(1989), *Nature's Capacities and Their Measurement*. Clarendon Press, Oxford.

Cohen, L. J., and Hesse, M. B. (eds.) (1980), *Applications of Inductive Logic*. Oxford University Press, New York.

Cohen, M. R., and Nagel, E. (1934), *An Introduction to Logic and Scientific Method*. Harcourt, Brace, New York.

Collier, J. (1983), "Frequency Dependent Causation: A Defense of Giere," *Philosophy of Science 50:* 618–25.

Davis, W. A. (1988), "Probabilistic Theories of Causation," in *Probability and Causality*, edited by James H. Fetzer. Reidel, Dordrecht, pp. 133–60.

DeFinetti, B. (1937), "Foresight: Its Logical Laws, Its Subjective Sources," translated by H. E. Kyburg, Jr., in Kyburg and Smokler (1964), pp. 93–158.

Dretske, F., and Snyder, A. (1972), "Causal Irregularity," *Philosophy of Science 39:* 69–71.

Dupré, J. (1984), "Probabilistic Causality Emancipated," in French, Uehling, and Wettstein (1984), pp. 169–75.

Dupré, J., and Cartwright, N. (1988), "Probability and Causality: Why Hume and Indeterminism Don't Mix," *Nous 22:* 521–36.

Edgeworth, F. Y. (1892), "Correlated Averages," *Philosophical Magazine 34:* 191–204.

(1910), "On the Application of the Calculus of Probability to Statistics," *International Statistical Institute Bulletin 18:* 505–36.

Eells, E. (1982), *Rational Decision and Causality*. Cambridge University Press, New York and Cambridge.

(1983), "Objective Probability Theory Theory," *Synthese 57:* 387–442.

(1986), "Probabilistic Causal Interaction," *Philosophy of Science 53:* 52–64.

(1987a), "Probabilistic Causality: Reply to John Dupré," *Philosophy of Science 54:* 105–14.

(1987b), "Cartwright and Otte on Simpson's Paradox," *Philosophy of Science 54:* 233–43.

(1988a), "Probabilistic Causal Interaction and Disjunctive Causal Factors," in *Probability and Causality*, edited by James H. Fetzer, Reidel, Dordrecht, pp. 189–209.

(1988b), "Probabilistic Causal Levels," in Skyrms and Harper (1988), pp. 109–33.

(1988c), "Eliminating Singular Causes: Reply to Nancy Cartwright," in Skyrms and Harper (1988), pp. 99–104.

404

Eells, E., and Sober, E. (1983), "Probabilistic Causality and the Question of Transitivity," *Philosophy of Science 50:* 35–57.

(1986), "Common Causes and Decision Theory," *Philosophy of Science 53:* 223–45.

(1987), "Old Problems for a New Theory: Mayo on Giere's Theory of Causation," *Philosophical Studies 52:* 291–307.

Einstein, A., Podolsky, B., and Rosen, N. (1935), "Can Quantum-Mechanical Description of Physical Reality be Considered Complete?," *Physical Review 47:* 777–80.

Fairley, W. B., and Mosteller, F. (eds.) (1977), *Statistics and Public Policy.* Addison-Wesley, Reading, MA.

Fetzer, J. H. (1971), "Dispositional Probabilities," in Buck and Cohen (1971), pp. 473–82.

(1977), "Reichenbach, Reference Classes, and Single Case 'Probabilities'," *Synthese 34:* 185–217; Errata, *37:* 113–14. Reprinted in Salmon (1979), pp. 187–219.

(1981), *Scientific Knowledge.* Reidel, Dordrecht.

(1986), "Methodological Individualism: Singular Causal Systems and Their Population Manifestations," *Synthese 68:* 99–128.

Fetzer, J. H., and Nute, D. E. (1979), "Syntax, Semantics, and Ontology: A Probabilistic Causal Calculus," *Synthese 40:* 453–95.

(1980), "A Probabilistic Causal Calculus: Conflicting Conceptions," *Synthese 44:* 241–6. Errata, *Synthese 48* (1981): 493.

Fisher, L., and Patil, K. (1974), "Matching and Unrelatedness," *American Journal of Epidemiology 100:* 347–9.

Fisher, R. A. (1959), *Smoking: The Cancer Controversy.* Oliver and Boyd, Edinburgh and London.

Forster, M. R. (1988), "Sober's Principle of Common Cause and the Problem of Comparing Incomplete Hypotheses," *Philosophy of Science 55:* 538–59.

French, P. A., Uehling, T. E., Jr., and Wettstein, H. K. (eds.) (1984), *Midwest Studies in Philosophy IX: Causation and Causal Theories.* University of Minnesota Press, Minneapolis.

Ghirardi, G. C., Rimini, A., and Weber, T. (1980), "A General Argument against Superluminal Transmission through the Quantum Mechanical Measurement Process," *Lettere al Nuovo Cimento 27:* 293–8.

Giere, R. N. (1973), "Objective Single-Case Probabilities and the Foundations of Statistics," in Suppes, Henkin, Joja, and Moisil (1973), pp. 467–83.

(1979), *Understanding Scientific Reasoning.* Holt, Reinhart and Winston, New York.

(1980), "Causal Systems and Statistical Hypotheses," in Cohen and Hesse (1980), pp. 251–70.

(1984a), *Understanding Scientific Reasoning,* 2nd edition. Holt, Reinhart and Winston, New York.

(1984b), "Causal Models with Frequency Dependence," *Journal of Philosophy 81:* 384–91.

Good, I. J. (1961–2), "A Causal Calculus I–II," *British Journal for the Philosophy of Science 11:* 305–18; *12:* 43–51; Errata and Corrigenda, *13:* 88. Reprinted in Good (1983), pp. 197–217.

(1980), "Some Comments on Probabilistic Causality," *Pacific Philosophical Quarterly 61:* 301–4.

(1983), *Good Thinking.* University of Minnesota Press, Minneapolis.

(1985), "Causal Propensity: A Review," in Asquith and Kitcher (1985), pp. 829–50.

Granger, C. (1969), "Investigating Causal Relations by Econometric Models and Cross Spectral Methods," *Econometrica 37:* 424–38.

(1980), "Testing for Causality: A Personal Viewpoint," *Journal of Economic Dynamics and Control 2:* 329–52.

Hacking, I. (1967), "Slightly More Realistic Personal Probability," *Philosophy of Science 34:* 311–25.

(1980), "Grounding Probabilities from Below," in Asquith and Giere (1980), pp. 110–16.

Hesslow, G. (1976), "Discussion: Two Notes on the Probabilistic Approach to Causality," *Philosophy of Science 43:* 290–2.

Horwich, P. (1987), *Asymmetries in Time.* MIT Press/Bradford Book, Cambridge, MA.

Humphreys, P. (1980), "Cutting the Causal Chain," *Pacific Philosophical Quarterly 61:* 305–14.

(1981), "Aleatory Explanations," *Synthese 48:* 225–32.

(1983), "Aleatory Explanations Expanded," in Asquith and Nickles (1983).

Jarrett, J. P. (1984), "On the Physical Significance of the Locality Conditions in the Bell Arguments," *Nous 18:* 569–89.

Jeffrey, R. C. (ed.) (1980), *Studies in Inductive Logic, II.* University of California Press, Berkeley and Los Angeles.

Jeffrey, R. C. (1983), *The Logic of Decision,* 2nd edition. University of Chicago Press, Chicago and London.

Kitcher, P. (1985), "Two Approaches to Explanation," *Journal of Philosophy 82:* 632–9.

Kline, A. D. (1980), "Are There Cases of Simultaneous Causation?," in Asquith and Giere (1980), pp. 292–301.

Kolmogorov, A. N. (1933), *Foundations of the Theory of Probability*. 2nd English edition (1956). Translated and edited by N. Morrison. Chelsea, New York.

Kyburg, H. E., Jr. (1974), "Propensities and Probabilities," *British Journal for the Philosophy of Science 25:* 359–75. Reprinted with minor changes in Toumela (1978), pp. 277–301.

(1978), "Subjective Probability: Criticisms, Reflections, and Problems," *Journal of Philosophical Logic 7:* 157–80.

Kyburg, H. E., Jr. and Smokler, H. E. (eds.) (1964), *Studies in Subjective Probability.* John Wiley and Sons, Inc., New York.

Laplace, Pierre Simon, Marquis de, (1819), *A Philosophical Essay on Probabilities* (1951), translated by F. W. Truscott and F. L. Emory. Dover, New York.

Lewis, D. (1973), "Causation," *Journal of Philosophy 70:* 556–67.

(1979), "Counterfactual Dependence and Time's Arrow," *Nous 13:* 455–76.

(1986), *Philosophical Papers, vol. 2.* Oxford University Press, London.

Mackie, J. L. (1974), *The Cement of the Universe.* Oxford University Press, London.

Mayo, D. (1986), "Understanding Frequency-Dependent Causation," *Philosophical Studies 49:* 109–24.

Mellor, D. H. (1978) (ed.), *The Foundations of Mathematics and other Essays: Essays by Frank Plumpton Ramsey.* Routledge and Kegan Paul, London.

Miettinen, O. S. (1970), "Matching and Design Efficiency in Retrospective Studies," *American Journal of Epidemiology 91:* 111–8.

(1974), "Confounding and Effect Modification," *American Journal of Epidemiology 100:* 350–3.

Miller, G. (1985), "Correlations and Giere's Theory of Causation," *Philosophy of Science 52:* 612–14.

Otte, R. (1981), "A Critique of Suppes' Theory of Probabilistic Causality," *Synthese 48:* 167–89.

(1985), "Probabilistic Causality and Simpson's Paradox," *Philosophy of Science 52:* 110–25.

Pearson, K. (1897), "Mathematical Contributions to the Theory of Evolution," *Proceedings of the Royal Society of London 60:* 489–503.

Popper, K. R. (1959), "The Propensity Interpretation of Probability," *British Journal for the Philosophy of Science 10:* 25–42.

Ramsey, F. P. (1926), "Truth and Probability." In his *The Foundations of Mathematics and other Logical Essays,* edited by R. B. Braithwaite, pp. 156–98 (1931), Routledge and Kegan Paul, London. Reprinted in Kyburg and Smokler (1964), pp. 61–92, and in Mellor (1978), pp. 58–100.

Reichenbach, H. (1935/1949), *The Theory of Probability*. University of California Press, Berkeley and Los Angeles. (First published as, *Wahrscheinlichkeitslehre*, Leiden).

(1956), *The Direction of Time*. University of California Press, Berkeley and Los Angeles.

Rosen, D. A. (1978), "In Defense of a Probabilistic Theory of Causality," *Philosophy of Science 45*: 604–13.

Russell, B. (1948), *Human Knowledge, Its Scope and Limits*. Simon and Schuster, New York.

Salmon, W. C. (1966), *The Foundations of Scientific Inference*. University of Pittsburgh Press, Pittsburgh.

(1971), *Statistical Explanation and Statistic Relevance*. University of Pittsburgh Press, Pittsburgh.

(1978), "Why Ask 'Why'?: An Inquiry Concerning Scientific Explanation," *Proceedings and Addresses of the American Philosophical Association*, vol. 5, no. 6: 683–705.

(1980), "Probabilistic Causality," *Pacific Philosophical Quarterly 61*: 50–74.

Salmon, W. C. (ed.) (1979), *Hans Reichenbach: Logical Empiricist*. Reidel, Dordrecht.

(1984), *Scientific Explanation and the Causal Structure of the World*. Princeton University Press, Princeton, NJ.

Savage, L. J. (1954), *The Foundations of Statistics*. 2nd edition, 1972. Dover, New York.

Schilpp, P. A. (ed.) (1974), *The Philosophy of Karl Popper*. Open Court, La Salle, IL.

Simon, H. (1954), "Spurious Correlation: A Causal Interpretation," *American Statistical Association Journal 49*: 467–79.

(1957), *Models of Man*. Wiley, New York.

Simpson, E. H. (1951), "The Interpretation of Interaction in Contingency Tables," *Journal of the Royal Statistical Society, Ser. B, 13*: 238–41.

Sklar, L. (1970), "Is Probability a Dispositional Property?," *Journal of Philosophy 67*: 355–66.

Skyrms, B. (1980), *Causal Necessity*. Yale University Press, New Haven and London.

(1984a), *Pragmatics and Empiricism*. Yale University Press, New Haven and London.

(1984b), "EPR: Lessons for Metaphysics." In French, Uehling, and Wettstein (1984), pp. 245–55.

(1988), "Probability and Causation," *Journal of Econometrics 39*: 53–68.

Skyrms, B., and Harper, W. L. (eds.) (1988), *Causation, Chance, and Credence*, vol. 1. Kluwer Academic Publishers, Dordrecht.

408

Sober, E. (1982), "Frequency Dependent Causation," *Journal of Philosophy* 79: 247–53.

(1984a), *The Nature of Selection*. MIT Press/Bradford Book, Cambridge, MA.

(1984b), "Common Cause Explanation," *Philosophy of Science 51:* 212–41.

(1985a), "Discussion: What Would Happen If Everyone Did It?," *Philosophy of Science 52:* 141–50.

(1985b), "Two Concepts of Cause." In Asquith and Kitcher (1985), pp. 405–24.

(1987a), "Parsimony, Likelihood, and the Principle of the Common Cause," *Philosophy of Science 54:* 465–9.

(1987b), "Explanation and Causation" (review article for Wesley Salmon's *Scientific Explanation and the Causal Structure of the World* (1984)), *British Journal for the Philosophy of Science 38:* 243–57.

(1988), "Apportioning Causal Responsibility," *Journal of Philosophy 85:* 303–18.

Suppes, P. (1970), *A Probabilistic Theory of Causality*. North Holland, Amsterdam.

(1974), "Popper's Analysis of Probability in Quantum Mechanics," in Schilpp (1974), pp. 760–74.

(1984), *Probabilistic Metaphysics*. Blackwell, Oxford.

Suppes, P., Henkin, L., Joja, A., and Moisil (eds.) (1973), *Logic, Methodology and Philosophy of Science IV.* North-Holland, Amsterdam and London.

Taylor, R. (1966), *Action and Purpose*. Prentice-Hall, Englewood Cliffs, NJ. Reprinted (1974), Humanities Press, New York.

Torretti, R. (1987), "Do Conjunctive Forks Always Point to a Common Cause?," *British Journal for the Philosophy of Science 38:* 384–7.

Toumela, R. (ed.) (1978), *Dispositions*. Reidel, Dordrecht.

van Fraassen, B. C. (1977a), "Relative Frequencies," *Synthese 34:* 133–66. Reprinted, in an expanded version, in Salmon (1979), pp. 129–67.

(1977b), "The Pragmatics of Explanation," *American Philosophical Quarterly 14:* 143–50.

(1980), *The Scientific Image*. Oxford University Press, Oxford.

(1982), "The Charybdis of Realism: Epistemological Implications of Bell's Inequality," *Synthese 52:* 25–38.

Venn, J. (1866), *The Logic of Chance*. 4th edition (1962). Chelsea, New York. (1st edition, London.)

von Mises, R. (1928), *Probability, Statistics and Truth*. 2nd English edition, (1957). Macmillan, New York.

(1964), *Mathematical Theory of Probability and Statistics* (H. Geiringer, ed.). Academic Press, New York.

von Wright, G. H. (1974), *Causality and Determinism*. Columbia University Press, New York.

Yule, G. U. (1910), "The Applications of the Methods of Correlation to Social and Economic Statistics," *International Statistical Institute Bulletin 18:* 537–51.

(1911), *An Introduction to the Theory of Statistics*. Griffen, London.

Index

411

412

positive causal factor, *see* causal factor, positive

principle of the common cause (Reichenbach's, stated), 60n.

probability
  actual frequency, 35n., 39ff
  axioms, 399
  classical, 35n.
  frequency, *see* probability: actual frequency, hypothetical frequency
  hypothetical frequency, 35n., 42ff.
  logical, 35n., 36ff.
  propensity, 35n., 42, 43ff.
  relative frequency, *see* probability: actual frequency, hypothetical frequency
  subjective, 35ff.

quantum mechanics, 17, 119ff., 283n.

Ramsey, F. P., 35n., 407
Reichenbach, H., 38, 60n., 82, 84, 107, 251n., 262, 268n., 269, 408
relativity, special theory of, 117n., 118ff., 240, 241, 278n.
revised context approach (to interaction), 130ff.
Rimini, A., 123n., 405
Rosen, D. A. 283ff., 408
Rosen, N., 117n., 118ff., 240, 278n., 405
Russell, B., 39n., 387–8, 408

Salmon, W. C., x, 1, 38n., 39n., 61n., 84, 107, 112ff., 118n., 126n., 148n., 233n., 262, 284, 367, 370n., 372ff., 391, 405, 408, 409
Savage, L. J., 35n., 408
Schilpp, P. A., 408, 409
screening off (definition of), 60n., 401
Simon, H., 59, 84, 408
Simpson, E. H., 2, 72, 408
Simpson's paradox, 2, 72
singly connected causal chain (definition of), 216
Sklar, L., 39n., 43n., 44, 408
Skyrms, B., x, 35n., 36n., 42, 43, 53, 59n., 67, 74, 84, 85, 87, 93, 95ff., 118n., 121, 125, 126n., 269, 404, 408, 409

Smokler, H. E., 404, 407, 408
Snyder, A., 17, 18, 281, 404
Sober, E., x, xi, xii, 61n., 85, 86n., 95, 96, 109n., 157n., 171, 176, 178n., 181, 193n., 194n., 211n., 213n., 216, 219, 220, 221, 251n., 272ff., 283, 381n., 386n., 405, 409
spacelike separated (definition of), 118n.
subsequent causal factor, *see* causal factor, subsequent
Suppes, P., x, 1, 39n., 59n., 84, 243, 251n., 262, 267ff., 405, 409

Taylor, R., 241, 409
timelike separated (definition of), 118n.
token uncaused (definition of), 319
Tolson, W., x, 310n., 383
Torretti, R., 61n., 409
Toumela, R., 407, 409
transitivity
  of property-level causation and causal chains (definitions of), 210–11
  of token-level probabilistic causation, 347ff.
Truscott, F. W., 407

Uehling, T. E., Jr., 404, 405, 408
unanimity
  context (definition of), 86
  of intermediaries (definitions of), 220–1
  of token causal intermediaries (definition of), 350
uncaused (token), *see* token uncaused

van Fraassen, B. C., 39n., 61n., 107n., 118n., 409
Venn, J., 38n., 409
von Mises, R., 38, 410
von Wright, G. H., 240, 410

Weber, T., 123n., 405
Wettstein, H. K., 404, 405, 408

Yaqub, A., xi, 347n.
Yule, G. U., 72n., 84, 410